Data Visualization in Society

Data Visualization in Society

Edited by
Martin Engebretsen and Helen Kennedy

Routledge
Taylor & Francis Group
LONDON AND NEW YORK

The publication of this book is made possible by a grant from the Research Council of Norway (grant number 25936).

First published in 2020 by Amsterdam University Press Ltd.

Published 2025 by Routledge
4 Park Square, Milton Park, Abingdon, Oxon OX14 4RN
605 Third Avenue, New York, NY 10158

Routledge is an imprint of the Taylor & Francis Group, an informa business

ISBN: 9789463722902 (pbk)
ISBN: 9781003693697 (ebk)
NUR 811

Cover illustration: Prisca Schmarsow of Eyedea Studio

Cover design: Coördesign, Leiden

DOI 10.5117/9789463722902

Table of Contents

List of tables

List of figures

Acknowledgements

We would like to thank the Research Council of Norway for funding INDVIL, a research project from which this edited collection emerged (Innovative Data Visualization and Visual-Numeric Literacy (https://indvil.org, grant number 259536). We thank our colleagues on INDVIL—Sara Brinch, Verena Lechner, Torgeir Uberg Nærland, Jill Walker Rettberg, Mikael Snaprud, Elise Seip Tønnessen, and Wibke Weber—for a stimulating and productive three-year collaboration. They are all authors of chapters in this book. We are grateful to all of the authors who have contributed to this collection, for their excellent contributions, their receptiveness to feedback and their patience during the long process of bringing the book to fruition. We are extremely grateful to the University of Agder for hosting the INDVIL project and for the administrative support provided to us. In June 2018 we held a seminar with many of the book's authors at the Metochi Study Centre on the island of Lesbos. We thank the staff at Metochi for the wonderful service they provided and the fantastic atmosphere which enabled us to work hard together on the book. We thank reviewers for taking the time to read our many chapters with care and for providing helpful feedback. We are grateful also to Alberto Cairo for his extremely generous foreword, and to esteemed data visualization scholars for endorsing this volume on its cover. We want to express our enormous gratitude to all of the brilliant data visualizers whose work inspires our own, especially Andy Kirk, for ongoing support and openness to dialogue. We could not have made it through the final stages of producing this book without the invaluable help of Robin Steedman, whose work as an editorial assistant has been faultless. Thank you Robin. We also thank Prisca Schmarsow of Eyedea Studio (https://eyedeastudio.london/) for lovely cover design and the staff at the Amsterdam University Press for their assistance along the way. Finally, Helen would like to thank Martin, and Martin would like to thank Helen, for the enjoyable and rewarding experience of editing this book together.

Foreword: The dawn of a philosophy of visualization

Alberto Cairo, Knight Chair at the University of Miami and author of How Charts Lie

Geographer John Pickles once wrote that 'GIS is a set of tools, technologies, approaches and ideas that are vitally embedded in broader transformations of science, society, and culture'. That's true of data visualization too, therefore the relevance of the book that you have in your hands, *Data Visualization in Society.*

I often joke—although I'm inclined to believe—that a field *X* reaches maturity when a parallel field of 'philosophy of *X*' springs into existence. That hasn't happened yet with data visualization, at least formally. Might we be on the path to it, though? I hope so. Some books have paved the way. Think of David J. Staley's *Computers, Visualization, and History*, Charles Kostelnick and Michael Hassett's *Shaping Information*, and Wolff-Michael Roth's *Toward an Anthropology of Graphing*, all from the early 2000s. Or, more recently, Orit Halpern's *Beautiful Data* (2014), Johanna Drucker's *Graphesis* (2014), R. J. Andrews's *Info We Trust* (2019), Sandra Rendgen and Julius Wiedemann's *History of Information Graphics* (2019), or the upcoming *Data Feminism* (2020), by Catherine D'Ignazio and Lauren Klein, who have also contributed to this volume.

Books like these prove that writing about visualization doesn't mean just thinking about how to *design* visualizations, but also about what visualization *is*, why it is the way it is—and what it *could* be. Data visualization is a technology—or set of technologies—and, like artefacts such as the clock, the compass, the abacus, or the map, it transforms the way we see and relate to reality. As Langdon Winner suggested in *The Whale and the Reactor* (1986), a foundational book in the phenomenological philosophy of technology, to create technologies doesn't consist just of crafting stuff; rather, when technologies come about 'new worlds are being made'. What 'new worlds' does visualization generate? That's a question for a potential philosophy of visualization.

A philosophy of visualization may derive themes, methodologies, and language from a wide range of disciplines: epistemology, sociology, semiotics, history, ethics, critical theory fields such as critical cartography, or from the philosophies of science, statistics, art, and—perhaps more strongly than any other—the philosophy of technology. Philosophers of visualization should reason about visualization's history, assumptions, conventions, practices, and impacts on individuals, cultures, and societies. They will combine the observational, descriptive, and hermeneutical—dealing with what currently exists and why—the normative—, asking what should or shouldn't exist or happen—and the critical—, challenging visualization's core tenets.

Data Visualization in Society is a collection of chapters by scholars and professionals who don't call themselves philosophers of visualization but who, in practice, operate as such. I see this book as a relevant step toward the possible inception of the philosophy of data visualization as a discipline. I hope it will serve as a starting point for many inquiries by other thinkers. This includes myself: I read all chapters with pleasure and took copious notes on the margins. I know these scribbles will later echo in my own work.

That's the virtue of the best philosophical writing: it doesn't aspire to settle matters outright, but to inspire further reflection. *Data Visualization in Society* may spur questions such as: Does visualization pretend to be 'objective', or is it just wrongly perceived as such? What does 'objective' mean in the first place? What is the influence of visualization on politics? Is numeracy—numerical literacy—enough to design or read visualizations? Doesn't the fact that a substantial portion of the public isn't numerate—or 'graphicate'—deepen existing inequalities and even create new ones? What do we mean when we say that a visualization is 'beautiful'? Is the goal of visualization to convey facts and data, or can it also spark profound emotional experiences? If so, how? And many more.

The variety of topics and approaches of the chapters in this book is astounding, but what most have in common is an open ending: they are links in a chain of reasoning—a dialogue—that extends from the distant past and that, conceivably, and with the contribution of a large critical mass of academics and practitioners of the craft, will continue beyond the foreseeable future. That's where you come in: does any of these chapters inspire you? Do you agree or disagree with it? Reason why. Argue. Establish a conversation with it. Write and publish, and be open to further responses and critiques. That's how philosophy begins.

1. Introduction: The relationships between graphs, charts, maps and meanings, feelings, engagements

Helen Kennedy and Martin Engebretsen

Today we are witnessing an increased use of data visualization in a range of domains and genres. In journalism, education, and public information as well as in workplaces, diverse forms of graphs, charts, and maps are used to explain, persuade, and tell stories. At best, visual representations of statistics and other, often quantitative data can convey complex facts and patterns quickly and effectively. At worst, they can appear confusing or manipulative. In an era in which more and more data are produced and circulated through online networks, and digital tools make visualization production increasingly accessible, it is important to study the conditions under which such visual texts are generated, disseminated and thought to benefit processes of sense-making, learning, and engaging.

Data visualization is not new. The graphical representation of numeric information has roots in early map-making, and grew in importance with the widespread use of data and statistics for planning and commerce in the nineteenth century (Friendly, 2008). Still, in our contemporary society, several factors contribute to give data visualization a social relevance on a scale we have not seen before. One of these factors, as Kennedy, Hill, Aiello, and Allen (2016b, p. 715) put it, is that '[...] data are becoming increasingly valued and relied upon, as they come to play an ever more important role in decision-making and knowledge about the world'.

In other words, more data are generated, gathered, stored, and made accessible than ever before. Data gathering takes place in many domains, often by law, including commerce, education, health, transport, and cultural and social life. These data offer insights into societal patterns otherwise invisible and unnoticed. Such documentation has been conducted for decades, but technological and other developments have led to its sharp

Engebretsen, M. and H. Kennedy (eds.), *Data Visualization in Society.*
Taylor & Francis Group, 2020
DOI 10.5117/9789463722902_CH01

increase, and data are now being gathered in huge volumes as a result of new techniques of measurement. These combined phenomena, sometimes called 'datafication' (Mayer-Schönberger & Cukier, 2013, p. 78) are understood as a transformation disrupting the social world in all its forms (Couldry, 2016).

Furthermore, to make data accessible to publics, rather than remaining a useful source only for experts and decision-makers, a range of actors have campaigned to open up public data, to make them reusable for a variety of activities and democratic purposes. Open data initiatives and related campaigning activities contribute to accelerate the spread of data visualization, which often serve as a main entry point to data for non-experts.

Another important driver in the spread of data visualizations is the development of related technology. New tools and techniques for harvesting, filtering, analysing, and visualizing data make these processes easier and cheaper. We are also witnessing new arenas for dissemination of and engagement with data, as data-based techniques are increasingly used for informative, persuasive, and rhetorical purposes in political campaigns, health communication, education, and in newsrooms, where new data visualization teams are being constructed, combining visual creativity with data science skills and other domain expertise (Engebretsen, Kennedy, & Weber, 2018).

As a result of these varied processes, data visualizations have innovative semiotic forms and result in novel types of communication and interactivity. This implies that their potential for meaning-making, for evoking emotions, democratic participation, and other forms of engagement is also in a state of transformation. So, while data visualizations have a growing importance in society, their novel forms and uses mean that our understanding of how they work as semiotic and aesthetic phenomena and how they support or hinder personal and social agency is also in flux. These transformations coexist with more familiar debates about whether data visualizations do 'good' or 'bad': do they promote understanding and engagement, as some commentators argue (e.g. Few, 2008 and Cairo, 2013), or do they do ideological work, privileging certain views of the world, as others claim (e.g. Barnhurst, 1994 and Latour, 1986; sources taken from Kennedy et al., 2016b)?

The phenomena described here, both the new uses of data visualization and the debates about them, form the focus of this book, which draws on a range of research and development projects to reflect on data visualization in society. The book addresses these questions:

- Where and how do citizens and publics engage with data visualizations, and for what reasons?
- In what new social and cultural contexts are data visualizations emerging, and to what ends?

- How do data visualizations create meanings in the various social and cultural arenas in which they appear, and what are their discursive roles and functions?
- How do data visualizations arouse feelings in their audiences, what kinds of emotional responses are activated, and to what ends?
- What does literacy mean when it comes to data visualization, and how can data visualization literacy be enhanced?
- What kinds of aesthetic characteristics do data visualizations have?
- What is the political significance of data visualization, and in what ways do data visualizations play a role in citizens' participation in democratic systems?

What do we mean by *data* and *data visualization*?

In a scientific context, data are generally understood to result from the generation, collection, observation, or registration of objects, events, or processes suitable to serve some analytical purpose. Similarly, in the context of data visualization, data can be anything that can be subjected to categorization, abstraction, and translation into graphical representation: persons, places, documents, relations, sentences, salaries, to mention some examples. A main distinction is between qualitative data and quantitative data. While qualitative data are valued for the uniqueness of each individual unit, be it a poem, a sentence, or an interview, quantitative data are valued for characteristics shared by all or many units in a dataset. It is their shared characteristics that make them objects for counting or measuring, and thus for numeric representation and statistical processing.

Both qualitative and quantitative data can be visualized. It is possible to visualize semantic structures in a novel, or networks of relationships between the works in an art collection, as seen, for example, in the work of Stefanie Posavec (http://stefanieposavec.com/). Most, but not all, of the contributions in this book focus on the visualization of quantitative data, for the reasons given above—that is, because their proliferation and increasing openness, and the enhanced availability of related tools, make them a socially and culturally significant phenomenon.

Numeric data can be structured or unstructured. Structured data have been subjected to statistical treatment and are typically represented as numbers in a table, with columns and rows presenting units and variables and numeric values positioned in cells. A common example is the datasets accessible from national statistics institutes (NSIs) which are often presented to the public in tabular form. Unstructured data have not been subjected to any statistical or structuring processing, and appear as 'raw' data in an analogue or digital register, until the data are structured by someone with an intention to use them for some specific purpose. An automatic registration of cars driving through a tollbooth is one example.

'Big data' is a fashionable concept, although its use is rarely accompanied by a shared understanding of what it means or how it differs from 'small' data. Big data have been said to be characterized by three Vs: volume, variety, and velocity. More recently, additional Vs have been proposed, such as variability and value (http://whatis.techtarget.com/definition/3Vs; see also Kitchin, 2014 for additions which don't begin with V). When we talk about datasets consisting of thousands of rows of data, or new streams of data created every second, we are talking about big data. Data harvested from a social media platform, or from the activities on the finance market, are some examples. But exactly when data become big is hard to define.

In the same way that it is hard to distinguish between big and other data, the differences between data visualization, information visualization, information graphics, and scientific visualization are also blurred. As Kennedy and Allen write, data visualization 'has data at its heart' (2016, p. 309), and it often uses abstract, geometrical forms to represent numeric values and relations. In contrast, an information graphic explains phenomena graphically but may contain no numeric data, or it presents data in charts alongside other illustrations, like photographs or drawings. Scientific visualization is a concept mostly used in highly specialized, expert-to-expert contexts, for example within medicine and biology. Here, visualizations are used to illuminate specific aspects of certain physical objects or processes, and may include simulations, drawings, or processing of magnetic resonance imaging (Ambrosio, 2015).

Data visualizations are a discursive resource used in the dissemination of statistical information and often numeric data. In this book, data visualizations are understood as graphical representations of data which are primarily, but not solely, numeric. What's more, they are abstractions and reductions of the world, the result of human choices, social conventions, and technological processes and affordances, relating to generating, filtering, analysing, selecting, visualizing, and presenting data. Data visualizations (also called *dataviz* or DV) are created to 'facilitate understanding', to use Kirk's term (2016, p. 19; see also Borgo et al., 2013; Cairo, 2013), but they can also facilitate other things, such as persuasion. Consequently, we understand data visualizations as cultural artefacts with distinct semiotic, aesthetic, and social affordances. There is, however, much more to data visualization than what can be captured in any simple definition, as will become evident throughout the book.

How can dataviz produce meanings, feelings, and engagements?

In this book, we relate the social power of data visualizations to their abilities to produce meanings, feelings, and engagements in their users and audiences. Processes of socially situated meaning-making are best described in the field of social semiotics, first developed by the Australian linguist Michael Halliday (1978), later adapted to visual and multimodal artefacts by Gunther Kress and Theo van Leeuwen (1996) and others. In social semiotic theory, the meaning of semiotic material (which can include words, images, colours, and more) can be traced in three different dimensions, each relating to an aspect of the situation of communication. These are:

1. The field (or topic) of discourse. How does the semiotic material represent the world or ideas about the world? This is known as *ideational meaning.*
2. The participants involved in the process of communication. How does the semiotic material reflect, establish, or change the social relations between the participants? This is known as *interpersonal meaning.*
3. The semiotic resources activated in the process. How do all the elements of the semiotic material unite in a textual whole? This is known as *compositional meaning.*

In many situations, the semiotic material in question will be identified as 'a text', such as a multimodal webpage with words, images, and colours organized in a specific user interface. In other contexts, meaning is made through semiotic resources not conventionally identified as texts, such as buildings, clothes, and sculptures. Such artefacts nonetheless carry meaning based on certain culturally and historically formed conventions. The artefacts that this book is concerned with, data visualizations, will normally be produced, distributed, and used in ways comparable to other multimodal and mediated text types.

Semiotic interpretation and aesthetic experience (that is, our sensory impressions, as well as judgements based on taste) go hand in hand in our encounters with texts and other cultural artefacts, and where one stops and the other begins is hard to identify. Our encounters with form, colour, and composition are informed by bodily experience as well as aesthetic judgement, and so the aesthetic (as well as the semiotic) aspects of data visualization need to be taken into account. Also relevant to a discussion of meaning in data visualization is the issue of 'knowledge regimes', or epistemology. What aspects of reality are privileged in a semiotic text based on visualized, numeric data? What kinds of truth are foregrounded, and what knowledge, values, and attitudes result? Data visualizations may seem to reflect reality in a more direct way than words because they are based on numbers, which seem trustworthy (Porter, 1995). But this does not mean that they are more true, in the sense that they offer a more objective representation of the world. This issue informs several contributions to this book.

Data visualizations thus create meanings through visual and other codes. But they also generate feelings, by which we mean the emotional responses that are connected to human encounters with data visualizations. Meanings and feelings are inseparable in our situated interactions with texts. They influence each other, and together they form our responses to the texts and artefacts with which we interact (Lemke, 2015). A recent study by Kennedy and Hill (2017) revealed that data visualizations awaken a wide range of feelings in people who engage with them, activated either by the textual content of the visualizations, contextual factors like users' earlier experiences (with visualizations, their subject matter, or other relevant phenomena), or by the physical and psycho-social situation of use. In the analysis of their research findings, Kennedy and Hill cite Jagger (1989) who argues that 'emotion can be understood as an "epistemic resource", a way of knowing that is valuable for building a critique of the world' and Damasio (2006), who argues that without emotions, 'the ability to make rational decisions is hampered' (Kennedy and Hill, 2017, p. 12). Emotions are vital components for understanding the social world, including data visualizations. As such, they are a central focus in this book.

Our emotional engagement with data visualizations is also closely connected to their aesthetic aspects. The forms, colours, and arrangements of data visualizations trigger our senses in particular ways. In turn, the interplay between the semiotic, meaning-making aspects of data visualizations, and the emotions they evoke is closely related to their ability to elicit social engagement. Here, the concept of *engagement* has several layers. It can refer to the actual interaction with a data visualization, being engaged *with it*, or to emotional and practical responses, getting engaged *by it*. It can also refer to broader audience responses, for example the ways in which data visualizations are mobilized to prompt political engagement. These three aspects of engagement are closely interrelated, as can be seen in several chapters in the book.

We understand 'engaging with' visual representations of data to refer to 'the processes of looking, reading, interpreting and thinking that take place when people cast their eyes on data visualisations and try to make sense of them' (Kennedy et al., 2016b). For people who are not experts in data visualizations but who encounter them with growing frequency, engaging with them is not straightforward. Without the right skills, the ability to participate in data-driven conversations and decision-making will be off limits to certain groups, existing uneven power relations will be reproduced and new, data-based ones will emerge. This has troubling implications for democracy, or for getting *engaged by* the world around us, including by visualized data.

The expansion of data visualization in society therefore requires a new kind of literacy if it is to enable citizens to act in informed and critical ways. It also requires the assessment of data visualization's role in democracy, and the reassessment of democratic theory in light of developments in data visualization. This means asking a range of questions about the relationship between data visualization and democracy. It also means considering the factors in visualization consumption and production processes that affect engagement, which might include factors which extend beyond textual and technical matters, such as class, gender, race, age, location, political outlook, and education of audience members. Some of the contributions in this collection address these issues.

Data visualization as discourse

This book is a contribution to multidisciplinary and multifaceted academic conversation concerning the forms, uses, and roles of data visualization in society. As a collection of chapters which study the conditions under which visualizations are generated, disseminated, and thought to benefit processes of learning, development, and participation, to reuse our own phrase from above, it belongs to the large and diverse field of discourse studies. Although the individual chapters derive from a range of perspectives, the tradition of discourse studies provides a framework. The book leans on a social semiotic understanding of *discourse*—as the situated application of semiotic resources (such as words and images) by human agents in order to construct and share ideas about the world and to perform social action (or make things happen) (Kress, 2010; van Leeuwen, 2005). The potential meanings carried by semiotic resources are dependent on both cultural conventions and the particular situations of use, including the background and motivations of the human participants, the media used to produce and distribute the messages, and the social practice of which the semiotic material is an integrated part. Discourse studies can offer nuanced analyses of the mediated processes of communication in which data visualizations are situated and also illuminate processes of social struggle and control.

A discourse studies approach combines the micro level with the macro level. It focuses on the relations between the specific structures and forms of the semiotic artefact on the one hand, and the social, technological, and cultural contexts which form it and are formed by it, on the other (Fairclough, 2010; Chouliaraki & Fairclough, 1999; van Leeuwen, 2005). The concept of discourse thus offers a theoretical and methodological framework for analysing data visualization in discrete social practices, like journalism, public information campaigning, or health communication. These relations between the micro and the macro, between texts and contexts, are apparent in all chapters of the book, although some focus more on the micro level, and others more on the macro level.

Discourse studies include a range of approaches, from those based on an analysis of how meanings are shaped and negotiated in specific social situations, to critical investigations of how words and images play a role in creating or opposing power structures and social inequalities. The latter approaches are often grouped under the term *critical discourse studies* (or CDA), which was originally theoretically and methodologically modelled by Norman Fairclough (2010). In several chapters in this book, similar critical approaches to the relationship between semiotic practices and social inequalities are used, although the authors do not necessarily all see themselves as discourse studies scholars. Rather, authors adopt such approaches from within a diverse range of disciplines, including gender studies, science and technology studies, (digital) media studies, critical cartography, design, art history, literacy studies, ICT, and the emerging field of data studies. Together, the chapters shine a spotlight on data visualization as an important instance of *text-in-society.*

How the book is organized and targeted

The book is organized into five sections. The first, called 'Framing Data Visualization', does the work of framing the contributions in the rest of the book, drawing on a range of conceptual and theoretical resources. The three chapters in this section sketch out three significant issues with which subsequent chapters engage: epistemology, semiotics, and politics respectively. In the first chapter in this section, 'Ways of knowing with data visualization', Jill Walker Rettberg explores the ways of knowing that have historically been privileged by different systems for gathering and visualizing data. Giorgia Aiello then maps out how the strategies deployed in a social semiotic approach can help us to understand data visualization *in society* in 'Inventorizing, situating, transforming: Social semiotics and data visualization'. In the final chapter in this section, Torgeir Nærland maps out perspectives from which we might approach analyses of data visualization's politics, in 'The political significance of data visualization: Four key perspectives'.

The second section of the book, 'Living and Working with Data Visualization', includes chapters which reflect on diverse experiences of and with data visualization in private and professional settings. In Chapter 5, 'Rain on your radar: Engaging with weather data visualizations as part of everyday routines', Eef Masson and Karin van Es explore uses and evaluations of uses of weather data visualizations in everyday life. This is followed by a chapter by Salla-Maaria Laaksonen and Juho Pääkkönen, which shifts the focus to working environments, and explores the uses of data visualizations in social media analytics companies, their role in knowledge claims, and the mechanisms by which they achieve credibility. The chapter is called 'Between automation and interpretation: Using data visualization in social media analytics companies'. Chapter 7, 'Accessibility of data visualizations: An overview of European statistics institutes', by Mikael Snaprud and Andrea Velazquez, uses multiple approaches to assess the extent to which dataviz shared by National Statistics Institutes (NSIs) are accessible to people with disabilities, and the extent of preparedness for compliance with new EU legislation on web accessibility of NSIs, which are both important characteristics of democratic societies. This is followed by a chapter which explores how data visualizations are evaluated, and whether approaches to evaluation which account for the sociocultural contexts of and influences on dataviz might be possible. This chapter, by Arran Ridley and Christopher Birchall, is called 'Evaluating data visualization: Broadening the measures of success.' The subsequent chapter, 'Approaching data visualizations as interfaces: An empirical demonstration of how data are imag(in)ed', by Daniela van Geenen and Maranke Wieringa focuses on the case of a specific data visualization produced by the authors, to show how visualization practices allow for interfacing with data and that a particular visualization provides only one perspective on data. In Chapter 10, 'Visualizing data: A lived experience', Jill Simpson draws on her own experience of producing a small-data hand-drawn visualization to explore questions of subjectivity, authenticity, and honesty in data visualization. This section ends with a chapter by Helen Kennedy, Wibke Weber, and Martin Engebretsen called 'Data visualization and transparency in the news', which explores the relationship between data visualization and the emerging journalistic norm of transparency.

The third section, 'Data visualization, learning, and literacy' includes four chapters which focus on the skills needed to engage with and make sense of data visualizations. In Chapter 12, 'What is visual-numeric literacy, and how does it work?', Elise Seip Tønnessen reports on research based on observations of Norwegian social science classrooms in which students sought to develop and deploy skills to make sense of visualizations in an educational setting. The setting of the next two chapters moves beyond educational institutions. In Catherine D'Ignazio and Rahul Bhargharva's chapter, 'Data visualization literacy: A feminist starting point', the authors introduce a starting point for teaching data visualization which is grounded in feminist theory, process, and design principles, to counter the problem of unequal human relations produced through data. This is followed by 'Is literacy what we need in an unequal data society?', in which Lulu Pinney unearths the different notions of power that are embedded in different uses of literacy across academic literature, policy, and practice in order to critically interrogate the usefulness of literacy as a term and concept. In the final chapter in this section, 'Multimodal academic argument in data visualization', Arlene Archer and Travis Noakes investigate students' semiotic and rhetorical strategies for making an argument with data visualization and their implications for teaching students to become critical citizens.

The fourth section of the book, called 'Data Visualization Semiotics and Aesthetics', includes contributions which focus on the semiotic, aesthetic, visual, and stylistic dimensions of data visualizations and the ways these intersect with social and cultural considerations. Chapter 16, 'What we talk about when we talk about beautiful data visualizations', by Sara Brinch, presents an analysis of what is regarded as beautiful within the field of data visualization design, and at the same time interrogates 'beautiful' as an ambivalent and contested concept. The next chapter, 'A multimodal perspective on data visualization', by Tuomo Hiippala, examines the multimodality of data visualizations, or how they combine multiple modes of expression, such as written language, photographs, diagrammatic elements, and illustrations. This is followed by a chapter by Wibke Weber, 'Exploring narrativity in data visualization in journalism', which explores how and when data visualizations tell stories and the narrative constituents in data visualization, in order to argue that understanding how data are transformed into visual stories is key to understanding how facts are shaped and communicated in society. Chapter 19, by Jonathan Gray, is called 'The data epic: Visualization practices for narrating life and death at a distance'. The chapter proposes the notion of the 'data epic' to explore the narrative and affective capacities of distance in the context of 'public data culture'. This is followed by a chapter by Verena Lechner which focuses on a very specific aspect of data visualization form, the line, a graphical element widely used in data visualizations to signal a connection between other visual elements. The chapter, 'What a line can say: Investigating the semiotic potential of the connecting line in data visualization', investigates the semiotic functions that connecting lines can have and how these functions can be related to variations in form. The final chapter in this section, 'Humanizing data through graphic visualization', by Aria Alamalhodaei, Alexandra Alberda, and Anna Feigenbaum, considers how the emergent areas of Graphic Medicine and Graphic Social Science deal with numeric data in ways that humanize data, encouraging empathy and connection in audiences. Data visualization could learn from these unconventional fields, the authors propose.

The contributions to the final section, entitled 'Data Visualization and Inequalities', focus on the political dimensions of the social and cultural embedding of data visualization. Chapter 22, 'Visualizing diversity: Data deficiencies and semiotic strategies', by John P. Wihbey, Sarah J. Jackson, Pedro M. Cruz, and Brooke Foucault Welles, explores the complicated dynamics that are inherent to the practice of data visualization involving issues of race and identity. The chapter focuses on data from the US Census and the profound questions that are raised as visual forms purport to represent groups, and showcases a visualization produced by the authors to address the challenges that they discuss. This is followed by a chapter by Rosemary Lucy Hill, 'What is at stake in data visualization? A feminist critique of the rhetorical power of data visualizations in the media'. This chapter argues that visualizations relating to abortion often tell a narrow story, remove contextual detail and omit questions important to women's health. The final three chapters of this section and of the book focus on maps as particular visualizations of data. Chapter 24, 'The power of visualization choices: Different images of patterns in space', by Britta Ricker, Menno-Jan Kraak, and Yuri Engelhardt, uses a dataset related to the United Nations Gender Inequality Index to demonstrate the numerous decisions that are made in the process of creating a map and the types of representations that result. The next chapter, by Anna Berti Suman, 'Making visible politically masked risks: Inspecting unconventional data visualization of the Southeast Asian haze', investigates the potential of data visualization in stimulating a socially and legally accountable governance of environmental risk affecting public health, focusing on mapping efforts of the Southeast Asian haze performed by environmental NGOs and civil society. Finally, Chapter 26, 'How interactive maps mobilize people in geoactivism', by Miren Gutiérrez, explores how maps are employed in activism to unleash sentiments, focusing on three examples and employing as a lens the emotional turn currently influencing geography.

We structure the book in this way with the aim of highlighting the major issues concerning researchers of data visualization in society; many of the chapters cover more than one of the issues that are named in section titles, of course. The book aims to be accessible to a broad audience interested in data visualization's increasing prominence and visibility and its social role. Chapters are written in an accessible style and are relatively short, including real-world examples. All chapters draw on original academic research, and many of them refer to specific visualization projects and practices. All contribute to academic and public conversation about data visualization in society.

References

Ambrosio, C. (2015). Objectivity and representative practices across artistic and scientific visualization. In: A. Carusi, A. S. Hoel, T. Webmoor, & S. Woolgar (Eds.), *Visualization in the age of computerization* (pp. 118-144). London: Routledge.

Borgo, R., Kehrer, J., Chung, D. H. S., Maguire, E., Laramee, Robert, S., Hauser, H., Ward, M., & Chen, M. (2013). Glyph-based visualization: Foundations, design guidelines, techniques and applications. *Eurographics State of the Art Reports*, 2013, 39-63. Retrieved from http://diglib.eg.org/EG/DL/conf/EG2013/stars/039-063.pdf

Barnhurst, K. G. (1994). *Seeing the newspaper.* New York: St. Martin's Press.

Cairo, A. (2013) *The functional art: An introduction to information graphics and visualization.* Berkeley, CA: New Riders.

Chouliaraki, L., & Fairclough, N. (1999). *Discourse in late modernity: Rethinking critical discourse analysis.* Edinburgh: Edinburgh University Press.

Couldry, N. (2016). Foreword. In: S. Kubitschko & A. Kaun (Eds.), *Innovative methods in media and communication research.* (pp. i-viii). Cham: Palgrave Macmillan.

Damasio, A. R. (2006). *Descartes' error: Emotion, rationality and the human brain.* London: Vintage.

Engebretsen, M., Kennedy, H., Weber, W. (2017). Visualization practices in Scandinavian newsrooms: A qualitative study. *21st International Conference Information Visualisation (IV)*, 297-300. http://doi.org/10.1109/iV.2017.54).

Fairclough, N. (2010). *Critical discourse analysis: The critical study of language.* London: Routledge.

Few, S. (2008, August). What ordinary people need most from information visualization today. *Perceptual Edge: Visual Business Intelligence Newsletter.* Retrieved from http://www.perceptualedge.com/articles/visual_business_intelligence/what_people_need_from_infovis.pdf

Friendly, M. (2008). A brief history of data visualization. In: C.-H. Chen, W. Härdle, & A. Unwin (Eds.), *Handbook of data visualization.* (pp. 15-56). Berlin: Springer.

Halliday, M. A. K. (1978). *Language as social semiotic: The social interpretation of language and meaning*. London: Arnold.
Jaggar, A. M. (1989). Love and knowledge: Emotion in feminist epistemology. *Inquiry, 32*(2), 151-176. https://doi.org/10.1080/00201748908602185
Kennedy, H., & Allen, W. (2016). Data visualisation as an emerging tool for online research. In: N. G. Fielding, R. M. Lee, & G. Blank (Eds.), *The Sage handbook of online research methods* (2nd ed.). (pp. 307-326). London: Sage.
Kennedy, H., & Hill, R. L. (2017). The feeling of numbers: Emotions in everyday engagements with data and their visualisation. *Sociology, 52*(4), 830-848. https://doi.org/10.1177/0038038516674675
Kennedy, H., Hill, R. L., Allen, W., & Kirk, A. (2016a). Engaging with (big) data visualizations: Factors that affect engagement and resulting new definitions of effectiveness. *First Monday, 21*(11). https://doi.org/10.5210/fm.v21i11.6389
Kennedy, H., Hill, R. L., Aiello, G., & Allen, W. (2016b). The work that visualisation conventions do. *Information, Communication and Society, 19*(6), 715-735. https://doi.org/10.1080/1369118X.2016.1153126
Kirk, A. (2016). *Data visualisation: A handbook for data driven design*. London: Sage.
Kitchin, R. (2014). *The data revolution: Big data, open data, data infrastructures and their consequences*. London: Sage Publications.
Kress, G., & van Leeuwen, T. (1996) *Reading images: The grammar of visual design*. London: Routledge.
Latour, B. (1986). Visualization and cognition: Drawing things together. *Knowledge and Society, 6*, 1-40.
Lemke, J. (2015). Feeling and meaning: A unitary bio-semiotic account. In: P. P. Trifonas (Ed.), *International handbook of semiotics*. (pp. 589-616). New York & London: Springer.
Mayer-Schönberger, V., & Cukier, K. (2013). *Big data: A revolution that will transform how we live, work, and think*. Boston, MA: Houghton Mifflin Harcourt.
Porter, T. M. (1995). *Trust in numbers: The pursuit of objectivity in science and public life*. Princeton: Princeton University Press.
van Leeuwen, T. (2005). *Introducing social semiotics*. London & New York: Routledge.

About the authors

Helen Kennedy is Professor of Digital Society at the University of Sheffield. Her research traverses digital landscapes and is currently focused on datafication in everyday life. Her Seeing Data (seeingdata.org) research into how non-experts relate to data visualizations provides the inspiration for many contributions in this book.

Martin Engebretsen is Professor of Language and Communication at the University of Agder, Norway, and director of the INDVIL project (indvil.org), which also provides the inspiration for this book. His research areas include text and discourse studies, multimodality, digital journalism and visual communication.

Section I

Framing data visualization

2. Ways of knowing with data visualizations

Jill Walker Rettberg

Abstract
Data visualizations combine numeric data with visual representation, and these modes allow them to express certain kinds of knowledge more easily than others. This chapter uses examples of historical data visualizations in order to examine what ways of knowing they privilege. What is the difference between the spatial organization of tools in prehistoric homes and a photograph or bar chart showing information about the same tools, in terms of the kinds of knowledge they enable? How do the systems for gathering and visualizing data during the 18th and 19th centuries shape our understanding of the world? How do data visualizations make us feel that they are objective? How do they shape our ideas of what is possible?

Keywords: Dataism; God trick; Desire for numbers; Correlation and causation; The sublime; Epistemology of data visualization

Introduction

Data visualizations combine at least two modes of representation: numerical data and visual diagrams. For a computer program to be able to process data, it has to be converted to numbers, to the zeros and ones of machine code. In addition, the data need to be visually organized, which often requires dividing them into discrete quantities where lines, size, spatial placement, and other visual elements show certain patterns in the data. Each of these two modes of expression, the numeric and the visual, carries its own affordances and constraints for what they can express.

This anthology has several chapters that use concrete examples to discuss how data visualizations can be biased in their representations of

Engebretsen, M. and H. Kennedy (eds.), *Data Visualization in Society*.
Taylor & Francis Group, 2020
DOI 10.5117/9789463722902_CH02

data (Ricker, Kraak, & Engelhardt, this volume; D'Ignazio & Bhargava, this volume) or how data visualizations can work against the typical abstraction they entail to include individuals' stories (Alamalhodaei, Alberda, & Feigenbaum, this volume). My emphasis in this chapter is on examining the underlying mechanisms of data visualizations as an assemblage of data and visualizations. My exploration sits alongside existing critical work on data visualizations in feminist scholarship (D'Ignazio & Klein, 2016; Hill, Kennedy, & Gerrard, 2016), in the digital humanities (Drucker, 2011, 2014; Gitelman, 2013), and in critical algorithm studies and other scholarship on the epistemological basis for algorithmic processing of big data (Eubanks, 2018; Gillespie & Seaver, 2015; Noble, 2018).

Visual organization

Organizing objects visually and spatially is something humans and our ancestors have done for a long time. In her essay 'Visualizing Thought', Barbara Tversky describes how hominins living three-quarters of a million years ago organized their tools and belongings in different areas of their home. She argues that this is the basic precursor to any kind of visualization: 'Perhaps the simplest way to use space to communicate is to arrange or rearrange things in it. An early process is grouping things in space using proximity, putting similar things in close proximity and farther from dissimilar things' (Tversky, 2010, p. 504). We might extend Tversky's line of reasoning to the modern domestic habit of keeping forks in one partition of a kitchen drawer and knives in another, and argue that this is a way of visually and spatially communicating information about the forks and knives.

The data visualizations we see on computer screens or printed pages, or even early markings on stones or in the sand, are one step removed from the phenomena they represent or organize. If we walk into somebody's kitchen and open a drawer, we *see* the knives and forks in the kitchen drawer, but we also experience them in space, and we can touch them and pick them up. Now, imagine a data visualization about kitchen utensils on a screen. It could be very simple, showing the number of knives and forks and other utensils in a kitchen, perhaps organized as a bar chart, perhaps using little pictures of forks stacked up in one bar and knives in another to show the relative quantities. Or imagine a photograph of the kitchen drawer, or an Instagram-style flat lay photograph of all the knives and forks neatly laid out on a table and photographed from above.

Once the knives and forks are transferred from spatially organized objects to a visual representation on a two-dimensional surface, our distance from them increases. We interpret them as separate from us. A photograph of the drawer might not encourage a great deal of analytical dissection of the image, but the neatly organized flat lay photograph and the bar graph prioritize an analytic approach to that which is represented.

In his influential book about the transition from oral to literate cultures, Walter Ong (1982) argues that a fundamental difference between orality and literacy is that the visual nature of writing leads to ideas of objectivity that are impossible in oral culture. When we speak to each other in a face-to-face conversation, we are immersed in the sound, and because the speakers are in the same physical space, face-to-face oral discourse tends to be situated and concrete. Writing, on the other hand, separates the knower from the known. There is a distance between reader and writer. 'Sight isolates', Ong writes, while 'sound incorporates. Whereas sight situates the observer outside what he views, at a distance, sound pours into the hearer' (1982, p. 45). A typical visual ideal is clarity and distinctness, a taking apart, Ong argues, whereas the auditory ideal, by contrast, is harmony, a putting together (p. 71). He writes: 'A sound-dominated verbal economy is consonant with aggregative (harmonizing) tendencies rather than with analytic, dissecting tendencies (which would come with the inscribed, visualized word: vision is a dissecting sense)' (p. 73).

Ong does not discuss visualizations or diagrams, but following his reasoning, we can see a similar transition from the spatial organization of objects to the visual representation of objects on a page or other flat surface. Think back to the drawer of knives and forks as a way of organizing data: when there are real knives and real forks, the human is able to pick up a knife or a fork, move them around, manipulate them. Touch, like sound, involves closeness and participation. But the moment we switch from a physical drawer to a visual representation of a drawer, we are placed outside the representation, as analytical observers who feel an objective distance from what is seen. At least, this is true if we follow Ong, and not all would: Jonathan Sterne, for instance, criticizes Ong's framework as too simple a binary, too closely based on theological distinctions about the meaning of 'the word', and too little grounded in existing anthropological research on oral and literate cultures (Sterne, 2011). But whether or not Ong's framework is too simplistic, the basic idea that visual representation can lead to a more analytic approach is also expressed by other scholars coming from very different angles. Tversky also emphasizes analysis as key in visualization, but for her it is the persistence of images, that is, that they are not fleeting like the spoken word, that allows the 'perceptual processes' to occur that are needed for 'understanding, inference, and insight':

> Because [images] persist, they can be subjected to myriad perceptual processes: Compare, contrast, assess similarity, distance, direction, shape, and size, reverse figure and ground, rotate, group and regroup; that is, they can be mentally assessed and rearranged in multiple ways that contribute to understanding, inference, and insight. (Tversky, 2010, p. 500)

Systematizing data

Importantly, not only the visual, but also the data themselves share much of this promise of analytical objectivity. Data visualization had a golden age in the nineteenth century, at the same time as nation states began large-scale collection of statistical data (Friendly, 2006). However, it began a few centuries earlier, at the same time as the scientific method was developing, and with it the idea that humans could precisely observe the world and use those observations to understand it. Seventeenth- and eighteenth-century Europe saw an increasing trend towards observation, measurement, and quantification, and different fields developed new ways of measuring and quantifying things that had not previously been seen as interesting. Some of these methods were technological. For instance the invention of the telescope allowed Galileo to make observations about the solar system that would not previously have been possible. In our time, the existence of precise sensors and of computers that can process massive amounts of data allows for certain types of measurement, analysis, and visualization that were not possible a few decades ago.

Social and organizational changes also led to new kinds of quantification. National registries became common during the nineteenth century, for instance, allowing for analysis of trends over time or the comparison of different regions. For example, the first centralized national system of crime reporting was instituted in France in 1825, and collected information about all charges made in French courts on a quarterly basis (Friendly, 2006, p. 25). More and more information was collected, and by the end of the nineteenth century the French police not only had detailed statistics about crimes, but also systems for documenting and identifying criminals and suspects using a system of 'anthropometrics', devised by Alphonse Bertillon and involving very specific measurements of body parts (Kember, 2014). Once one has such a system, once it is possible to gather data that appears to give us knowledge, we end up with what Helen Kennedy calls a 'desire for numbers' that can lead to a lack of critical reflection about what those numbers mean and whether we truly need them (2016, p. 51).

This sense that systematized data have authority is an important aspect of the rhetorical power of data visualizations. While Ong and Tversky emphasized the *visual* as allowing for an analytical and perhaps objective stance, many have argued that it is the *data* themselves, or the quantitative nature of data visualizations, that lend them this sense of authority.

A perception of objectivity

According to Anthony McCosker and Rowan Wilken (2014), data visualizations often offer a 'fantasy of knowing' or of 'total knowledge', or in Donna Haraway's words, they claim to present a 'God's eye view' (Haraway, 1988, p. 581). The use of a data visualization in a newspaper article or a corporate report carries with it a rhetorical weight: the simple presence of the data visualization seems to state 'Look, we have data. This is true' (see Tal, Aner, & Wansink, 2016 on data visualization's association with truthfulness).

José van Dijck uses the term dataism to describe the ideology of big data, which is characterized by 'a widespread belief in the objective quantification and potential tracking of all kinds of human behavior and sociality through online media technologies' (2014, p. 198). Epistemologically, data visualizations build upon this trust in data.

We can trace many histories of society's growing trust in numbers. The registration of data about crimes and criminals mentioned above tells of one such history, which can be traced forwards to today's bodycams, surveillance, and biometrics (Gates, 2011). Another, parallel history is that of the transition from midwives and their home-based care of mothers and infants to the increasing medicalization of prenatal care. This story can be told as a transfer of power from women to men, but it can also be seen as a transfer of trust from humans to machines, as the increasing institutionalization of prenatal and infant care included a radical growth in the use of technology to monitor growth and health (Oppenheimer, 2013). Today, iPhone apps connect to digital scales that generate daily data visualizations of a baby's weight (Rettberg, 2014, p. 67) and smart socks generate continuous visualizations of a baby's heartbeat (Leaver, 2017).

The management of birth is one thread in this story of numbers. Another thread is the management, or perhaps rather the marketing, of instruments of death, as told by Donald Mackenzie in *Inventing Accuracy: A Historical*

Sociology of Nuclear Missile Guidance (1993). Or we might consider the prevention of life, a thread of the story told by Michelle Murphy in *The Economization of Life* (2017), where she discusses how demographic models comparing population size and financial growth created programmes intended to improve the future economies of developing countries through extensive birth control and abortion programmes.

The average as norm

Displaying data visually rather than as a table of numbers is a powerful method for finding patterns in the data. Some patterns recur in many different datasets, such as the bell-shaped curve seen in Figure 2.1, a graph showing the heights of Belgian men, which follows what is mathematically known as a normal distribution. Writing in the 1860s, Adolphe Quetelet interpreted this recurrence as evidence of a fundamental social law, and defined the central portion of the curve as 'normal', with those outside the normal zone seen as aberrations (1997). Sekula explains that '[t]hus conceived, the "average man" constituted an ideal, not only of social health, but of social stability and of beauty' (1986, p. 22). Quetelet's work leaned heavily upon data visualizations. He first showed his data in the form of a table, then showed it visualized, drawing conclusions from the patterns that became apparent when the numbers were shown as curves on an x- and y-axis.

The power of visualizations to show averages and patterns contributed to the nineteenth-century privileging of the 'norm', or as Lennard Davis argues, a 'generalized notion of the normal as an imperative', where 'the average then paradoxically becomes a kind of ideal, a position to be wished' (Davis, 2013, p. 2). This privileging of the average is a marked break from earlier traditions that saw the ideal body, represented for instance in paintings of Venus, as something 'mytho-poetical', a 'divine body' that is 'not attainable by a human' (Davis, 2013, p. 2).

As it turns out, the average human doesn't exist. Yes, that even curve shape shown in Figure 2.1 does show up again and again when you measure almost any aspect of humans—or of most things, really. But that doesn't mean that any individual human is 'average'. In her book *Technically Wrong* (2017), Sara Wachter-Boettcher tells the story of how the adjustable seatbelt was designed. Prior to its invention, the air force planned to design cockpits that fit 'the average pilot'—but they discovered that *none* of their pilots were of average size in all the ten dimensions they measured, such as height,

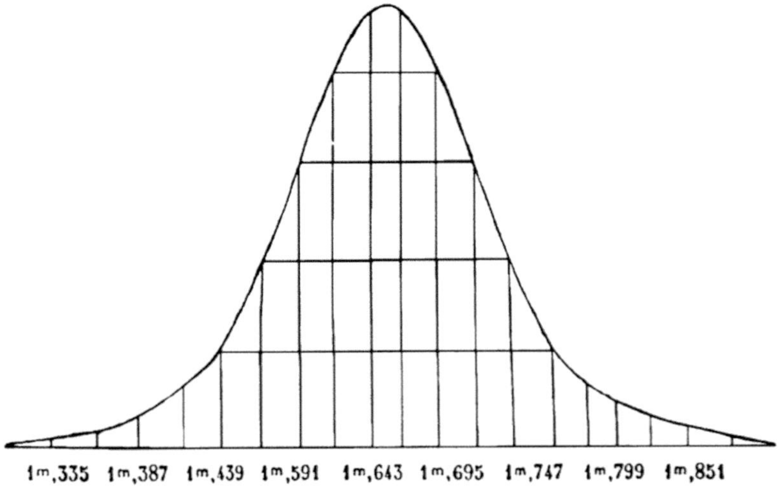

Figure 2.1. The height of Belgians from 18 to 20 years. Reprinted from *Physique sociale ou Essai sur le développement des facultés de l'homme* (p. 355), by A. Quetelet, 1997 [1869], Brussels: Académie Royale de Belgique. Copyright 1997 by Académie Royale de Belgique. Reprinted with permission.

wrist circumference, and shoulder width. Wachter-Boettcher uses this point to argue that it's important to design technology that fits people at each extreme rather than for the average person, as the air force did by creating adjustable seats and seat belts (Wachter-Boettcher, 2017). The idea of 'the average' may be encouraged by data visualizations, but that doesn't mean that it's necessarily the most useful way of viewing the data.

Correlation is easier than causation

Computers are extremely good at finding correlations. In fact, this is one of the mainstays of current models of deep machine learning, where software is fed 'big data' and works through it to find patterns. By analysing historical data, computers can find patterns that allow them to predict future behaviour. Often these predictions are eerily accurate. In some tests, AI systems do a better job at medical diagnosis than human doctors (Olson, 2018). It is wise to remember, though, that many stakeholders have a strong financial interest in convincing the general public that AI is efficient, perhaps more efficient than humans, and AI's ability to make accurate predictions is often overstated.

Visualizations of data also prioritize correlation over causation. They show patterns and relative size or position, but it is more difficult to show causality. Viktor Mayer-Schönberger and Kenneth Cukier argue in their book *Big Data* (2013) that we no longer need causality. If we can predict how likely patients are to take their medicine based on their car insurance payment history, why would we want or need to know *why* they don't take their medicine, Mayer-Schönberger and Cukier ask. Obviously their payment history doesn't *cause* their tendency to take or not take medicines as prescribed. But it no longer matters. Causality for them is simply 'human intuiting' that doesn't deepen our insight, it is merely a 'cognitive shortcut that gives us the illusion of insight but in reality leaves us in the dark about the world around us' (2013, p. 64). Others are more concerned about the downplay of causality, as Wendy Chun writes: 'Big data [...] offers a form of cognitive mapping that allegedly sees all, by ignoring causes' (2017, p. 56).

Different forms of representation emphasize different relationships and patterns. Quetelet's data visualizations contributed to the idea of the average as something to be sought after, whereas earlier forms of representation, such as paintings, were well-suited to representing ideal beauty as something beyond human perfection. Data visualizations prioritize correlation. Narrative, by contrast, is a form of representation that often emphasizes causal connections. Narratives organize events in time. Some also provide causal connections between the events. E. M. Forster argues that such connections separate a story, which is just events in time ('and then, and then'), from a plot, which adds causality. '"The king died and then the queen died," is a story. "The king died, then the queen died of grief" is a plot,' Forster wrote (1949, p. 82). Roland Barthes, on the other hand, argued that 'the mainspring of narrative' is the reader's assumption that an event that happens *after* another event is *caused by* the first event, meaning that 'narrative would be a systematic application of the logical fallacy [...] *post hoc, ergo propter hoc*' (1977, p. 94). Causation is not always evident, but different forms of representation emphasize causation or correlation in different ways. Visualizations do not usually portray narratives, although this is certainly possible, as discussed by Wibke Weber and others in this volume.

One important extension of the correlation/causation binary is that the algorithmic processing of data that lies behind data visualizations often use *proxies*. Often, we cannot measure the things we are really interested in, so we find something that we can measure and that we assume has a direct relationship to the thing we actually want to understand. For instance, we don't have a way of directly measuring human emotions. Yet developers, or at least the marketers of their products, appear confident that using machine vision algorithms to analyse facial expressions can tell us that somebody is 99% angry and 0.5% sad, for instance. In this case, the facial expressions are *proxies* that are presumed to correlate perfectly with a person's emotions, although this assumption builds upon psychological theories that were arguably outdated decades ago (Bjørnsten & Zacher Sørensen, 2017). We measure what we *can* measure and make claims based on that.

In other cases, maybe data *could* have been measured, but they were not, and so the datasets are incomplete. Machine learning can find correlations that appear to be valid in imperfect datasets. A useful example, discussed by economist Sendhil Mullainathan and medical researcher Ziad Obermeyer, demonstrates how machine learning in healthcare, despite excelling at 'predicting outcomes *y* based on inputs *x*', can lead to misleading or biased predictions. This is because inputs such as medical records and insurance claim data suffer from 'large and systematic mismeasurement' (2017, p. 476). They give the example of predictors for having a stroke. It is often difficult to tell if patients arriving at a hospital are at risk of having a stroke, so a team used machine learning to analyse historical patient data in order to find factors in their medical history that are predictors of likelihood of having a stroke. On the surface, such a 'prediction problem' doesn't need to prove causal connections, since the goal is simply to plan for a more efficient use of resources, allocating more resources to patients with a higher risk of having a stroke. But although the machine learning algorithm had a lot of patient data, it did not have all the necessary data, because a lot of information about patients does not end up in their medical journals. The algorithm found that statistically valid predictors for having a stroke included having been treated for a minor injury due to a fall, or for acute sinusitis, or having had a scan for colon cancer. Upon closer inspection, human researchers found that the minor injuries and scans were in fact *proxies* for patients who were likely to go to the doctor for relatively minor issues. These patients were more likely than the general population to have a stroke *diagnosed* by a doctor, though not necessarily more likely to actually have a stroke, as many strokes are not diagnosed. Such skewed data can easily end up in well-intended data visualizations.

Phantasmagrams and affect

Sometimes, data visualizations are used to make predictive claims or arguments that can shape our understanding of the world. This can happen in a conceptual manner, as when Quetelet used data visualizations to develop the idea of the average as ideal, or in a more concrete way, as when a data series shows an increase or a decrease and the visualization suggests that this trend will continue into the future. This predictive use of data visualization is becoming more automated in systems such as those offered by Palantir and other companies for risk prediction. For instance, in predictive policing, police departments have a live map of their district with percentages and colour codes showing places where there is a high risk of certain crimes occurring, based on data analysis of past crimes as well as data such as local weather reports and the school calendar. When data visualizations make claims about the future, they can also affect the future, and we should be wary of how they do so.

Michelle Murphy has used the term *phantasmagram* to describe the way that 20th-century economic and demographic models became not just descriptions of how the world works, but projections that lived lives of their own. She compares them to the phantasmagoria of the nineteenth century, 'ghostly simulations made by whirling magic lanterns that stimulated fright and awe' (Murphy, 2017, p. 53). She argues that demographic models are phantasmagrams, models that created new ways of seeing the world:

> Through the work of Keynes and other similarly minded macroeconomists, the national economy was explicated as a new aggregate kind, a collective blur of activity that nonetheless could be modeled as a set of predictable correlations, tendencies, forces, and rates representable in equations and graphs. When interest rates go up, investment goes down, employment drops, output falls. With equations and diagrams, mathematical modelling in the 1930s performatively discerned 'the economy' as a constellation of such interrelationships within a closed system whose boundary was the nation-state. (Murphy, 2017, p. 18)

The very idea of it being possible to measure the entire economic performance of a country as its Gross Domestic Product (GDP) is a phantasmagram, Murphy argues, which will always leave things out (unpaid labour, for instance) and miscount other components. GDP is an example of 'quantitative practices that are enriched with affect, propagate imaginaries, lure feeling, and hence have supernatural effects in surplus of their rational precepts' (2017, p. 24). The success of such a model lay 'not in its empirical veracity but in the way it gave form to a technocratic dream of a national macroeconomy that could be fostered, directed, and triggered by rearranging reproduction en mass,' Murphy argues (2017, p. 51).

We are used to thinking of quantitative models or visualizations as objective and rational. This is what José van Dijck calls dataism, as noted above. For Murphy to instead highlight the *affect* and even the *sublime* of economic models (2017, pp. 9, 23) is a very different approach that may seem at odds with our everyday experience of models, graphs, and other data visualizations. Perhaps, as Helen Kennedy and Rosemary Hill note, it is the very combination of the 'statistical *and* visual' in data visualizations that leads to their emotional impact (2017, p. 831). Kennedy and Hill discuss a range of emotional responses that participants in their focus groups expressed when looking at data visualizations. Here, I will discuss the sense of the sublime that Murphy touches upon.

The sublime is an old concept, used first by Longinus around 2000 years ago. For Longinus, the sublime was a rhetorical technique used in a speech to 'overcome the rational powers' of an audience (Longinus, 100 CE/1935). While Longinus theorized the sublime as a rhetorical technique for influencing people, Kant saw it as a human response to grandeur in art or nature. His concept of the mathematical sublime is awakened in us when we sense something that is absolutely large: it isn't of a specific size, it is great without comparison, so we can't grasp it mathematically. The sublime, for Kant, lies not in the object but in our experience of it. If you gaze at the night skies, or consider undying love or loyalty, then you may experience the sublime. Combining Kant and Murphy's ideas, then, we might say that the vastness of the idea of GDP, of being able to compute and visualize all the economy of all the world, also awakens this sense of the sublime.

The pleasure of the sublime, Kant writes, lies in the sense that our mind is broadened by this experience of the infinite that allows us to 'pass beyond the narrow confines of sensibility' (2007, p. 256). This sounds close to the reaction that designer Jer Thorp says he aims for when he designs a data visualization: 'First, it needs to be visually pleasing. I want people to say 'Oooh…!' when they turn the page to it. Once they're hooked, though, I want them to learn something—the 'Aaah!' moment' (2010). The initial pleasure should give way to rational understanding. Although some have criticized the obsession with the visual beauty of data visualizations (McCosker & Wilken, 2014), Kant's idea of the sublime as something that can lead to a deeper understanding can also be seen as aligned with ideas of embodied knowledge and the role of emotions and the senses in knowledge. Affect and emotion can offer us kinds of knowledge that are not directly accessible through purely rational analysis.

Data visualizations are not simply visual and they are not simply quantitative. They are a form of communication that emphasizes data. Sometimes it is the very fact that they present reality as understandable and predictable through data models that makes data visualizations so convincing.

References

Barthes, R. (1977). *Image music text.* (S. Heath, Trans.). London: Fontana Press.

Bjørnsten, T. B., & Zacher Sørensen, M.-M. (2017). Uncertainties of facial emotion recognition technologies and the automation of emotional labour. *Digital Creativity, 28*(4), 297-307. https://doi.org/10.1080/14626268.2017.1383271

Chun, W. H. K. (2017). *Updating to remain the same: Habitual new media.* Cambridge, MA: MIT Press.

Davis, L. J. (2013). Introduction: Disability, normality, and power. In: L. J. Davis (Ed.), *The disability studies reader.* (pp. 1-17). Florence, KY: Taylor & Francis.

D'Ignazio, C., & Klein, L. (2016, October). *Feminist Data Visualization.* Paper presented at and published in the workshop proceedings from the Workshop on Visualization for the Digital Humanities at IEEE VIS Conference, Baltimore, Maryland.

Drucker, J. (2011). Humanities approaches to graphical display. *Digital Humanities Quarterly, 5*(1). Retrieved June 12, 2019 from http://www.digitalhumanities.org/dhq/vol/5/1/000091/000091.html

Drucker, J. (2014). *Graphesis: Visual forms of knowledge production.* Cambridge, MA: Harvard University Press.

Eubanks, V. (2018). *Automating inequality: How high-tech tools profile, police, and punish the poor.* New York: St. Martin's Press.

Forster, E. M. (1949 [orig. 1927]). *Aspects of the novel* (Pocket ed.). London: Edward Arnold & Co.

Friendly, M. (2008). A brief history of data visualization. In: C.-H. Chen, W. Härdle, & A. Unwin (Eds.), *Handbook of data visualization.* (pp. 15-56). Berlin: Springer.

Gates, K. (2011). *Our biometric future: Facial recognition technology and the culture of surveillance.* New York: NYU Press.

Gillespie, T., & Seaver, N. (2015, November 5). Critical Algorithm Studies: a Reading List [Blog post]. Retrieved from https://socialmediacollective.org/reading-lists/critical-algorithm-studies/

Gitelman, L. (Ed.). (2013). *Raw data is an oxymoron.* Cambridge, MA: MIT Press.

Haraway, D. (1988). Situated knowledges: The science question in feminism and the privilege of partial perspective. *Feminist Studies, 14*(3), 575-599. https://doi.org/10.2307/3178066

Hill, R. L., Kennedy, H., & Gerrard, Y. (2016). Visualizing junk: Big data visualizations and the need for feminist data studies. *Journal of Communication Inquiry, 40*(4), 331-350. https://doi.org/10.1177/0196859916666041

Kant, I. (2007). *Critique of judgement* (J. C. Meredith, Trans.; N. Walker, Ed.). Oxford & New York: Oxford University Press.

Kember, S. (2014). Face recognition and the emergence of smart photography. *Journal of Visual Culture, 13*(2), 182-199. https://doi.org/10.1177/1470412914541767

Kennedy, H. (2016). *Post, mine, repeat: Social media data mining becomes ordinary.* London: Palgrave Macmillan Limited.

Kennedy, H., & Hill, R. L. (2017). The feeling of numbers: Emotions in everyday engagements with data and their visualisation. *Sociology, 52*(4), 830-848. https://doi.org/10.1177/0038038516674675

Leaver, T. (2017). Intimate surveillance: Normalizing parental monitoring and mediation of infants online. *Social Media + Society, 3*(2). https://doi.org/10.1177/2056305117707192

Longinus. (1935 [ca. 100 CE]). *On the sublime* (W. R. Roberts, Trans.) Cambridge: Cambridge University Press.

Mackenzie, D. (1993). *Inventing accuracy: A historical sociology of nuclear missile guidance.* Cambridge, MA: MIT Press.

Mayer-Schönberger, V., & Cukier, K. (2013). *Big data: A revolution that will transform how we live, work, and think.* Boston, MA: Houghton Mifflin Harcourt.

McCosker, A., & Wilken, R. (2014). Rethinking 'big data' as visual knowledge: The sublime and the diagrammatic in data visualisation. *Visual Studies, 29*(2), 155-164. https://doi.org/10.1080/1472586X.2014.887268

Mullainathan, S., & Obermeyer, Z. (2017). Does machine learning automate moral hazard and error? *American Economic Review, 107*(5), 476-480. https://doi.org/10.1257/aer.p20171084

Murphy, M. (2017). *The economization of life.* Durham, NC: Duke University Press.

Noble, S. U. (2018). *Algorithms of oppression: How search engines reinforce racism.* New York: NYU Press.

Olson, P. (2018, June 28). This AI just beat human doctors on a clinical exam. *Forbes.* Retrieved from https://www.forbes.com/sites/parmyolson/2018/06/28/ai-doctors-exam-babylon-health/

O'Neil, C. (2017). *Weapons of math destruction: How big data increases inequality and threatens democracy.* London: Penguin.

Ong, W. J. (1982). *Orality and literacy: The technologizing of the word.* London: Routledge.

Oppenheimer, G. (2013). Medical care of infants in the United States: A review of recent work. In: R. Greene (Ed.), *History of medicine.* (pp. 103-128). Florence, KY: Taylor and Francis.

Quetelet, A. (1997). *Physique sociale ou Essai sur le développement des facultés de l'homme* [*1869*]. (É. Vilquin & J.-P. Sanderson, Eds.). Brussels: Académie Royale de Belgique.

Rettberg, J. W. (2014). *Seeing ourselves through technology: How we use selfies, blogs and wearable devices to see and shape ourselves.* Basingstoke: Palgrave.

Sekula, A. (1986). The body and the archive. *October, 39*, 3-64. https://doi.org/10.2307/778312

Sterne, J. (2011). The theology of sound: A critique of orality. *Canadian Journal of Communication, 36*(2). https://doi.org/10.22230/cjc.2011v36n2a2223

Tal, A., & Wansink, B. (2016). Blinded with science: Trivial graphs and formulas increase ad persuasiveness and belief in product efficacy. *Public Understanding of Science*, 25(1), 117-125.

Thorp, J. (2010, July 12). Wired UK, Barabási Lab and Big Data [Blog post]. Retrieved from http://blog.blprnt.com/blog/blprnt/wired-uk-barabasi-lab-and-big-data

Tversky, B. (2010). Visualizing thought. *Topics in Cognitive Science, 3*(3), 499-535. https://doi.org/10.1111/j.1756-8765.2010.01113.x

van Dijck, J. (2014). Datafication, dataism and dataveillance: Big Data between scientific paradigm and ideology. *Surveillance & Society, 12*(2), 197-208. http://doi.org/10.24908/ss.v12i2.4776

Wachter-Boettcher, S. (2017). *Technically wrong: Sexist apps, biased algorithms, and other threats of toxic tech.* New York: Norton.

About the author

Jill Walker Rettberg is Professor of Digital Culture at the University of Bergen, and principal investigator of an ERC (European Research Council) project that explores how machine vision and new algorithmic images are affecting us as a society and as individuals. She is the author of *Blogging* (Polity, 2014) and *Seeing Ourselves Through Technology* (Palgrave, 2014).

3. Inventorizing, situating, transforming: Social semiotics and data visualization

Giorgia Aiello

Abstract

This chapter is an overview of social semiotics as a productive framework for research on data visualization. It provides conceptual instruments that can be used to explore the relationship between the formal properties of data visualization and the meanings and practices that these may promote or hinder among users. In particular, the chapter argues that a social semiotic framework can be used to *inventorize*, *situate*, and *transform* visualization resources. Overall, it links descriptive, interpretive, and critical objectives to generate a framework aimed at understanding how data visualization 'works' from a formal standpoint, what meanings are consistently associated with particular semiotic resources, and how both key semiotic 'rules' and dominant meanings may be questioned and changed.

Keywords: Social semiotics; Data visualization; Semiotic resources; Visualization design

Introduction

This chapter is a focused critical overview of social semiotics as a productive framework for research on data visualization. It aims to provide conceptual instruments that can be used to explore the relationship between the formal properties of data visualization and the kinds of responses, engagements, and practices that these may promote or hinder among users. Over the last decade or so, and in the wake of digitalization and datafication, data visualization has emerged rapidly as what Engebretsen and Weber (2017) have defined as a 'super-genre' that is used to accomplish a wide variety of communicative tasks across an increasing number of professional and

Engebretsen, M. and H. Kennedy (eds.), *Data Visualization in Society*.
Taylor & Francis Group, 2020
DOI 10.5117/9789463722902_CH03

institutional communities of users. Beer and Burrows (2013) highlight that, as a whole, we have witnessed the rise of a 'visualization of culture' and a 'culture of visualization' across spheres of social activity and cultural production, as 'there are not many things that have yet to be visualized and archived' (p. 62). Different kinds of data visualization have become privileged *signs* to mark the rationality of particular processes and promote specific attitudes towards various aspects of everyday life, ranging from policymaking to personal productivity. As Ledin and Machin (2018) point out, often diagrams, charts, and other types of visualization are used not only to illustrate how things are but also, above all, 'how things should be done' (p. 335).

Precisely because of the increasing social significance of this phenomenon, there is a growing body of academic literature centred on critical, practical, and combined approaches to the formal and overall aesthetic qualities of data visualization. Generally speaking, these approaches offer very useful insights to examine data visualization design from an ideological, professional, or praxis-based standpoint. On the one hand, it has become increasingly urgent to examine what Kennedy and Hill (2017) define as the 'visual sensibilities' (p. 2) that are at work in the ways in which ordinary people respond culturally and engage emotionally with data and their visualizations. On the other hand, professional and institutional uses of data visualization techniques must be examined in the light of their underlying histories, conventions, and changes over time and across contexts. For these reasons, a detailed appraisal of data visualization's main semiotic resources, or its tools for meaning-making, is key to empirical research in this field. Unlike other currently more widespread approaches to data visualization research rooted in cultural and social theory, a social semiotic approach focuses keenly on the formal properties of visualizations together with their semiotic and social affordances.

As I will explain later, a social semiotic approach entails a systematic mapping of semiotic resources together with an empirical if not ethnographic investigation of how such resources came to be the way they are, how they are used or understood by a variety of individuals and groups of people, and how they are shaped by dominant practices and regulated by given institutions. It is in this sense that social semiotics is inherently critical, as it relates texts to contexts to reflect on the social and political implications of meaning-making. However, social semioticians are concerned not only with the *politics* but also with the *potentials* of semiosis. Ultimately, one of the major aims of social semiotics is to contribute to semiotic innovation, or envision ways in which the 'rules' of sign-making may be broken or changed (van Leeuwen, 2005). This matters because semiotic innovation can contribute to engendering social change. In this chapter, I therefore argue that a social semiotic framework of this kind can and ought to be extended further to *inventorize, situate,* and *transform* the semiotic resources associated with data visualization.

To explore relevant conceptual tools that are central to social semiotics as a mode of inquiry, then, the chapter begins with a broad discussion of the methodological dimensions of social semiotics, together with an initial discussion of existing scholarship in this area. I then delve into three main theoretical and analytical areas. First, I outline some of the major sources and methods that we can harness to begin inventorizing data visualization resources. In doing so, I review a selection of analyses of relevant multimodal semiotic artefacts and technologies such as diagrams (Ledin & Machin, 2016a and 2016b; Bateman et al., 2017), infographics (Bateman et al., 2017; Amit-Danhi & Shifman, 2018), and PowerPoint (Djonov & van Leeuwen, 2013; Zhao et al., 2014). Second, I explain how data visualization resources can be situated in their contexts, particularly through historical and ethnographic approaches. Finally, I advance the idea that social semiotics can contribute to transforming data visualization resources. The overall aim here is to link descriptive, interpretive, and critical objectives to generate a framework aimed at understanding how data visualization 'works' from a formal standpoint, what meanings are consistently associated with particular semiotic resources, and how both key semiotic 'rules' and dominant meanings may be questioned if not changed.

Why social semiotics?

Critiques of data visualization often focus on the truth-making claims and related epistemological implications of its design (see Halpern, 2015). For example, recently Gray et al. (2016) explored some of the ways in which data visualization's 'ways of seeing' and 'ways of knowing' can be understood in relation to 'the aesthetics, cultures, values, ideals and practices associated with their production' (p. 294). When it comes to research on the visual and multimodal detail of data visualization, there is still a predominance of practice-based research. Edward Tufte's groundbreaking work on the design norms underlying the visual display of information has been both widely criticized and surpassed by technological and cultural changes in how visualizations are both produced and used (see Tufte, 1983, 1997). Levels of interest in research on the 'good practices' of data visualization design have grown among praxis-oriented thinkers. In his data visualization handbook, for example, Andy Kirk (2016) offers guidance on the development of design solutions across the 'five layers of the visualisation design anatomy' (p. 145), which he defines as data representation, interactivity, annotation, colour, and composition.

Synthesizing critical and practice-based approaches, Catherine D'Ignazio and Lauren Klein (2016) have claimed that theories from the humanities can be used to inform and change visualization design. Their contribution is largely focused on ensuring that the design process is inclusive and pluralistic at all stages, from the selection of data sources and representational strategies to the ways in which design teams are composed and the insights and experiences of end users are taken into account. And because they speak as part of a science and technology studies debate on dominant epistemological perspectives and power relations in data visualization design, D'Ignazio and Klein also primarily focus on structure and practice rather than form and meaning.

In the collaborative study on visualization conventions that I conducted with a group of researchers led by Helen Kennedy, we laid the foundations for a social semiotic approach to research on data visualization, with the explicit aim to understand how power works through some of the key semiotic resources found across visualizations (Kennedy et al., 2016). Likewise, Ledin and Machin (2018) propose a general framework for the study of 'data presentation' as a *semiotic material*, or a particular form of communication set apart by unique affordances and canons of use. In turn, Engebretsen and Weber (2017) highlight that data visualization is multimodal, as it is usually enacted as a deployment of multiple graphic modes including, for example, 'typography, layout, maps, diagrams, and drawings' (p. 279) together with colour as 'an integrated component in all the other ones' (p. 279). As they explain, in digital media data visualizations 'can be static and monologic, but they can also be dynamic and dialogic' (p. 289), they can be more or less explorative or open to interpretation, and they can be both pictorial or non-pictorial, with building blocks like photographs, illustrations, geometric shapes, and abstract motifs being equally available to visualization designers. This is important work, but nonetheless, there is little systematic research that combines *both* how data visualization design works semiotically *and* the politics and potentials of this semiotic work in relation to specific contexts and for particular groups of people.

As a methodology that is highly akin to critical discourse analysis, social semiotics is interested in what Caldas-Coulthard and van Leeuwen (2003) define as 'the processes and products of discourse' (p. 3), or both sign-making practices and their concrete outcomes together with their underlying 'ways of knowing' and implications for our 'ways of seeing'. In this sense, social semiotics is not merely a method or collection of methods, but rather a theoretical approach to empirical research. Like critical discourse analysis, and through a Foucauldian lens, social semiotics considers language and sign-making more broadly as key to the reproduction or transformation of social structures. However, social semiotics is also interested in how language and other modes of communication, particularly visuals, work together to make meaning.

Social semiotics originates from a synthesis of structuralist semiotics and Halliday's (1978, 1985) systemic functional linguistics. Social semiotics is functionalist in that it considers all sign-making as having been developed to perform specific actions, or *semiotic work* (Hodge & Kress, 1988). Just like semiotics, it is also concerned with the internal structures of texts and, increasingly, also of other semiotic artefacts (e.g. architecture) and semiotic technologies (e.g. PowerPoint). Unlike traditional semiotics as well as other textual methodologies, social semiotics places emphasis on 'how people make signs in the context of interpersonal and institutional power relations to achieve specific aims' (MODE, 2012). In doing so, social semiotics therefore posits that the physiological and technological means (e.g. sound or imagery) that we use to communicate are to be examined as semiotic *resources* which can be, and in fact most often are, actively mobilized to achieve political, economic, and ideological ends.

This dynamic approach to defining key concepts extends to the notion of meaning, which is not fixed, and where semiotic resources 'have a meaning potential, based on their past uses, and a set of affordances based on their possible uses' (van Leeuwen, 2005, p. 285). The nature of such meaning potentials depends on concrete uses of semiotic resources in specific social contexts where their uses are governed by what van Leeuwen (2005) calls 'semiotic regimes'. In other words, sign-making is regulated through social practices and guided by authority, expertise, or simple conformity in particular contexts. Hence, social semiotics is also able to account both for top-down power and bottom-up polysemy in relation to the uses of semiotic resources.

As I mentioned in the introduction, then, the critical aims of social semiotics are inherent in its approach to examining sign-making, which is always both descriptive and interpretive. Combining a systematic appraisal of semiotic repertoires with an understanding of how their meaning potentials are established over time and in context enables the analyst to understand how semiotic resources are shaped by power relations and, in turn, also 'who made the rules and how and why they might be changed' (Jewitt & Oyama, 2001, p. 135). In his book-length introduction to social semiotics, Theo van Leeuwen (2005) explains that (social) semioticians do three main things. First, they 'collect, document and systematically catalogue semiotic resources—including their history' (p. 3). Second, social semioticians 'investigate how these resources are used in specific historical, cultural and institutional contexts, and how people talk about them in these contexts—plan them, teach them, justify them, critique them, etc.' (p. 3). Finally, they also 'contribute to the discovery and development of new semiotic resources and new uses of existing semiotic resources' (p. 3).

To this three-pronged definition, I would also add that social semiotics extends Roland Barthes's original, though unfinished agenda in *Mythologies*, where he emphasized the need to create 'an appropriate method of detailed analysis' (Barthes, 1972, p. 9) to reveal and undermine the meanings established and perpetuated by the bourgeoise, which he defined as 'the essential enemy' (p. 9). While Barthes's definition of power and the status quo was specific to his time and intellectual background, social semiotics can still be seen as a way to carry out Barthes's *semioclasm*, or a radical attack on the naturalization of signs followed by a more democratic redefinition of what widely shared semiotic practices may look like (Aiello, 2006).

With its ability to link texts with contexts, semiotic production with social action, and meaning with power, social semiotics is an especially congenial framework for research on data visualization. For these reasons, here I propose that a social semiotic framework should be used systematically to *inventorize*, *situate*, and finally also *transform* the semiotic resources of data visualization as a multimodal 'super-genre' in its own right.

Inventorizing data visualization resources

As a first step in our social semiotic approach, we must therefore begin by *inventorizing* the semiotic resources that are typical of data visualization across media and contexts. As van Leeuwen (2005) explains, '[t]o make an inventory we first need a collection' (p. 6). In other words, we must identify and catalogue resources that are representative of data visualization as a whole. This is a particularly challenging task, both because uses of data visualization cut across a vast range of social spheres, and because the existing empirical base to systematically describe key data visualization resources is still thin.

To begin building an inventory of data visualization resources and their possible combinations, we can draw from existing social semiotic and multimodal studies of data visualization and of related semiotic objects. In the study led by Helen Kennedy mentioned earlier, we identify four key data visualization conventions, namely two-dimensional viewpoints, clean layouts, geometric shapes and lines, and the inclusion of data sources (Kennedy et al., 2016). By the same token, in their recent book on visual analysis, Ledin and Machin (2018) examine different types of 'data presentation' through a social semiotic lens, including lists, bullet points, line graphs, bar charts, and flow charts. In this analysis, they identify a set of semiotic resources, namely paradigms, spatialization, vertical and horizontal orientation, graphic shapes and icons, temporality, and causality. Similar analyses of related semiotic objects like diagrams, infographics, and PowerPoint can also be useful in building an inventory of data visualization resources. This is not only because some of these are used in data visualization (e.g. diagrams) or are, at times, confused with data visualizations (e.g. infographics), but also because these analyses offer a discussion of findings and concepts that are useful for a social semiotic analysis of data visualization. What diagrams, infographics and PowerPoint have in common with data visualization is that they are all often used to relay 'hard' facts and key strategic points, usually with the aim to maximize an organization's outputs and increase its competitiveness.

Research on diagrams has focused both on the features of diagrams as semiotic objects in their own right (Ledin & Machin, 2016a) and on the existence of a 'diagrammatic mode', which can manifest itself both independently (e.g. through charts, graphs, and schematic drawings, or 'self-standing' diagrams) but also in combination with other semiotic modes. Bateman et al. (2017) explain that the diagrammatic mode can work together with other modes so as to 'form composite units' (p. 279) that are often set apart by the 'stacking' of elements such as labels and connecting lines over illustrations, maps, or photographs. They argue that information graphics are the resulting 'composite' mode, as these provide the 'glue' to the 'rhetorical relations between contributions from an equally wide range of semiotic modes' (p. 294). In providing this rhetorical cohesion, information graphics rely not only on diagrammatic elements, but also and perhaps most importantly on layout space as a semiotic resource in its own right. Amit-Danhi and Shifman (2018) highlight that the composite nature of digital infographics is also increasingly mobilized to 'embed a rhetoric of participation' (p. 15), for example by letting users choose layouts and selections of data.

Along the same lines, Theo van Leeuwen's collaborative work on PowerPoint highlights the increasing importance of semiotic resources that are typical of visual design, rather than traditional media as such, in everyday communication—such as typography, layout, colour, and texture (Djonov & van Leeuwen, 2013). In doing so, it focuses on inventorizing the resources that the software itself makes available by design, for example by privileging certain resources and uses over others in its interface or help menu. This work contributes an understanding of the relationship between software and their uses, thus moving away from the notion of 'text' to investigate the relationship between semiotic technologies and semiotic practices (Zhao, Djonov, & van Leeuwen, 2014).

In addition to findings from existing analyses, practice-oriented publications like Andy Kirk's data visualization design handbook or Alberto Cairo's guide to information graphics and data visualization can offer a good starting point for the development of an inventory of the modes and resources that are used by designers themselves for the creation of 'good' visualizations (Kirk, 2016; Cairo, 2013). Finally, it is foremost through extensive empirical data collection both from a variety of media (e.g. news media, school textbooks, government websites) and in the field (i.e. through contact with designers, media professionals, and ordinary users) that we can build a systematic inventory of data visualization resources.

This first step of the social semiotic approach may be interpreted as an attempt to outline a 'grammar' of data visualization design, or what Machin (2007) defines as a 'lexicon of elements that can be chosen to create meaning in combinations' and 'a finite system of rules' (p. 185) for their combination. However, it would be problematic to think of such an inventory as a grammar, in that our goal here is not so much to understand how data visualization is and ought to be done, but rather what its major resources are, and how these are mobilized in particular contexts and for specific purposes (see Engebretsen & Weber, 2017).

Situating data visualization resources

Precisely for this reason, the next step of our social semiotic framework entails an attempt to *situate* data visualization resources in their social and cultural contexts. As Jewitt et al. (2016) explain, one of the main aims of social semiotics is 'to understand the social dimensions of meaning, its production, interpretation and circulation, and its implications' (p. 58). Both historical and ethnographic methods are often invoked as key to a social semiotic understanding of meaning-making. Cultural and social histories of a variety of resources—like, for example, colour—are used productively to locate their origins, understand the material, cultural, and political forces that shaped them, and trace their changes over time (see, for example, the history of the colour blue by Michel Pastoureau, 2001). However, fieldwork, and ethnographic research in particular, has often remained an ideal among social semioticians. One exception is my own work, in which

I have adopted a multi-sited ethnographic approach to investigate the practices, motivations, and outputs of image-makers like photographers and graphic designers (Aiello, 2012a, 2012b). As Marcus (1995) writes, when the object of ethnographic investigation is in 'the realm of discourse and modes of thought, then the circulation of signs, symbols, and metaphors guides the design of ethnography' (p. 108). Because of this focus on the social lives of signs, rather than of particular sites or communities, a social semiotic approach will entail a focus on data visualization as it is produced and used across different social and geographical locales.

This said, there is also much to be learned from existing and ongoing ethnographic studies of particular sites and settings in which data visualization is produced, used, or consumed. Alongside Helen Kennedy's collaborative work on designers' intentions and ordinary people's responses with regard to data visualization, there is also a growing body of work on the production and uses of data visualizations in newsrooms (see Engebretsen et al., 2018). In this regard, a social semiotic approach to data visualization can also benefit from sociological research on digital and data journalism, in that it offers detailed accounts of the material resources, skills, and tools that are available to those who make decisions about data visualizations across news media (Fink & Anderson, 2015). This said, when interviewing participants, it is important that researchers ask questions not so much about the intentions, motivations, feelings, and overall actions of participants in relation to data visualization, but more specifically about how they use or interpret particular semiotic resources. This can be done through elicitation or reconstructive methods, where participants are asked to comment on particular texts (in this case, specific visualizations) that the researcher shares with them or asks them to share during the interview. Ultimately, asking questions about 'the set of semiotic choices that typify a given context' (van Leeuwen, 2005, p. 14) contributes both to understanding the context itself and the reasons why specific semiotic resources come to be the way they are. In situating visualization resources in their contexts, particularly through ethnographic fieldwork, researchers will often also come across 'new' resources, which will thus go to enrich and extend their initial inventory.

Transforming data visualization resources

The knowledge generated through the descriptive and interpretive stages of the social semiotic approach to data visualization leads to an understanding of visualization resources as part of broader cultural processes and power relations. A third and final stage in this framework focuses both on the *politics* and *potentials* of data visualization. Major semiotic resources and their combinations can be *transformed* to break away from dominant 'visual sensibilities' and therefore also promote particular forms of social action and social change. As I highlighted earlier in the chapter, the goal of social semiotics is to interrogate as well as redefine sign-making. This is not considered to be a neutral process, but rather as having both power-laden origins and powerful implications.

It can therefore be useful to combine both critical and creative ends to understand how data visualization may be both part of what Fairclough (1995) has termed the 'technologization of discourse' and what van Leeuwen (2008) more recently defined as 'the new writing', or the new dominant language of multimodal communication. On the one hand, data visualization may be seen as part of a powerful impetus towards the standardization of semiotic resources for 'the engineering of social change' (Fairclough, 1995, p. 3). In other words, broader shifts in discursive practices are often aimed at changing the ways in which given institutions—e.g. news media, universities, and governments—and publics think and act in relation to particular issues. For example, Fairclough (1992, 1996) focused extensively on how language was used to promote and normalize both marketization and managerialism in public institutions like schools, universities, and hospitals. Through an analysis of how data visualization resources may be increasingly codified within and across institutions, and how such processes of semiotic codification may be tied to broader structures of power, we can begin to provide an evidence-based, sustained critique of the politics of data visualization. In this regard, for example, Ledin and Machin (2016a, 2016b, 2018) are currently building a body of work on how the discourses of performance management and marketized steering are recontextualized into increasingly ubiquitous 'strategic diagrams'. These are used to translate values like competitiveness and accountability 'into graphic shapes' with 'a clear logic of cause and effect' (Ledin & Machin, 2016a, p. 323).

On the other hand, data visualization ought to be approached as evolving, rather than fixed or unchangeable. According to van Leeuwen (2008), in 'the new writing' the distinction between different semiotic modes is increasingly blurred and, in fact, their relationships are always expressed visually—for example, through layout and 'cohesive uses of colour, typography and other stylistic elements' (p. 132). Across types of media (e.g. websites, newspapers and magazines, institutional documents, and PowerPoint presentations), imagery now tends to be actively combined with writing and other semiotic resources. Hence, writing or images alone are no longer the most authoritative sources of information and persuasion in isolation from one another. Unlike the 'old writing', then, this new 'language' is grounded in principles of visual design (rather than image or word alone) that used to be relegated to professional niches such as web and graphic design. What this means is that data visualization is part and parcel of much broader semiotic practices that are increasingly shaped by normative discourses found in style manuals and formal teaching in art and design schools, but that are also learned through 'best practice' (or approaches that are widely accepted and prescribed as being most effective and sound) and built into semiotic technologies like office software.

These normative discourses regulate the uses of particular semiotic resources and users' competencies in spite of 'all-too-easy affirmations of boundless choice and endless creative opportunity' (van Leeuwen, 2008, p. 135). This said, van Leeuwen (2008) also exhorts students and scholars of visual communication to investigate how these new 'languages' work in practice, to understand what they can and cannot do, and assess how homogenous or varied their applications and uses are in different contexts. For example, in addition to outlining guidelines for 'good' data visualization design, data visualization designers and their students can use a social semiotic approach to examine the histories of particular semiotic resources (e.g. colour, but also shape or layout) as well as understand how these may be used in different social and cultural contexts. Likewise, praxis-oriented scholars of data visualization may want to shift their attention from the broader power structures and work practices that shape data visualization design to include considerations about the ways in which key semiotic resources are used and interpreted by specific groups of people. In both cases, a social semiotic approach may offer an enriched outlook on how data visualization design 'works' in society—thus yielding practical insights into how to adjust and indeed also transform key formal characteristics for purposes like inclusion and equality.

Conclusion

Research on data visualization in society can benefit greatly from approaches that examine the formal—that is, both visual and multimodal—characteristics of visualization design in relation to their implications for how we 'make sense' of the knowledges, facts, and perspectives communicated by data visualizations. A social semiotic framework contributes to a systematic investigation of semiotic resources like colour and layout, for example, together with visualization elements like graphs and charts. However, this is an approach that doesn't stop at a description—however systematic and comprehensive—of form. Instead, it links form to context to understand how semiotic resources work in practice, what they mean and do in everyday life, and ultimately also how they might be changed to do good, or at least do better. By inventorizing and situating data visualization resources, we can build evidence aimed at engaging with the politics and potentials of increasingly dominant, transversal uses of data visualization. In this way, we can also contribute to transforming a range of semiotic practices related to the production and uses of data visualization in everyday life.

References

Aiello, G. (2006). Theoretical advances in critical visual analysis: Perception, ideology, mythologies and social semiotics. *Journal of Visual Literacy, 26*(2), 89-102. https://doi.org/10.1080/23796529.2006.11674635

Aiello, G. (2012a). The 'other' Europeans: The semiotic imperative of style in *Euro Visions* by Magnum Photos. *Visual Communication, 11*(1), 49-77. https://doi.org/10.1177/1470357211424686

Aiello, G. (2012b). All Tögethé® now: The recontextualization of branding and the stylization of diversity in EU public communication. *Social Semiotics, 22*(4), 459-477. https://doi.org/10.1080/10350330.2012.693291

Amit-Danhi, E. R., & Shifman, L. (2018). Digital political infographics: A rhetorical palette of an emergent genre. *New Media & Society, 20*(10), 3540-3559. https://doi.org/10.1177/1461444817750565.

Barthes, R. (1972). *Mythologies*. New York: Hill and Wang.

Bateman, J., Wildfeuer, J., & Hiippala, T. (2017). *Multimodality: Foundations, research and analysis—A problem-oriented introduction*. Berlin: De Gruyter Mouton.

Beer, D., & Burrows, R. (2013). Popular culture, digital archives and the new social life of data. *Theory, Culture & Society, 30*(4), 47-71. https://doi.org/10.1177/0263276413476542

Cairo, A. (2013). *The functional art: An introduction to information graphics and visualization*. Berkeley, CA: New Riders.

Caldas-Coulthard, C. R., & van Leeuwen, T. (2003). Critical social semiotics: Introduction. *Social Semiotics, 13*(1), 3-4. https://doi.org/10.1080/1035033032000133481

D'Ignazio, C., & Klein, L. (2016, October). *Feminist Data Visualization*. Paper presented at and published in the workshop proceedings from the Workshop on Visualization for the Digital Humanities at IEEE VIS Conference, Baltimore, Maryland.

Djonov, E., & van Leeuwen, T. (2013). Between the grid and composition: Layout in PowerPoint's design and use. *Semiotica, 197*, 1-34. https://doi.org/10.1515/sem-2013-0078

Engebretsen, M., Kennedy, H., & Weber, W. (2018). Data visualization in Scandinavian newsrooms. *Nordicom Review*, 39(2).

Engebretsen, M., & Weber, W. (2017). Graphic modes: The visual representation of data. In: C. Cotter and D. Perrin (Eds.), *The Routledge handbook of language and media*. (pp. 277-295). London: Routledge.

Fairclough, N. (1992). *Discourse and social change*. Cambridge: Polity.

Fairclough, N. (1995). *Critical discourse analysis*. Harlow: Longman.

Fairclough, N. (1996). Technologisation of discourse. In: C. R. Caldas-Coulthard & M. Coulthard (Eds.), *Texts and practices: Readings in critical discourse analysis*. (pp. 71-83). London: Routledge.

Fink, K., & Anderson, C. W. (2015). Data journalism in the United States. *Journalism Studies, 16*(4), 467-481. https://doi.org/10.1080/1461670X.2014.939852

Gray, J., Bounegru, L., Milan, S., & Ciuccarelli, P. (2016). Ways of seeing data: Toward a critical literacy for data visualizations as research objects and research devices. In: S. Kubitschko & A. Kaun (Eds.), *Innovative methods in media and communication research*. (pp. 227-251). Cham: Palgrave Macmillan.

Halliday, M. A. K. (1978). *Language as social semiotic: The social interpretation of language and meaning*. London: Arnold.

Halliday, M. A. K. (1985). *An introduction to functional grammar*. London: Arnold.

Halpern, O. (2015). *Beautiful data: A history of vision and reason since 1945*. Durham: Duke University Press.

Hodge, R., & Kress, G. (1988). *Social semiotics*. Cambridge: Polity.

Jewitt, C., Bezemer, J., & O'Halloran K. (2016). *Introducing multimodality*. London: Routledge.

Jewitt, C., & Oyama, R. (2001). Visual meaning: A social semiotic approach. In: T. van Leeuwen & C. Jewitt (Eds.), *Handbook of visual analysis*. (pp. 134-156). London: Sage.

Kennedy, H., Hill, R. L., Aiello, G., & Allen, W. (2016). The work that visualisation conventions do. *Information, Communication and Society, 19*(6), 715-735. https://doi.org/10.1080/1369118X.2016.1153126

Kennedy, H., & Hill, R. L. (2017). The feeling of numbers: Emotions in everyday engagements with data and their visualisation. *Sociology, 52*(4), 830-848. https://doi.org/10.1177/0038038516674675

Kirk, A. (2016). *Data visualisation: A handbook for data driven design*. London: Sage.

Kress, G., & van Leeuwen, T. (2006). *Reading images: The grammar of visual design* (2nd ed.). London: Routledge.

Ledin, P., & Machin, D. (2016a). Strategic diagrams and the technologization of culture. *Journal of Language and Politics, 15*(3), 322-336. https://doi.org/10.1075/jlp.15.3.06led
Ledin, P., & Machin, D. (2016b). The evolution of performance management discourse in corporate strategy diagrams for public institutions. *Discourse, Context & Media, 13*(Part B), 122-131. https://doi.org/10.1016/j.dcm.2016.05.004
Ledin, P., & Machin, D. (2018). *Doing visual analysis: From theory to practice*. London: Sage.
Machin, D. (2007). *Introduction to multimodal analysis*. London: Bloomsbury.
MODE (2012). Social semiotics. Retrieved November 27, 2018 from https://multimodalityglossary.wordpress.com/social-semiotics/
Pastoureau, M. (2001). *Blue: The history of a color*. Princeton, NJ: Princeton University Press.
Thurlow, C., & Aiello, G. (2007). National pride, global capital: A social semiotic analysis of transnational visual branding in the airline industry. *Visual Communication, 6*(3), 305-344. https://doi.org/10.1177/1470357207081002
Tufte, E. R. (1983). *The visual display of quantitative information*. Cheshire: Graphics Press.
Tufte, E. R. (1997). *Visual explanations: Images and quantities, evidence and narrative*. Cheshire: Graphics Press.
van Leeuwen, T. (2005). *Introducing social semiotics*. London & New York: Routledge.
van Leeuwen, T. (2008). New forms of writing, new visual competencies. *Visual Studies, 23*(2), 130-135. https://doi.org/10.1080/14725860802276263
Zhao, S., Djonov, E., & van Leeuwen, T. (2014). Semiotic technology and practice: A multimodal social semiotic approach to PowerPoint. *Text & Talk, 34*(3), 349-375.

About the author

Giorgia Aiello is Associate Professor in the School of Media and Communication at the University of Leeds, United Kingdom. Giorgia's research focuses on the politics and potentials of visuality, materiality, and multimodality. Her book *Visual Communication: Understanding Images in Media Culture* (with Katy Parry) was published by SAGE at the end of 2019.

4. The political significance of data visualization: Four key perspectives

Torgeir Uberg Nærland

Abstract

Practitioners and scholars alike assume that data visualization can have political significance—as vehicle for progressive change, manipulation, or maintaining the status quo. There are, however, a variety of ways in which we can think of data visualization as politically significant. These perspectives imply differing notions of both 'politics' and 'significance'. Drawing upon political and social theory, this chapter identifies and outlines four key perspectives: data visualization and 1) public deliberation, 2) ideology, 3) citizenship, and 4) as a political-administrative steering tool. The aim of this chapter is thus to provide a framework that helps clarify the various contexts, processes, and capacities through which data visualizations attain political significance.

Keywords: Data visualization; Politics; Democracy; Citizenship; Ideology

Introduction

Data and their visualizations are becoming increasingly important in a variety of domains of Western societies (van Dijck, 2014; Couldry & Hepp, 2016). Kennedy, Hill, Aiello, and Allen (2016, p. 715) comment that '[...] data are becoming increasingly valued and relied upon, as they come to play an ever more important role in decision-making and knowledge about the world'. Through their encoding, circulation, and uptake in private, public as well as institutional contexts, data visualizations operate in real-world contexts where politics and power are at work. As such, they have potential significance as instruments or sites for change.

Engebretsen, M. and H. Kennedy (eds.), *Data Visualization in Society*.
Taylor & Francis Group, 2020
DOI 10.5117/9789463722902_CH04

However, where this significance resides, and how data visualizations become *politically* significant, is both a matter of context and of how we define 'politics'. This chapter outlines what are argued to be the four most important perspectives through which we can think of data visualizations as politically significant. In doing so, it outlines the contexts these perspectives belong to and the notion of politics these contexts imply.

Before we can outline these perspectives, it is necessary to make some general demarcations about what we mean when we talk about political significance in the context of data visualizations. First, 'politics' may be understood in narrow terms, as the workings of political parties, processes, and institutions. Politics may also be understood in a wider sense, as the struggle for power more broadly, as this struggle takes place both in the private as well as the cultural sphere, and by symbolic as well as material means. The perspectives outlined in this chapter span from narrow to wide understandings of politics. As will be outlined, data visualizations may assume direct significance as part of the decision-making processes in political institutions. Or, as will be highlighted, data visualizations may assume less direct yet critical significance as a resource for citizenship and participation, or as part of ideological struggles.

Second, we need to clarify what we mean by significance. This chapter is concerned with the effects the circulations of data visualizations have on society. The chapter premises that for data visualizations to be politically significant, they need to be engaged with in real life contexts—be it in institutional or informal contexts. They need to be connected to processes of change, or the maintaining of the status quo.

In the following, four perspectives are presented. These perspectives do not exhaust the possible ways through which data visualizations may lead to change in the world. Rather, they synthesize what is argued to be the main variants. Moreover, there may be considerable overlap between the perspectives presented, and they are indeed interlinked. Yet, these perspectives are not reducible to each other.

Data visualization and public deliberation

Data visualization and public deliberation is a perspective capturing scenarios where data visualizations enter public discussion concerning matters that are contested or that need collective resolution. This perspective is closely affiliated with what has come to be known as a deliberative model of democracy (Habermas, 1994). This model presupposes a well-functioning public sphere, a space where different actors—private, institutional, political—come together to discuss matters of collective importance. Such space may be of a physical nature, as in the case of pubs or assembly halls. But when we today talk about the public sphere, we primarily mean the symbolic space facilitated by the media. It is a normative model inasmuch as it posits that deliberation—the public exchange of arguments—should be rational, and the public opinion that arises from deliberation should form the basis for legitimate political decision-making.

A deliberative model of democracy emphasizes the importance of citizen participation and the public exchange of rational arguments. Crucially, the discussion that takes place in the public sphere is, ideally, connected to actual decision-making. The public sphere should be the mediating space between private persons, civil society, and political decision-makers (Habermas, 2006). The core idea is that political decisions should be grounded in public opinion, not only on electoral results and on the negotiations among elite actors.

From this perspective, data visualizations become significant as part of public discourse. Consider for instance how visualizations of carbon emissions data frequently are employed in public discussions about transport policy. Data visualizations may here function as support for an argument, or as arguments in their own right. As part of public deliberation, data visualizations contribute to the formation of public opinion about contested matters, to which decision- and policymakers ideally should be attentive. The public circulation of data visualizations may also inform decision-makers directly, and people's voting preferences. In addition to how voters make tactical decisions based on visualizations of parties' electoral performances, visualizations of the different parties' stances on key political issues—from climate to immigration policy—inform voters' party preferences.

However, the questions of *who* engages with data visualizations and *where* in the public sphere engagement occurs are important for their bearings on political decision-making. As such, the political theorist Nancy Fraser (1992) introduces a clarifying distinction between 'weak' and 'strong' publics. Weak publics, according to Fraser (1992, p. 134), are those publics 'whose deliberative practice consists exclusively of opinion formation and does not also encompass decision making'. Weak publics would typically include non-elite citizens and media audiences. Strong publics, by contrast, are those 'whose discourse encompasses both opinion formation and decision making' (p. 134) and may include politicians and bureaucrats. Consequently, the direct impact of data visualizations on political decisions will be stronger when they engage strong publics.

Further, the public sphere may facilitate interventions or 'watchdogging' where critical issues are brought into public attention by means of data visualizations. Whereas this is an institutional responsibility of journalism, Suman (this volume) illustrates how also actors from civil society (in her case Greenpeace) can spread visualizations of political interest (haze reports) in the public sphere, to ensure greater accountability in the government's handling of such problems and also to mobilize critical publics.

Data visualizations and ideology

The perspective of *Data visualizations and ideology* captures the ways in which data visualizations privilege certain views of the world, and through dissemination and audience engagement thus work as manifestations or carriers of ideology. From this perspective, data visualizations are integral to the production of meanings, signs, and values in social life, and, according to Marxian thought, a vehicle for the legitimation of the ideas of a particular group or class. Through their dissemination, data visualizations thus may work in the service of particular ideologies—be it for change or for preserving the status quo. Several chapters in this book highlight how data visualizations are not innocent or neutral representations of facts, but are indeed promoting a certain view of the world or establishing a certain kind of epistemology. Hill (this volume) for instance, shows how data visualizations of abortion work to naturalize limitations on access to reproductive healthcare.

From an ideological perspective, data visualizations primarily have *pre*-political significance, rather than direct bearings upon politics (understood in a narrow sense). Data visualizations can contribute to naturalize or challenge certain broad worldviews. Consider for instance how data visualizations can frame socio-economic disparities as dramatic and critical, or conversely, as natural and inevitable. Such worldviews promoted through data visualizations may in turn be highly significant in legitimating or challenging the priorities of political bodies or actors, or in informing voting preferences.

This perspective contrasts to that of deliberation. A deliberative perspective presupposes that data visualizations form part of the exchange of arguments open to validation or critique. In contrast, data visualizations seen from an ideological perspective, work to conceal or naturalise propositions that are nonetheless laden with a particular view of the world.

David Beer (2016) presents a compelling example of how data visualizations do ideological work, or, more precisely, how the visualization of metrics naturalize and augment neoliberal ideology. Beer argues that the pervasiveness of metrics and their circulations produce certain kinds of knowledge ('regimes of truth'). Key to his argument is that the increasing pervasiveness of metrics in social and political life generates a kind of 'numerical thinking' in which (self-)measurement and competition are internalised as self-evident values. In turn, these values are contingent with the modus operandi of neoliberalism where competition and free markets sit at the core. Data visualization is thus a key vehicle for the promotion of numerical thinking. As Beer writes (2016, p. 114), 'How metrics look and how they are visualized can dictate their impact. In each case, these metrics have the capacity to create realities'. Whereas this dynamic does not impact directly on the workings of politics in a narrow sense, the advancements of metrics through visualizations can be seen to pave the way for neoliberal governance.

Beer's account, however, brings into attention an important distinction between data visualizations *as* ideology and data visualizations as *carriers of* ideology. Beer is exemplifying the former. For Beer, who is inspired by the discourse theory of Michel Foucault, data visualizations discursively constitute the trust in numbers that is at the heart of neoliberal ideology. In its forms and modes of production, data visualization here *embodies* the very logic of neoliberalism. In contrast, we can think of data visualization as a tool for symbolic *representation* of issues of ideological significance. From this latter perspective, ideology is not contained in the form of data visualizations themselves, but is a matter of what is represented and how. Consider for instance how visualizations of a country's socio-economic performance may highlight data indicating either commercial growth or the redistribution of resources. Whereas the former can be seen to promote a view of the world where market liberalism is natural and desirable, the latter could be seen to promote social democracy or socialism.

Either way, data visualizations may be used instrumentally by various actors to support or promote particular worldviews. In such scenarios, the public sphere can be seen as a 'battlefield' for conflicting or competing ideologies. A number of examples of how data visualizations can be used to challenge certain worldviews are contained within this book. One is offered by the contribution by Ricker, Kraak, and Engelhardt (this volume), calling for a feminist cartography. These scholars argue that the production and dissemination of maps attentive to gender issues and needs are important to challenge patriarchal ideology.

Data visualizations and citizenship

Data visualizations and citizenship is a perspective emphasizing the different ways in which data visualization can enable people to function as citizens. It does not capture direct impacts of data visualizations on political processes or decision-making. Rather, the political significance here resides in how data visualization may foster engagement with these processes and political participation more broadly.

This perspective lends itself to participatory models of democracy, most notably what is known as the republican and deliberative models of democracy (Held, 2006). These posit that democratic citizenship is not confined to the act of voting, and that broad citizen participation and engagement constitute the core of democratic politics. It is important to note, though, that in the same way that data visualization works as a resource for informed and critical citizenship, it may also work as a tool for misinformation and manipulation, and consequently contribute to the erosion of informed and critical citizenship.

An obvious capacity through which data visualization may enable citizens is by providing them with information and with tools for making sense of complicated political issues. It may enable citizens to take part in political will and opinion formation as well as to form informed party preferences. Moreover, data visualizations may also provide valuable input to the everyday and informal discussions among 'ordinary people', sitting at the core of deliberative models of democracy.

Coleman and Moss's (2016) study of televised election debates and their audiences offers one example of how data visualization may work to promote informed and critical citizenship. In the context of television debates, they identify data visualization as a key sense-making technology through which viewers can be addressed in an inclusive manner by politicians, as well as a tool for citizens to understand and evaluate claims made by politicians and parties.

Moreover, given open data sources and rising levels of technical literacy, the production and dissemination of data visualizations *by* citizens or activists constitutes a bottom-up form of civic engagement in itself. Such a bottom-up perspective is highlighted by D'Ignazio & Bhargava (this volume). These contributors argue that the diffusion of visual-numerical literacy is critical for enabling non-elite members of society to produce their own counter-hegemonic data visualizations. Similarly, Pinney (this volume) highlights the importance of data literacy for participation in today's datafied society.

So far in my treatment of data visualizations' relevance for citizenship, I have highlighted what could roughly be labelled as 'cognitive' dimensions. However, people's engagement with political causes or issues, and inclination to participate more generally, is a question about more than rational judgements and uptake of factual information. It is also a matter of feelings and belongings. Civic engagement hinges on sympathies, antipathies, identifications, passions, and so on. In order to be motivated to act as a citizen, one needs to feel as part of the community that makes up the polity (Kymlicka, 2015; Dahlgren, 2002). Or conversely, feelings of being excluded from community may also motivate political engagement, or political struggle for inclusion more generally. These affective and affinitive aspects of citizenship imply a significant role for data visualizations.

For one, data visualization may spur emotional engagement around certain causes. As shown by Kennedy and Hill (2017), emotions are an integral part of the experiences people have when encountering different aspects of data visualizations, including the data themselves and the subject matter of the visualizations. This point is also made by Gutiérrez (this volume). In the critical context of how industrial countries exploit developing countries' natural resources, she highlights the potential of affective data visualization for mobilizing people to become political activists.

Moreover, data visualizations may play an important role in democratic inclusion. Democratic inclusiveness is not only a matter of who gets to speak or vote. It also concerns whether people feel they are represented in and part of a community or not. As argued by the political philosopher Charles Taylor (1994), the recognition—the positive affirmation—of people's presence is a key motivating force for participation in society. Elsewhere, I have also argued (2017) that media constantly mirror back images of their audiences, who in turn interpret and reflect upon these images. Media representations thus constitute an increasingly important source for recognition. Crucially, data visualizations also bring representations of identities and perspectives into the public sphere, which are engaged with by members of the public. As Kennedy and Hill point out:

> Although more abstract than other visual forms, data visualizations act as media images as well as representations of data, and as such they have the potential to evoke empathy, pity, sorrow, shame and other emotions. (2017, p. 14)

Consequently, audiences may feel recognized, misrecognized, or not recognized at all by data visualizations (see Wihbey, Jackson, Cruz, & Foucault Welles, this volume). Such visualization-enhanced recognition may in turn be important for people's sense of belonging to a community, and in turn their motivations for civic engagement.

A very basic example is national weather forecasts. These visualizations involve a simple form of recognition: the acknowledgement of peripheral parts of a country and the people who live in them as part of the country. Another example is political maps. Political maps involve the recognition of sovereign territories as nation states, and the non-recognition of others—often those with unfulfilled claims to sovereignty. What is included and what is not in data visualizations—who is given *visibility* through data visualizations—thus emerges as a condition for recognition. This capacity of data visualizations to give visibility to groups or persons is addressed in this book in Alamalhodaei, Alberda, and Feigenbaum's chapter, which calls for more 'humanized 'data visualizations. Similarly, Gray's chapter (this volume) highlights the narrative and affective capacities of data visualizations, which in turn may enable visibility and recognition of persons or groups.

Yet another example is visualizations of crime statistics and how these routinely present specific immigrant groups as perpetrators of crime. Seen from the perspective of persons belonging to these specific immigrant groups, these visualizations may form part of an overall negative media framing that, for them, is experienced as a misrecognition of their presence in society, and as being counted as a burden rather than a resource. As a source for (mis)recognition, data visualizations thus may contribute, positively or negatively, to people's sense of being accepted, and consequently, their motivation for civic participation.

Data visualizations as political-administrative steering tool

The perspective of *Data visualizations as political-administrative steering tool* captures scenarios where data visualization is used instrumentally to guide policy or decision-making. It is thus a perspective in which data visualization is assumed to have a strong and direct link to politics. In contrast to the other perspectives, the significance of data visualizations here does not necessarily depend on their circulation in the public sphere or their uptake by non-expert citizens. Rather, the perspective assumes a trajectory directly from experts to policymakers or between other elite actors, who are often connected to scientific, economic, and political institutions. I will illustrate this perspective using an example from the field of global climate policy.[1]

1 In researching this perspective, I consulted Eilif Ursin Reed who is a communication adviser at CICERO Center for International Climate Research, and to whom I am grateful for comments and advice.

Here, I zoom in on how visualizations of climate data inform how The United Nations Framework Convention on Climate Change (UNFCCC) sets its climate policy goals. A main focus of this political body is to set the temperature target; the maximum allowable warming to avoid dangerous anthropogenic interference in the climate. The UNFCCC regularly commissions scientific reports on which to base policymaking. These reports are commissioned from The Intergovernmental Panel on Climate Change (IPCC), a scientific body consisting of thousands of scientist and other experts. As part of these lengthy reports the IPPC produces a short version of the report, called *Summary for Policymakers*, which addresses policymakers directly. Among other things, this summary presents research-based scenarios guiding policymakers, who also finally approve the summary. These summaries routinely contain data visualizations.

For instance, a regular staple in these summaries has been the visualization feature called the 'burning ember' (see *New York Times*, 2009). In the form of coloured bar graphs, the 'burning ember' visualizes risks (the redder, the more critical) associated with different scenarios of increased global mean temperatures. As such, the 'burning ember' provides an example of how data visualizations address 'strong' publics, whose discourse encompasses both opinion formation and decision-making. It is thus also an example of how data visualizations may attain very concrete and manifest political significance as aids for political decisions. However, the inclusion of the 'burning ember' has been criticized precisely for being too instructive. Rather than merely visualizing problems—which is the mandate of the scientists—it is criticized for employing visual rhetoric that command certain responses (see, for instance, Mahoney & Hulme, 2012).

Conclusion: Avenues for further research

This chapter has outlined four important perspectives through which we can think of data visualizations as politically significant. Moreover, it has attempted to clarify the contexts where data visualizations become politically significant, and the notions of politics implied by these contexts. Each of these perspectives implies different avenues for research. In the following, I will briefly point to some of the most important of these.

The ways in which data visualizations form part of public deliberations actualizes questions about the argumentative and rhetorical nature of such visualizations. Do data visualizations, as they appear in public debate, work to clarify or conceal arguments? Do they lay themselves open to (in) validation? And how are rhetorical devices used to convince? Such questions are important to answer in order to attain a more critical understanding of how data visualizations contribute to public and political discourse—or more generally; to manipulative or argumentative public spheres.

Likewise, there is a need for empirical research into how data visualizations textually promote ideology, and how citizens' worldviews are shaped or negotiated in their encounters with data visualizations. A further step would be to explore empirically, and in more detail, how the ideological work done by data visualizations connects to or prepares the ground for political agendas.

Moreover, there is a need for a clearer understanding of how the expansion of data visualization affects people's ability to function as citizens. Through which capacities and in which contexts do data visualizations work as a resource for citizenship, and when do they not? In particular, the affective and affinitive dimensions of how people engage with data visualizations warrant further research. When and how do data visualizations engender feelings of being recognized among audiences, and how may such feelings contribute to audiences' civic affinities?

Lastly, there is need for more empirical research into when and how data visualizations are used instrumentally as an aid in political or administrative decision-making processes. Such endeavours would enable insight into some of the very concrete and manifest ways in which data visualization affects politics. This would require investigations into the specific contexts where decision-making takes place, be they political, administrative, or legal bodies.

Some of this much-needed research is underway, and can be found in the chapters in this book.

References

Beer, D. (2016). *Metric power.* York: Palgrave Macmillan.

Coleman, S., & Moss, G. (2016). Rethinking election debates: What citizens are entitled to expect. *The International Journal of Press/Politics, 21*(1), 3-24. https://doi.org/10.1177/1940161215609732

Couldry, N., & Hepp, A. (2016). *The mediated construction of reality.* Cambridge: Polity Press.

Dahlgren, P. (2002). In search of the talkative public: Media, deliberative democracy and civic culture. *Javnost: The Public, 9*(3), 5-26.

Fraser, N. (1992). Rethinking the public sphere: A contribution to the critique of actually existing democracy. In: C. Calhoun (Ed.), *Habermas and the public sphere.* (pp. 109-43). Cambridge: MIT Press.

Habermas, J. (1994). Three normative models of democracy. *Constellations, 1*(1), 1-10. https://doi.org/10.1111/j.1467-8675.1994.tb00001.x

Habermas, J. (2006). Political communication in media society: Does democracy still enjoy an epistemic dimension? The impact of normative theory on empirical research. *Communication Theory, 16*(4), 411-426. https://doi.org/10.1111/j.1468-2885.2006.00280.x

Held, D. (2006). *Models of democracy* (3rd ed.). Cambridge and Malden, MA: Polity Press.

Kennedy, H., & Hill, R. L. (2017). The feeling of numbers: Emotions in everyday engagements with data and their visualisation. *Sociology, 52*(4), 830-848. https://doi.org/10.1177/0038038516674675

Kennedy, H., Hill, R. L., Aiello, G., & Allen, W. (2016). The work that visualisation conventions do. *Information, Communication and Society, 19*(6), 715-735. https://doi.org/10.1080/1369118X.2016.1153126

Kymlicka, W. (2015). Solidarity in diverse societies: Beyond neoliberal multiculturalism and welfare chauvinism. *Comparative Migration Studies, 3*(17). https://doi.org/10.1186/s40878-015-0017-4

Mahoney, M., & Hulme, M. (2012). The colour of risk: An exploration of the IPCC's 'Burning Embers' diagram. *Spontaneous Generations: A Journal for the History and Philosophy of Science, 6*(1), 75-89. https://doi.org/10.4245/sponge.v6i1.16075

Nærland, T. U. (2017). Altogether now? Symbolic recognition, musical media events and the forging of civic bonds among minority youth in Norway. *European Journal of Cultural Studies.* Advance online publication. https://doi.org/10.1177/1367549417719013

New York Times (2009) [Digital image]. Retrieved August 28, 2018 from https://archive.nytimes.com/www.nytimes.com/imagepages/2009/02/23/science/dotearth_large.html

Taylor, C. (1994). The politics of recognition. In: A. Gutmann, A. Appiah, & C. Taylor (Eds.), *Multiculturalism: Examining the politics of recognition.* (pp. 25-73). Princeton, NJ: Princeton University Press.

van Dijk, J. (2014). Datafication, dataism and dataveillance: Big data between scientific paradigm and ideology. *Surveillance & Society, 12*(2), 197-208. http://doi.org/10.24908/ss.v12i2.4776

About the author

Torgeir Uberg Nærland is a researcher at the Department of Information Science and Media Studies at the University of Bergen. Focusing on textual, receptive, and institutional dimensions of expressive culture, his research explores the intersections between aesthetics and politics.

Section II

Living and working with data visualization

5. Rain on your radar: Engaging with weather data visualizations as part of everyday routines

Eef Masson and Karin van Es

Abstract

This chapter discusses visualizations of weather data, used to communicate short-term precipitation predictions to lay audiences. Focusing on the example of *Buienradar*, a popular Dutch weather forecast website and app, it investigates how people engage with such representations on a daily basis, how they interpret them, and how their readings of them affect their actions and decisions, shaping their day-to-day routines. The research is based on semi-structured interviews with users with different demographic profiles. Aside from establishing usage patterns or preferences and readerly strategies, the chapter also considers people's own evaluations of their conduct in relation to the *Buienradar* service, and more broadly, their reflections on the significance of weather data visualizations to their lives.

Keywords: Weather data; Data visualization; Data usage; Readerly strategies; Daily life; Routines

Introduction

In late August of 2017, the spokesperson for a Dutch association of campsite owners criticized *Buienradar*, an often-used weather forecast website and app, for the financial setbacks its members had incurred over the course of the summer. In an interview with a local newspaper, he posited a causal relation between patrons' use of the service and cancellations received in the week prior to their stay (Baard & Hellegers, 2017). The news report

Engebretsen, M. and H. Kennedy (eds.), *Data Visualization in Society*.
Taylor & Francis Group, 2020
DOI 10.5117/9789463722902_CH05

suggests that he primarily blamed the weather service itself, as a source of misleading information. But his statements also betray frustration with the customers, for blindly trusting the overly cautious predictions made.

While this position may sound extreme, it does build on widespread assumptions about how people today access, and act upon, information about the weather, as obtained via a range of (often digital) media. In 2001, the media scholar Marita Sturken already observed that the weather 'is no longer something one goes outside to register, that one experiences on the ground and in the flesh. It has become, rather, a technological experience, seen from satellites and endlessly monitored on television and the Internet' (2001, p. 161). But the above anecdote also invokes associations with the sort of (humorous) comments, proliferating online, that suggest that people these days would rather believe what their weather apps tell them than to trust their own senses.

Buienradar, the main target of the campsite owner's frustrations, is something of a household name in the Netherlands. Launched in 2006, it was the first service in the country to make use of data from KNMI, the national weather office, in order to visualize, in rather distinctive ways, both recent and current rainfall, as based on precipitation detections, and projections for future rainfall. Its present default view has two key elements (see Figures 5.1 and 5.2 below). On the one hand, the actual *buienradar*, literally 'shower radar': a map of the Netherlands showing rain clouds in different colours, denoting the amount of rain (in mm/h) observed or predicted, traversing the territory in small increments. And on the other, a so-called *regengrafiek* or 'rain chart': a line graph showing the amount of rain per temporal unit for a given place. In addition, the platform also provides information and predictions on a range of other weather phenomena, in different forms and for different time frames.

Informal exchanges with users suggest that *Buienradar*'s data visualizations, or readings thereof, affect how they live their lives on a daily basis. But the sorts of actions and decisions mentioned are generally more mundane than those alluded to in the anecdote above. In addition, such conversations reveal that we do not actually *know* very much about how readings of weather visualizations precisely take shape. Nor, for that matter, about how such representations, with all the epistemic power they wield and the interpretive pitfalls they present (cf. Kessler & Schäfer, 2018; Smith, 2018), get navigated on a daily basis, as part of the routines of people's everyday lives.

In recent years, data scholars have deplored the dearth of empirical study into how people encounter, use, and reflect on data on a daily basis (e.g. Couldry & Powell, 2014, p. 2; Michael & Lupton, 2016, p. 110; Pink, Sumartojo,

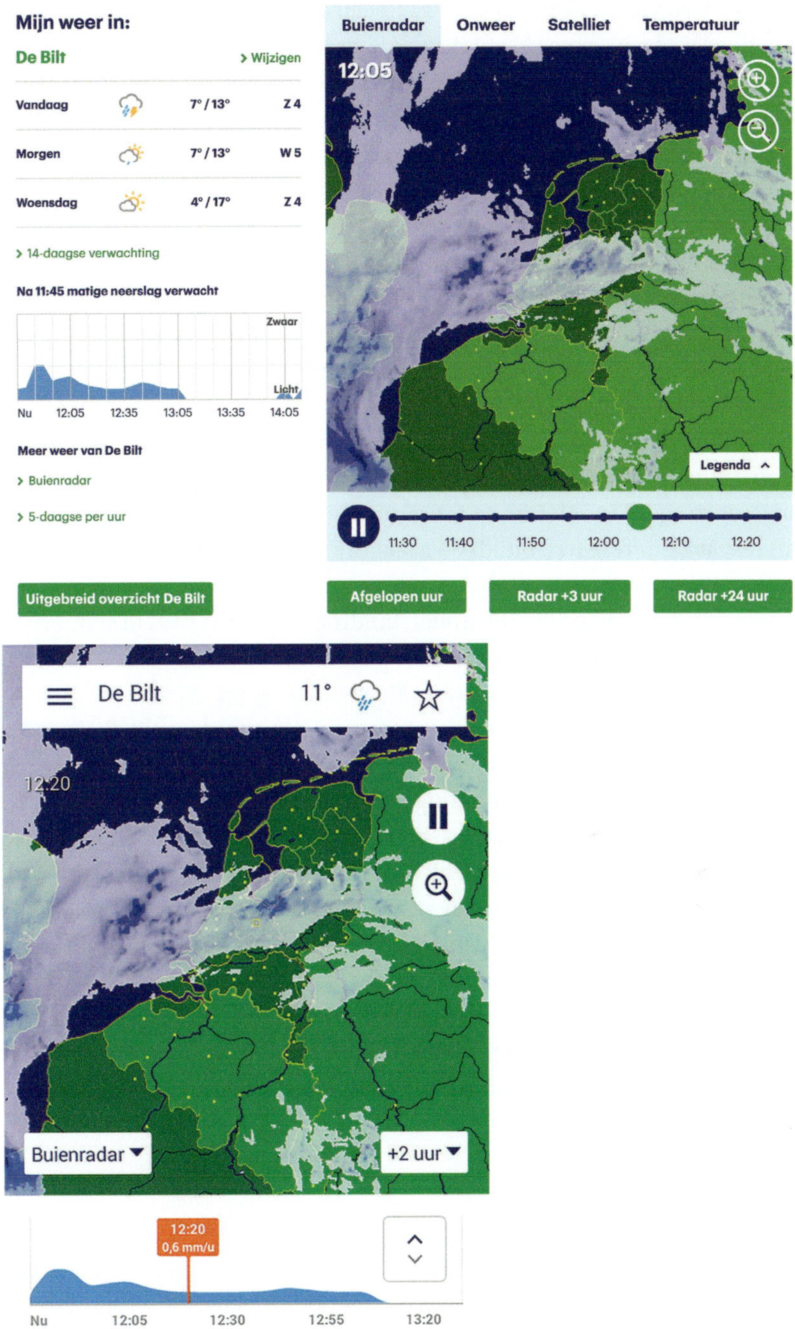

Figures 5.1 and 5.2. Default views for the *Buienradar* website and (Android) app for Monday April 30, 2018 at around 11:35 a.m. CET. Screenshots by Eef Masson, used under quotation exception. Copyright 2006-20 by RTL Nederland.

Lupton, & Heyes La Bond, 2017, p. 2). Specifically, they have identified the experiences of non-experts and the relations between data use and everyday activities as 'critical absences' in research so far (Kennedy, 2018, p. 19). With this chapter, we want to make a preliminary contribution to the shared attempt—among others through this volume—to start a scholarly debate on the topic.

In doing so, we position ourselves on the intersection of two types of research. On the one hand, we want to build on previous studies of the ways in which *data usage is integrated into daily life*. With the spread of consumer digital media, there is a renewed interest in how media employment relates to 'everyday temporalities, materialities and routine' (Pink & Leder Mackley, 2013, p. 680). Here, we focus specifically on interactions with data *visualizations*. On the other hand, we also want to learn more about how people concretely *read and understand such visualizations* (Ruckenstein, 2014), once again in relation to the situations of which their use is part. In this respect, our research builds on a lengthy tradition of reception research. This tradition, we argue, retains its relevance in the digital age—especially insofar as it considers how the understanding of texts as sites of semiosis is affected by their various 'contexts', for instance technological or social (see Livingstone & Das, 2013, pp. 105-106; Mathieu, 2015, pp. 16, 19).

In the opening sections of the chapter, we briefly introduce the *Buienradar* service and explain how we conducted our exploratory empirical research into people's use and understanding of the visualizations it provides. Next, we discuss our results. We focus, first, on what we learned about how people commonly *use Buienradar*, and which views or settings they prefer, and why. Then, we relate how they actually *read* them. Here, we consider questions both about the relations they establish between data, their representation, and acts of interpretation, and about the readerly strategies they apply. Finally, we look at how users *act upon* their readings and integrate them into their everyday routines, concluding with a section on the *broader significance* of the *Buienradar* visualizations to their lives.

Buienradar: Some background

Buienradar was developed by three Dutch siblings, but inspired by a practice observed on American television (e.g. Galasz, 2014): the broadcasting of short-term precipitation projections based on public, radar-generated weather data. Initially, it exclusively provided precipitation information, based on data obtained from KNMI; later, it broadened its scope to other atmospheric conditions such as temperature, relying also on additional sources. In 2011, the company was bought by the commercial broadcaster RTL, which now operates the website and apps. Use of the service has always been free, with revenue coming from advertisements.

In promotional texts, *Buienradar* defines itself primarily in the following terms: as a platform for (precise) information about precipitation, in visual form, at very short notice. Its creators claim that in launching the service, they appealed to a desire among audiences for forecasts that were both easier and quicker to read, and more unambiguous than those offered through other channels. Users of weather media, they argue, felt hampered by the 'intervention' of experts. On the one hand, because they craved precision and certainty rather than nuance and cautiousness; on the other, because they were rarely interested in how predictions came about. The initiators expected that in providing 'direct' access to weather data, the service would enable the user to take a meteorologist's place, seeing 'at a glance' what was to happen at specific points in time (e.g. Ermstrang, 2011).

Despite increased competition, but also critique from weather experts (critique variously concerning the implausibility of very precise precipitation predictions, or the flaws of the particular technology for data collection that the service capitalizes on; see e.g. Galasz, 2014; Elegeert, 2015; van Leur, n.d.), *Buienradar* remains highly popular in the Netherlands. In February of 2018, the website and app together reached 3.8 million local users (Verenigde Internet Exploitanten, n.d.)—almost 25 percent of the population over the age of six. But their cultural significance arguably reaches much further, as our interviews suggest that the name 'Buienradar' is sometimes used eponymously for similar services.

Methodological considerations

In light of our wish to gain preliminary insight into how people understand weather data visualizations in relation to the specifics of their everyday lives, we chose to conduct a series of interviews as a basis for our observations. This way, we were best able consider the mutually productive relation between the two, taking into account that daily routines do not merely 'accommodate' for interactions with data, but also shape those interactions, and vice versa (e.g. Pink et al., 2017). This method also has the advantage that it allows us insight not only into people's understandings of weather data visualizations and their experiences of living with them, but also into how they personally assess them. Such reflection by users is of interest here, because it is informative of how they personally gauge the importance of visualizations and because it sheds light on their own perceptions of the issues such representations raise and the pleasures they provide.

Our findings are based on semi-structured interviews with sixteen users of the *Buienradar* website and app. In selecting respondents, mostly from our personal network, we sought to consider the diversity of actual experiences among a range of people. This resulted in variations in age (with participants between 25 and 71, more or less evenly spread across the decades) and gender (eight men and eight women), family structure (people living alone or with a partner, versus members of families with children) and occupation (salaried versus self-employed, and within different sectors). Arguably, our sample is somewhat biased in terms of educational level, in that most of the people we interviewed have completed some form of further education (vocational or academic). Also, for practicality's sake, all interview subjects have been recruited from the Randstad area of the Netherlands (the megalopolis comprising the country's largest cities), where we live and work. Most of the interviews lasted between ten and twenty minutes, and they followed roughly the same pattern.

Usage patterns and preferences

Most of our interviewees regularly access information about the weather; two thirds do so at least once a day. About half of them rely for this purpose on the general news media: broadcasts on radio or television or (online) news publications. Oftentimes, they do not actively seek out such information, but encounter it as part of their daily routines in media consumption. Those who go looking for forecasts tend to prefer specialist websites or apps (sometimes as pre-installed on their devices). Overall, source selection is quite arbitrary: respondents often alternate between services, and 'googling for the weather' is common, especially in looking for longer-range predictions (e.g. prior to holiday travel).

If we compare forecasts in the mainstream media and on general weather sites with those provided by *Buienradar* and similar services, more distinct user patterns emerge. 'Traditional' forecasts, as we know them from newspapers and TV, tend to focus on averages for the day and week, and mostly feature still or animated maps and tables with icons and numerical information (see Figure 5.3). Generally speaking, people opt for *Buienradar* when they are looking specifically for predictions of rainfall (as opposed to other weather conditions) that are also more precise—both in terms of

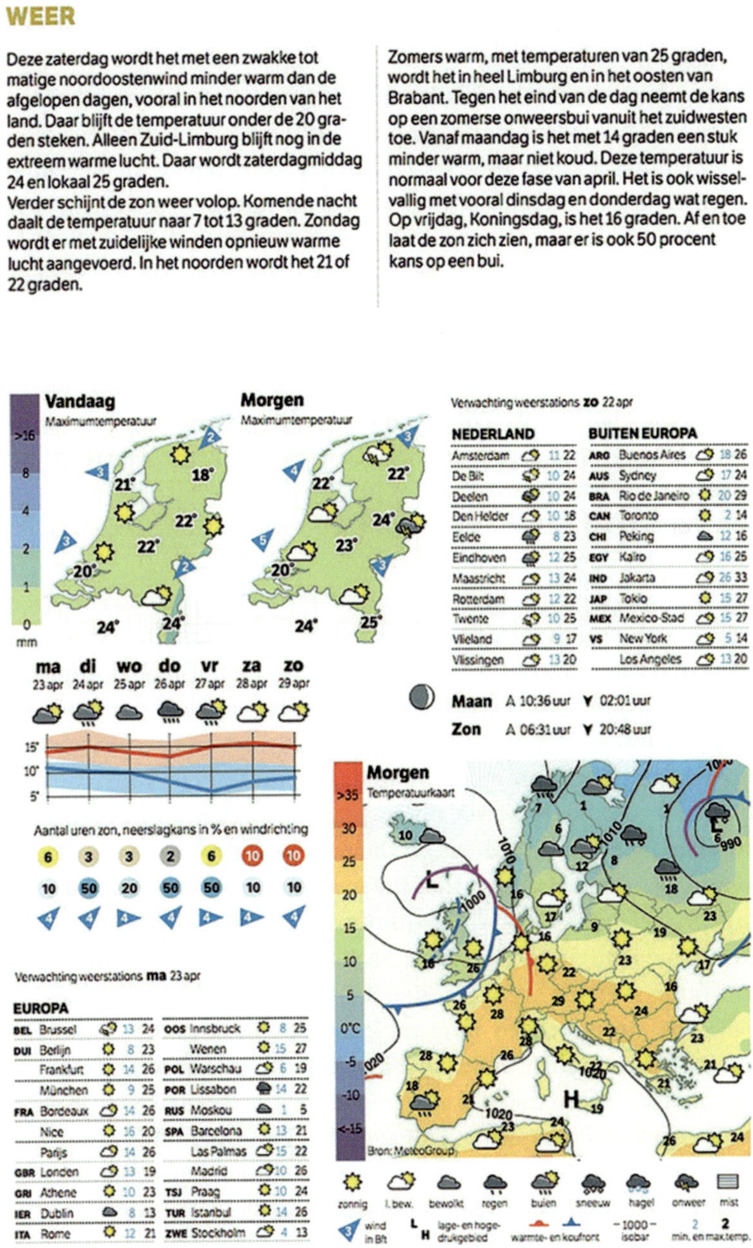

WEER

Deze zaterdag wordt het met een zwakke tot matige noordoostenwind minder warm dan de afgelopen dagen, vooral in het noorden van het land. Daar blijft de temperatuur onder de 20 graden steken. Alleen Zuid-Limburg blijft nog in de extreem warme lucht. Daar wordt zaterdagmiddag 24 en lokaal 25 graden.
Verder schijnt de zon weer volop. Komende nacht daalt de temperatuur naar 7 tot 13 graden. Zondag wordt er met zuidelijke winden opnieuw warme lucht aangevoerd. In het noorden wordt het 21 of 22 graden.

Zomers warm, met temperaturen van 25 graden, wordt het in heel Limburg en in het oosten van Brabant. Tegen het eind van de dag neemt de kans op een zomerse onweersbui vanuit het zuidwesten toe. Vanaf maandag is het met 14 graden een stuk minder warm, maar niet koud. Deze temperatuur is normaal voor deze fase van april. Het is ook wisselvallig met vooral dinsdag en donderdag wat regen. Op vrijdag, Koningsdag, is het 16 graden. Af en toe laat de zon zich zien, maar er is ook 50 procent kans op een bui.

Verwachting weerstations **zo** 22 apr

NEDERLAND

Amsterdam	11	22
De Bilt	10	24
Deelen	10	24
Den Helder	10	18
Eelde	8	23
Eindhoven	12	25
Maastricht	13	24
Rotterdam	12	22
Twente	10	25
Vlieland	9	17
Vlissingen	13	20

BUITEN EUROPA

ARG	Buenos Aires	18	26
AUS	Sydney	17	24
BRA	Rio de Janeiro	20	29
CAN	Toronto	2	14
CHI	Peking	12	16
EGY	Kairo	16	25
IND	Jakarta	26	33
JAP	Tokio	15	27
MEX	Mexico-Stad	15	27
VS	New York	5	14
	Los Angeles	13	20

Maan A 10:36 uur ⋎ 02:01 uur
Zon A 06:31 uur ⋎ 20:48 uur

Verwachting weerstations **ma** 23 apr

EUROPA

BEL	Brussel	13	24	OOS	Innsbruck	8	25
DUI	Berlijn	8	23		Wenen	15	27
	Frankfurt	14	26	POL	Warschau	6	19
	München	9	25	POR	Lissabon	14	22
FRA	Bordeaux	14	26	RUS	Moskou	1	5
	Nice	16	20	SPA	Barcelona	13	21
	Parijs	14	26		Las Palmas	15	22
GBR	Londen	13	19		Madrid	10	26
GRI	Athene	10	23	TSJ	Praag	10	24
IER	Dublin	8	13	TUR	Istanbul	14	26
ITA	Rome	12	21	ZWE	Stockholm	4	13

Figure 5.3. Weather report with textual and graphic elements in *NRC Handelsblad* (a Dutch national newspaper) for the weekend of April 21 and 22, 2018. Screenshot by Eef Masson, used under quotation exception. Copyright 2018 by *NRC Handelsblad*.

when and where the rain will fall. As a rule, moreover, they are interested in short-term predictions (that is, information concerning the next one to three hours).

Respondents tend to use *Buienradar* as they are about to undertake an activity that involves leaving the house, often for a journey somewhere. Overwhelmingly, interviewees establish a relation here with bicycling—a highly common means of transportation in the Randstad area. Other activities that prompt them to consult the service range from such day-to-day pursuits as walking the dog or hanging the laundry to dry, to sports practice at different levels of expertise. While some users check *Buienradar* as a matter of habit, others do so only if it is either (heavily) raining already, or if they have reason to believe that it might. In other words, people are motivated to access the platform by a desire to know if they may 'get wet'—often in hopes that they can adapt their plans so as to avoid it. In this respect, the intensive use of weather apps seems to have engendered a shift in terms of how weather forecasts are commonly used (cf. for instance Lazo, Morss, & Demuth, 2009, p. 792).

Our conversations also reveal strong but diverging preferences for specific *Buienradar* functionalities and types of visualization. In addition, they suggest that users, over time, develop their own habits in navigating them. As regards preferences, our respondents roughly divide into three groups, based on whether they are interested primarily, or even exclusively, in the aforementioned 'shower radar' (map representation) or 'rain chart' (line graph), or a combination. A majority prefer the geographical representation, focusing in their readings on the relation between current location (sometimes set to default, so that the map shows only a select part of the country; see Figure 5.4) and the timing of a given stage in the animation of rainclouds moving over it. Others, however, radically prefer the line graph, often with the argument that it is 'clearer' or that it provides 'more specific' or 'more detailed' information (either in terms of location, or in the sense that rainfall is more precisely quantified). For yet another group of respondents, use of the map and graph forms part of a two-stage process, whereby the graph is consulted for additional information.

Aside from the map to chart navigation, common interactions with the default view involve zooming in on the map, and specific ways of toggling between the one- and three-hour views (on the website) or moving one's cursor between different projection times (all media). Those who navigate beyond the initial map and chart (roughly half of our respondents) tend to do so only incidentally, and often in search of other kinds of information than about precipitation.

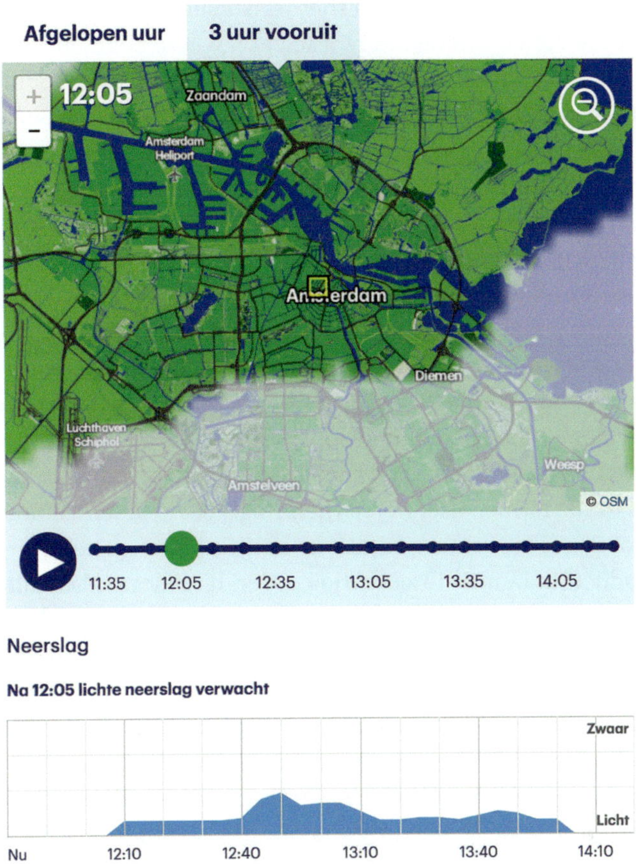

Figure 5.4. Shower radar and rain graph visualizations on the *Buienradar* website, set to Amsterdam, for Monday April 30, 2018, 11:35 a.m. CET. Screenshot by Eef Masson, used under quotation exception. Copyright 2006-20 by RTL Nederland.

Buienradar readings: Data and visualization, prediction and interpretation

Each interview began with a request to explain to a hypothetical interlocutor what *Buienradar* is. In retrospect, the answers given are quite revealing of people's understanding of the service. Most characterizations focused either on the predictive aspect of the information provided or on the fact that it is rendered in a primarily visual form. In some cases, respondents highlighted precisely the combination of those features. A couple of interviewees also named a specific *type* of representation, usually 'map' (a choice suggesting the close association of *Buienradar* with the geographic view, in the common perception). A few even used such terms as 'photographic' or 'radar', alluding to the specific imaging technologies that they (mistakenly) assumed were used to produce the views. Interestingly, a select few also characterized it as a platform enabling the consultation of weather 'data'; however, only one used the term 'visualization' (importantly, someone with a professional interest in the topic). In other words, respondents tended to be acutely aware of the fact that what *Buienradar* offers are representations—even if they did not actually conceptualize them as data visualizations and were in doubt as to which other label to use instead.

Furthermore, formulations used throughout the interviews attest to diverse, and in some respects contradictory, assumptions about the role of data and interpretation, both in the information provided and in the way *Buienradar* presents it. This diversity manifests the most clearly if we separate claims along those two dimensions: statements about weather information, specifically prediction, and about weather (data) visualization.

With respect to the former, users overwhelmingly seem to understand information that concerns future conditions as interpretive, and by implication, as products of human intervention. Overall, they are also quite permissive here in matters of accuracy: since the weather is hard, perhaps impossible, to predict, it is not at all odd that forecasts are not always 'right'.

However, readings of the *Buienradar* views as *representations of* data or information reveal very different assumptions about what exactly it is users are presented with. Formulations that show awareness of the representational status of the shower radar, rain chart, or any of the other visualizations provided still attest to an understanding of their relation to reality as barely mediated. Telling in this context was the use, during interviews, of such terms as 'photographic' or 'radar' (the latter likely prompted by the tool's own name). Aside from the fact that such choices in wording attest to an at best rudimentary understanding of the relation between weather data and their registration, as well as their visualization, they also suggest that interviewees infer a direct, indexical relation between what they see on *Buienradar*, and 'the world out there'. Moreover, interpretations were often phrased in terms suggestive of the visualizations' presumed objectivity and evidentiary power. In this respect, they align with the service's self-promotion as one that provides direct access to 'raw data', eliminating in the process any form of human 'meddling'.

Evidently, the dimensions of weather prediction and data visualization, in the respondents' accounts, cannot always be disentangled. Even so, the interview results attest to these users' desire to also consider the merits of the representation as such; for example, in comments on the clarity of maps, charts, or tables. Once again, this suggests that they have an eye for what data visualizations do—even when they obscure their own status as representations.

Buienradar readings: Interpretive strategies

Aside from navigational habits, the users we interviewed also displayed personalized readerly routines. During the interviews, we asked them to vocalize their thought process as they contemplated the different visualizations. In doing so, we realized that their interpretations came about in intuitive ways and were often based on information once verified but then modified as part of individualized reading strategies. In many cases, for instance, interpretations of the map visualization accounted for the colour of animated clouds. However, while the map's legend is quite unequivocal about how those colours are encoded, the interviewees' readings of them were highly diverse. Many understood them in terms of rain intensity ('how heavily it will rain'), an interpretation that ties in quite closely with their actual coding in terms of precipitation volumes. Others, however, did not take the colours to carry any meaning at all. And some respondents, including some true *Buienradar* aficionados, associated them with rather more complex or encompassing atmospheric conditions (for instance, 'red' as taken to denote 'thunder' or 'stormy weather'). These last examples suggest that our users, even if they built in their interpretations on what they had previously heard or read about the codes deployed, would oftentimes add to or tweak the information obtained.

However, such reading habits do not necessarily derive from limited engagement with the site or app, or the visualizations specifically. In this respect also, our data show considerable differences between interview subjects, who may be roughly divided into two groups based on the expectations they have from the service. Those in the first group tend to avoid information and representations that present some sort of an interpretational hurdle, for instance because they require non-standard knowledge. One map user for instance complained that the rain graph mistakenly presumes that the user understands what it means to confront a specific amount of precipitation (in mm). Respondents in this group therefore also applied simplifying reading strategies (e.g. interpreting the line in said graph as indicative of 'rain' or 'no rain', rather than a certain measure of precipitation). Another interviewee had difficulties interpreting tables with probability figures (presumably, a common issue in the reception of weather forecasts;

see Gigerenzer, Hertwig, Van Den Broek, Fasolo, & Katsikopoulos, 2005), which he therefore avoided. A select few also expressed a preference for simplicity in visual design, objecting to the *Buienradar* (desktop) site's overly cluttered interface or overload of information.

A second group of respondents, by contrast, seemed prepared to engage much more deeply with the service's representations. They made elaborate studies of their preferred weather visualizations, or took a comparative approach, contrasting the different visualizations amongst each other and even with those on other sites or apps. Some did so from a critical impulse, suspicious of either weather prediction or visualization practice. Others took this approach because they needed to very precisely plan (recurrent) activities that were weather-dependent, such as outdoor sports. These accounts suggest that in accessing the service, both factions tried to 'penetrate its underlying system', so as to be able to see more clearly what the prediction and/or visualization algorithms actually do. Arguably, they thus attest to a drive to 'take the forecasters' place'—but with a different motivation than *Buienradar*'s initiators anticipated. Here, the perceived problem is not one of specialist intervention (which it is, for *some* users!) but rather, that the access to data that *Buienradar* provides is not quite 'direct' *enough*—in spite of the makers' pledges. For this group, unimpeded access is in the interest of a more nuanced understanding of the reality the data reference, and presumably, the data's representation blocks this reality from view.

Buienradar in the routines of everyday life

Many of our participants who access *Buienradar* prior to open-air activity take practical decisions based on their readings of the visualizations encountered. They use them for instance in determining what to wear or how to dress the children, whether to take further protective devices such as umbrellas, or even—if they have the choice—which means of transportation to choose. Prior research suggests that such decision-taking habits are common for forecasts across the board, regardless of the media or representations involved (cf. Lazo et al., 2009, p. 792). A difference, however, is that *Buienradar* users sometimes also delay their plans, or even cancel them, quoting the very precise information the service provides. While obviously more common in people who have more of a hand in how they organize their days, such behaviour was widespread amongst our participants.

Lazo, Morss, and Demuth (2009) observe that weather forecasts, as forms of communication, are 'part of the infrastructure of our lives and livelihoods' (2009, p. 795). The *Buienradar* case confirms this, to the extent even that accessing the site or app has become profoundly entwined with people's daily routines and patterns of behaviour. We distinguished earlier between those people who check the service when prompted by current weather conditions and those who do it as a force of habit. For the latter, the act of checking becomes inextricably interwoven with moments of departure. One account further suggested that such behaviour may be engrained in the social conduct of (specific) collectives. The respondent in question—someone in his early thirties—related that during outings with groups of peers, whenever plans were being made to move from one location to another, one person would always check *Buienradar*.

At times, the habitual use of such weather services may even become a routine in itself, functioning as a propeller in (re)shaping the flow of everyday life (cf. Nansen, Arnold, Gibbs, & Davis, 2009). One interviewee, a homemaker, explicitly assigned the service a role in setting up her day, but also claimed that the mere act of accessing the site helped her give her life substance. Arguably, this is only possible because *Buienradar* provides a continuous stream of perpetually updated information—much like other contemporary (social) media do.

Summing up: *Buienradar*'s significance to people's lives

At the beginning of this chapter, we referenced some sources that observe a widespread blind trust in information about the weather as presented by such services as *Buienradar*. Our own account suggests that users do indeed take the platform's visualizations very seriously, in that they consult them repeatedly and act upon how they read them. However, they seem to do so *in spite of* a profound scepticism towards the information the platform provides. In the context of our conversations, such mistrust often derived from awareness of the fundamental unpredictability of atmospheric conditions, informed by a diverse body of (lay) knowledge about the limits of weather forecasting. But in a select few cases, interviewees also attributed it to the intricacies of data visualization (unsurprisingly, mostly respondents engaged in study or professional activities that presuppose a certain interest in such matters). For example, a couple of users argued that the *Buienradar* maps and charts were (necessarily) selective in what they show, and one person suspected that they might actually be misleading.

Another was even prompted by the interview to wonder about which data models were used, and how this affected what she saw. Yet as a rule, such understandings did not seem to prevent the speakers from relying on the service. With reference to Sturken, we therefore conclude that today still, there is a widespread yearning for an experience of control through monitoring—even of something as fundamentally uncontrollable as the weather (2001, pp. 162, 165).

In light of the above, it is hardly surprising that people gave rather ambiguous answers to questions about *Buienradar*'s importance to their lives. On the one hand, they found the service very useful. Some argued that while they previously did 'just as well without', not having it would require an adjustment—and a far-reaching reorganization of their daily routines. A few respondents actually found this scenario appealing, as they realized that *Buienradar*'s use profoundly impacted on the rhythms of their personal lives, or even, on people's sense of self-reliance. But on the other hand, they also took care to put the service's importance into perspective, pointing among others to the banality of the information provided and the availability of practical solutions and precautions. Overall, their behaviours supported the sincerity of their claims. For instance, several respondents related that they decided at some point to remove the (storage-consuming) *Buienradar* app from their phones, opting instead for the mobile site, because other functionalities were more crucial to their lives (a navigation tool, for instance, or more space for pictures).

Presumptions of 'blind trust' in the face of technology are further undercut by people's profound awareness of their own habits as users, and above all, by their preparedness to reflect on them. Many interviewees volunteered to comment—albeit sometimes with shame or in self-mockery—on the paradoxical aspects of their behaviour: the apparently inverse relation between how they act upon *Buienradar* information, and a fundamental suspiciousness towards what the service does (predicting) or how it does it (selectively visualizing extrapolated data). Some also showed awareness of the social conditioning of their conduct, and of the relation between the platform's economics (e.g. its use of adverts) and their dependence on it. A few even expressed appreciation of the pitfalls of an increasingly datafied existence—either for political reasons (e.g. in light of data collection and privacy-related issues) or social ones (as in the comments on self-reliance). This strengthens us in our conviction, inspired by Couldry, Fotopoulou, and Dickens (2016), that we cannot reduce the users of data visualizations to actors without agency, and should be alert also to signs of reflexiveness.

Presumably, one reason why such critical attitudes are easily overlooked is that they may actually coincide with (intensive forms of) use. And even, we would like to add, with *enjoyment* of such use. Several of our respondents access *Buienradar* also because they derive some form of pleasure from engaging with its visualizations. For example, one Amsterdam resident explained that she finds the default map view the more attractive one, because it not only shows what is going to happen in her current location, but also 'how a rain shower develops' as it passes east over the country, which she finds 'fun to watch'. While she also derives pleasure from studying physical indicators of atmospheric conditions or developments—for instance, the movement of a real-life flag or vane—there is an added appeal to weather observation via the shower radar. This suggests in turn that *Buienradar*'s use for monitoring the weather is about more than just 'mastering' one's experiential world: it is also about engaging (in the process) with the latest technologies, and the particular gratification this provides.

Acknowledgements

Many thanks to all our interviewees for sharing their experiences; to Pauline van Romondt Vis for her valuable suggestions on conducting user research; and to the reviewers of this text for their editorial input.

References

Baard, L., Hellegers, B. (2017, August 29). Buienradar oorzaak matig zomerseizoen Twentse campings. *Tubantia*. Retrieved from https://www.tubantia.nl/enschede/buienradar-oorzaak-matig-zomerseizoen-twentse-campings~ac5e2182/

Couldry, N., Powell, A. (2014). Big data from the bottom up. *Big Data & Society, 1*(2), 1-5. https://doi.org/10.1177/2053951714539277

Couldry, N., Fotopoulou, A., Dickens, L. (2016). Real social analytics: A contribution towards a phenomenology of a digital world. *The British Journal of Sociology, 67*(1), 118-137. https://doi.org/10.1111/1468-4446.12183

Elegeert, T. (2015, May 8). Hoe betrouwbaar is Buienradar eigenlijk? *HP/De Tijd*. Retrieved from https://www.hpdetijd.nl/2015-05-08/hoe-betrouwbaar-is-buienradar/

Ermstrang, A. (2011, January 5). In één oogopslag. *Terdege*, pp. 17-18. [Retrieved with Digibron.]

Galasz, P. (2014, July 12). Het regent buienradars, maar kloppen ze ook? *De Morgen*. Retrieved from https://www.demorgen.be/wetenschap/het-regent-buienradars-maar-kloppen-ze-ook-b9a55f42/

Gigerenzer, G., Hertwig, R., Van Den Broek, E., Fasolo, B., Katsikopoulos, K. V. (2005). 'A 30% chance of rain tomorrow': How does the public understand probabilistic weather forecasts? *Risk Analysis, 25*(3), 623-629. https://doi.org/10.1111/j.1539-6924.2005.00608.x

Kennedy, H. (2018). Living with data: Aligning data studies and data activism through a focus on everyday experiences of datafication. *Krisis, 2018*(1), 18-30. http://krisis.eu/living-with-data/

Kessler, F., Schäfer, M. T. (2018). Trust in techno-images: Early media collections as precursors of big data. *TMG—Journal for Media History, 21*(2), 52-66. http://doi.org/10.18146/2213-7653.2018.366

Lazo, J. K., Morss, R. E., Demuth, J. L. (2009). 300 million served: Sources, perceptions, uses, and values of weather forecasts. *Bulletin of the American Meteorological Society, 90*(6), 785-98. https://doi.org/10.1175/2008BAMS2604.1

Livingstone, S., Das, R. (2013). The end of audiences?: Theoretical echoes of reception amid the uncertainties of use. In: J. Hartley, J. Burgess, and A. Bruns (Eds.), *A companion to new media dynamics*. (pp. 104-121). Oxford: Wiley-Blackwell.

Mathieu, D. (2015). The continued relevance of reception analysis in the age of social media. *Trípodos, 2015*(36), 13-34. Retrieved October 9, 2018 from http://www.tripodos.com/index.php/Facultat_Comunicacio_Blanquerna/article/view/240.

Michael, M., Lupton, D. (2016). Toward a manifesto for the 'public understanding of big data'. *Public Understanding of Science, 25*(1), 104-116. http://dx.doi.org/10.1177/0963662515609005

Nansen, B., Arnold, M., Gibbs, M. R., Davis, H. (2009). Domestic orchestration: Rhythms in the mediated home. *Time & Society, 18*(2-3), 181-207. https://doi.org/10.1177/0961463X09338082

Pink, S., Leder Mackley, K. (2013). Saturated and situated: Expanding the meaning of media in the routines of everyday life. *Media, Culture & Society, 35*(6), 677-691. http://dx.doi.org/10.1177/0163443713491298

Pink, S., Sumartojo, S., Lupton, D., Heyes La Bond, C. (2017). Mundane data: The routines, contingencies and accomplishments of digital living. *Big Data & Society, 4*(1), 1-12. http://dx.doi.org/10.1177/2053951717700924

Ruckenstein, M. (2014). Visualized and interacted life: Personal analytics and engagements with data doubles. *Societies, 4*(1), 68-84. https://doi.org/doi:10.3390/soc4010068

Smith, G. J. D. (2018). Data doxa: The affective consequences of data practices. *Big Data & Society, 5*(1), 1-15. https://doi.org/10.1177/2053951717751551

Sturken, M. (2001). Desiring the weather: El Niño, the media, and California identity. *Public Culture, 13*(2), 161-189.
van Leur, H. (n.d.). Pluimen en apps [Blog post]. Retrieved October 9, 2018 from http://www.helgavanleur.nl/Columns/19/pluimen-en-apps
Verenigde Internet Exploitanten (VINEX). (n.d.). *NOBO top 50 feb 2018*. Retrieved October 9, 2018 from http://www.vinex.nl/resultaten/archief/

About the authors

Eef Masson is an Assistant Professor of Media Studies at the University of Amsterdam, Netherlands. She has published on non-fiction and non-theatrical film, media archives, museum media, and the use of digital tools for media (history) research. Recently, she has collaborated on projects on visual analysis and data visualization, specifically for purposes of archival access.

Karin van Es is an Assistant Professor of Media and Culture Studies at Utrecht University, Netherlands, and coordinator of the Datafied Society research platform. Co-editor of the volume *The Datafied Society* (AUP, 2017), she focuses in her current work on public service television and public values in the digital age.

6. Between automation and interpretation: Using data visualization in social media analytics companies

Salla-Maaria Laaksonen and Juho Pääkkönen

Abstract
This chapter explores the use of data visualizations in social media analytics companies. Drawing on a dataset of ethnographic field notes and thematic interviews in four Finnish social media analytics companies, we argue that data visualizations are crucially involved in how analytics-based knowledge claims become accepted by companies and their clients. Basing on previous research on visualizations in organizations and as a representational practice, we explore their role in social media analytics. We identify three practices of using visualizations, which we have named have simple-boxing, flatter-boxing, and pretty-boxing. We argue that these practices enable analysts to achieve the simultaneous aims of producing credible and valuable analytics in a context marked by high business promises.

Keywords: Visualization in data analytics; Analytics as business; Automated analytics; Interpretation; Visual analytics; Epistemology of data analytics

Introduction

The abundance of social media data and the development of computational methods have led to the birth of new business opportunities, including the growing field of data analytics. One nascent area inside this field is social media analytics, or refining and processing data generated by human behaviour on social media, with the aim of transferring them to

Engebretsen, M. and H. Kennedy (eds.), *Data Visualization in Society*.
Taylor & Francis Group, 2020
DOI 10.5117/9789463722902_CH06

business knowledge. David Beer (2017) describes data analytics companies as new powerful *data intermediaries* who build infrastructure to shape the circulation of social data and construct narratives of omnipotent and intelligent calculative technologies to reach objective, correct solutions and decisions in organizations. Beer (2017, p. 466) posits that the 'data analytics industry is powerful in shaping what is said, made visible or known through data', and calls for more studies explicating the ways in which this industry cultivates visions of data-led thinking. At present, we lack an adequate account of knowledge production in data analytics, although the field's societal implications are proliferating (cf. Beer & Burrows, 2013). This issue is particularly pressing for novel contexts of analytics, such as social media analytics, which are currently becoming established (Kennedy, 2016).

An essential part of making sense of digital datasets is the use of visualizations—such as regular line, bar, or pie charts, or more sophisticated algorithmic visualizations, such as clustering diagrams or network graphs (Kennedy & Hill, 2018). Visualization techniques also play an increasingly important role in organizations, which utilize visual representations of data in both their internal and outward communication and negotiations (Quattrone et al., 2012; Halpern, 2014). While the most obvious aim of data visualizations is to communicate information, they can also shape the actions and understandings of their users (cf. Beer & Burrows, 2013; Kennedy, Hill, Aiello, & Allen, 2016). This means that they are devices which bear material agency in their immediate purposes and contexts of use (cf. Leonardi, 2011) and work to construct conceptions of both data and analytics.

In this chapter, we explore processes of knowledge production in data analytics, by investigating the use of data visualization in the business of social media analytics. We approach visualizations as visual representations of data, the use of which is intertwined with epistemic conceptions that guide how analytics are conducted, both as business and as knowledge production. Thus we are interested in the conceptions of what constitutes knowledge in analytics, and the role visualizations play in the process. Drawing on ethnographic field notes and thematic interviews in four Finnish social media analytics companies, we argue that data visualizations are crucially involved in how analytics-based knowledge claims become accepted by companies and their clients. We explore their role in this process, and the associated epistemic conceptions concerning social media data and its analysis.

To begin, and to formulate our research question, we first introduce social media analytics as a research context, and discuss previous literature on the uses of visualization beyond social media analytics. Then, in the subsequent sections, we present our empirical material and findings.

Social media analytics as a research context

Visuals have achieved a dominant position in contemporary organizations (e.g. Sørensen, 2014). Organizational, economic, and political life is represented and enacted by visualizations such as flowcharts and diagrams (Locke & Lowe, 2012), infographics (Amit-Danhi & Shifman, 2018), or network visualizations (Venturini, Bounegru, Jacomy, & Gray, 2017), which have reached an almost paradigmatic status in social media visualizations. In the business context, visualizations are strongly marked with practical value evaluations: they make sense of complex data, translating them to usable or valuable information (Halpern, 2014). Being valuable means giving insights to the current state of affairs but also includes a prophetic dimension by enabling a sight to the future (Beer, 2017). Visualizations, hence, are technologies that find value which it is impossible for an unaided human eye to locate.

Visualization in social media analytics is marked by two specificities: consulting business and the nature of data. First, while statistical methods and visualization practices are old conventions, the data on which the analyses are conducted are of a novel nature: the validity and generalizability of social media data have not been demonstrated or commonly constructed. Many hype-generating narratives evolve around the business insights that can be extracted from social media analytics, but practices of using these novel data sources are yet to be firmly established.

Second, social media analytics are situated in a business-to-business (B2B) context, where companies' business proposals are coined around the notion of selling data, which also means they need to construct value for those data and repeat the hype narrative highlighting its importance, omnipotence and visionary capabilities (Beer, 2017). This places visualizations in a new context, where their role and function is defined in B2B negotiations. Analytics companies are simultaneously required to produce credible analytics and justified knowledge claims in a nascent area, while meeting hype-inflated expectations of business value. These two aims are potentially in conflict, for instance when selling analytics demands that results are presented in an appealing form, which might not represent the analytics process accurately.

Data visualizations as revelation and persuasion

As noted above, in organizations, visualizations are an established form of knowledge production. They are representations that have been established as useful and credible means of depicting information, making them intertwined with practices and structures of institutional politics and legitimacy (Scott & Orlikowski, 2012; Lynch & Woolgar, 1988). The role of visualizations is further emphasized through the prevalent process of datafication, whereby increasingly many aspects of human life are quantified and captured in a form amenable for computational analysis (e.g. van Dijck, 2014). Data visualization is essentially a way to access and make sense of data, and thus the increasing pace of datafication leads to a growing importance of visualization methods (Kennedy & Hill, 2018).

Previous research has discussed visualization in data analytics as a representational practice that enables one to *reveal* hitherto hidden, or otherwise inaccessible, information in data (Coopmans, 2014; Halpern, 2014). As such, the commercial attraction of visualization stems from the promise of yielding insight into data, portrayed as a vital strategic resource for business (Coopmans, 2014). However, data analysis promises unforeseen increases in business value only for those who are in possession of the required visualization tools and skills to use them—a conception dubbed as *artful revelation* by Catelijne Coopmans (2014). Underlying the notion of artful revelation is the idea that visual representations of data enable viewers to see for themselves, or witness patterns in data first-hand, thus lending credibility to the produced knowledge. The idea of first-hand witnessing as a source of credibility has its roots in the history of scientific experiments and publishing (Shapin, 1984). Further, in the context of visualizations, credibility through witnessing is linked to the ideal of 'mechanical objectivity' (Daston & Galison, 1992), according to which visual representations should strive to depict patterns in data truthfully, free from the biasing influence of subjective interests and aesthetic judgements (Frow, 2014; see also Kennedy et al., 2016).

However, digital visualization techniques can also allow images to be manipulated without being restricted by the underlying data, to make them aesthetically pleasing, or to highlight certain selected aspects and downplay others (Frow, 2014). Consequently, as Emma Frow (2014) has argued in the context of scientific publishing, the issue of credibility in digital visualization concerns the skills of those producing the visualizations. Visualizations that look carefully prepared and aesthetically pleasing are easily 'interpreted by readers as reflecting skill and expertise on the part of the author' (p. 258). Thus, credible revelation in digital data visualization is as much a matter of skilfully selecting and portraying the right aspects in data, as one of depicting objectively existing patterns.

In this vein, critical accounts portray visualizations as doing *persuasive work*, that is, presenting particular viewpoints as more acceptable or valuable than others (see Kennedy et al., 2016 for a summary). Kennedy et al. (2016) have shown that the persuasive powers of visualization can be linked to prevalent visualization conventions, which imbue visualizations with an 'aura of objectivity' (p. 723), making them seem transparent and factual. However, as Frow (2014) argues, perceptions of the objectivity of digital visualizations—and thus the issue of their credibility—are intertwined with assessments of the skill of the visualizers on the part of the visualizations' viewers. Consequently, skilful use and aesthetic aspects are central to the persuasive work of visualizations.

Acknowledging the specificities of social media analytics as a research context, as well as the previous research on the purposes of visualizations in society, we formulate our research question as follows: *How are data visualizations used in social media analytics, and how are the credibility and value of analytics constructed through visualization processes?*

Data and method

Our empirical study is based on ethnographic field notes and thematic interviews collected in four companies that analyse social media data as a part of their business endeavours. In total six person across the four companies were interviewed. The interviewees were all in positions of management or middle-management, although, in the smaller companies these are in practice also operative employees. The companies vary in size and stage: Three are mid or early-stage start-ups, with less than fifteen employees. The fourth company is an established firm, the main business of which is surveys, but which aims to expand to social media data. The main product of one of the smaller companies is network analyses and visualizations. Another smaller company focuses mainly on distributing and cleaning social media datasets instead of doing in-depth analyses themselves, but they offer some basic visualizations and summaries.

We approached the interview transcripts and ethnographic notes qualitatively, focusing on the parts where the material concerns visualization. Using Atlas.TI and an inductive approach, we trace the different functions and meanings associated with visualizations, and the ways in which they are used to reveal information and to persuade. Both authors first coded the material independently using a grounded approach (cf. Glaser & Strauss, 1967), after which we met to discuss and refine the coding, and to group our codings into larger categories and themes. After reclassifying the material, we ended up with six themes: interestingness, objectivity, credibility, communication to lay people, aesthetics, and customer apprehension. Next, the passages under these codes were reread and reorganized to chart out, first, the central role visualizations play in imbuing social media analytics with credibility and value, and second, the practices of generating visualizations, through which this role is enacted.

Credibility and value in social media analytics

In data analytics, the belief that data constitute a baseline of truth is prevalent (Ruppert, Isin, & Bigo, 2017). However, in our material, we observed that social media analytics are characterized by the conception that social media data are *messy* and *heterogeneous*; it is a form of data that is unreliable compared to traditional, institutionalized data forms which have been accepted as solid evidence for phenomena. Despite this starting point, our case companies were preoccupied with the idea that there is something essential hidden in the data: it is a matter of revelation and presentation to find it. The analysts expressed that the credibility of analysis is based, first, on the data containing patterns, which can be uncovered and utilized in a reliable manner, and second, the practice of using automated means to produce results that are free from subjective bias. Consequently, the results of analysis and measurements were deemed credible when they were grounded in, or driven by, data and analysed using objective means. This notion relies on the idea of quantification as a source of trust in contexts where expertise has not been demonstrated (Porter, 1995; Halpern, 2014; Elish & boyd, 2018).

In a context marked by the idea of messy data, visualizations are used for imbuing analytics with credibility and value. They are used to build coherence and to reveal hidden structures and patterns of that which *is* in the data. Thus, epistemically visualizations not only describe, but bring out the essential, show patterns in data in a manner that does not distort their depiction, letting the data drive the visualization. Automation plays a crucial role in the process, as interpretation takes place as an interactive process between the machine and the human:

> We would need something programmatic to browse through that stuff and condense it, and to categorize, classify and present it. And then the researcher could sort of see that infographic, in quotation marks, and detect the point they want to take up, what I want to bring out and make a point, that this information could be valuable.

Credibility through automation, then, is seen as a prerequisite for valuable, marketable analyses. However, although automation and data-drivenness are associated with credible analytics, visualizations are a means through which patterns in data are ultimately revealed, in the sense of making them visible, that is, amenable for assessment and examination. Without visualization one has to trust the algorithm, but visualizations let the interpreter do the interpretations. This is a form of credibility produced through witnessing, which is specific to visual depiction of patterns.

Despite being based on algorithmic data-driven analytics, data visualization is also an acquired skill when it comes to discovering patterns in data. In Halpern's (2014, p. 22) words, visualizing is 'making something that is out of sensory recognition relatable to the human being'. Thus, visualization expertise is relevant for assessing the credibility of analytics, as the results of algorithmic analysis can be hard to decipher without the aid of visual representations. Further, demonstrating expertise in visualization to clients also lends value to analytics, as will become apparent below. Next, we will scrutinize three practices through which human skills are involved in the process of creating credibility and value in social media data visualization.

Simple-boxing to conceal complexity

The consulting context is marked with an unequal distribution of knowledge: the customers are buying expertise from the consultant. This generates a need to communicate potentially complex information to the customers in an understandable delivery format, which implies simplification. The necessity of simplicity also affects visualization choices and preferences; in practice, it leads good visualization practice to be a customer-driven concept within the companies. Good visualizations are easily comprehended and effectively work as descriptions of the situation or, preferably, decision-making devices for the customers. This means that visualizations need to show, in one glimpse, understandable depictions of reality, stripped of its messiness and complexities. This is a process referred to as 'simple-boxing' by one of our informants, during which details of the complex reality become hidden.

> So we will probably put some colour symbols there like red and green or something according to the sentiment. So that in a glimpse you could see if there's a lot of red or blue or green.

In this vein, as a part of its simple-boxing process, the network analysis company has developed heuristics considering the amount of nodes that can be reasonably presented in a network graph so that it remains understandable to the customer. The company has learned by practice to filter out information that is deemed unnecessary and irrelevant based on expertise accumulated in business negotiations. This process is also affected by the presentation technologies and conventions institutionalized in business practices. One of our case companies explained how they select the dimensions of a statistical model so that they can be fitted and beautifully visualized on a single PowerPoint slide. They have developed an understanding that neither a salesperson nor the customer can handle more dimensions.

As a way of communicating complex information, simple-boxing also serves to enable assessments of the credibility and value of analytics. The following quote from the firm specializing in survey research demonstrates how communicating information effectively means revealing it by means of data visualization.

> We have in fact thought about the issue of how to get to raw data in such a way [...] that we could somehow define it more strictly, standardize in a way the output [...] to visualize it to the researcher so that it would be easier for them to [...] recognize whether there is anything meaningful there, versus just getting 50,000 lines of some text.

Simple-boxing thus is a way of crafting visualizations that enables the analysts to comprehend complex information and communicate it effectively to different parties, including clients, consultants selling the analytics, and colleagues. Nevertheless, simple, comprehensible visualizations are regarded as representing patterns in data truthfully, provided they are based on analyses that are judged to be reliable. For instance, one analyst suggested that a visualization of clustering results truthfully depicts the underlying patterns in data. This was despite a somewhat subjective choice of the number of clustering dimensions displayed in the results. Thus, simple-boxing with visualizations need not imply unrealistic representation, provided that the depicted results are produced through credible analytics.

Flatter-boxing to highlight the interesting

Apart from working as devices that reveal essential information, our informants also identify a need for visualizations to point out what is *interesting* in the data.

Revealing the interesting is a practice thought to be dependent on the skills of the analyst, which involves finding the potential structure and patterns, and picking correct ways to visualize them. For example, both generating and interpreting visualizations are referred to as craftsmanship in the network analyses consulting company—an expert can immediately see if there is a structure in a network visualization. As the following quote highlights, the represented patterns might not constitute an 'absolute truth', and thus their status is dependent on the conceptions and craftsmanship of the analyst.

> Absolute truths are hard to formulate about [network visualizations]. When you have, over the years, done quite many of those [...] you start to form some sort of a picture about what the network might tell, kind of a carpenter's feel, that you learn to recognize different kinds of wood merely by smell or touch.

One popular way of building structure and pointing out the interesting from social media data is by making lists and rankings. This is a practice postulated by the business context: in a situation of constantly striving for profit and competition any information that creates order becomes valuable (cf. Halpern, 2014). For instance, one of our case companies publishes lists of actors' performance on social media and uses various visualizations to track their relative positions over time. These processes require visualizations to depict patterns in data that are deemed to be interesting, such as displaying differences or orderings among the measured units, 'so that they can constantly observe on one screen if they are going down or up'.

Thus, visualizations generate value for analytics by highlighting differences. What is crucial for this kind of revelation is that patterns are discovered that match the analyst's conceptions of valuable information, with less emphasis being based on reliable and data-driven analysis processes. Hence, the skills of the human analyst are essential for finding and depicting the interesting. We have named this practice flatter-boxing. In line with the notion of artful revelation (Coopmans, 2014) the use of visualization skills, or craftsmanship learned through experience, is a condition for the production of valuable analytics that reveal interesting patterns in data.

Pretty-boxing to induce marvel and convey expertise

Apart from being simple and interesting, visualizations also need to be attractive (cf. Brinch, this volume). Visualizations are representational devices that both communicate expertise and build trust. This effect is largely gained through aesthetics; visualizations need to look beautiful but also difficult to create (cf. Coopmans, 2014; Frow, 2014). This is particularly the case with network analyses, which are easy to sell merely because they generate what the consultant calls a 'wow effect':

> It's always like wow, because you get fancy, pretty visualizations, so that it's an essential part of the network analysis that you generate the visualization. People often equate [the analysis and the visualization], [...] people are satisfied since they get a view of what is really happening, that it's not just a stream of messages, but there is a structure.

Compared to topic detection methods, another computational text analysis method with less clear visualization practices, the interviewee thought that networks are easier to interpret for people with no technical background. At the same time, the interviewee acknowledges that the customer can concoct a story based on network diagrams; they are like Rorschach tests that support many versions of the truth, when sometimes no truth actually exists.

Thus, aesthetically pleasing visualizations can give an illusory feeling of understanding. As Kennedy et al. (2016) have argued, the persuasive work done by visualizations is in part due to prevalent visualization conventions. Our evidence conforms to the notion that aesthetic considerations are involved in assessing analytics (cf. Frow, 2014). Through pretty-boxing analytics with beautiful visual renderings, analysts can convey expertise, which simultaneously builds credibility and value. However, this persuasiveness of visual presentation is considered dangerous in cases where the reliability of analytics has not been demonstrated. As the analysts of the survey analytics firm expressed in the interview, the results of even epistemically dubious analytics can be successfully marketed, given appropriate presentation.

Discussion: When and how visualizations work

The evaluation of visualizations among our informants is functionally oriented: visualizations should *work*, or manage to do what is expected in a business context. When it comes to social media data, the notion of the data being messy, heterogeneous, and complex is a narrative which is part of the business offering of these companies. In our case companies, data visualizations are used to translate knowledge extracted by various algorithms to applicable business knowledge. In order to be applicable, the produced analytics are required to be simultaneously credible and valuable. These aims cannot be straightforwardly accomplished by applying seemingly objective algorithmic procedures to extract information from data. In addition to appearances of objectivity, successful consulting also requires that the product is relevant and interesting for the customer, and that the customer is able to understand and benefit from the results.

In our research, the translating role that visualizations play between humans and technology in the context of social media analytics becomes evident. As explained above, visualizations are regarded as tools which can hide complexity and represent data beautifully, potentially glossing over multifaceted decisions made during data analysis. As Coopmans (2014) argues, data visualization is a representational practice which enables analysts to discover valuable information in data and convey their expertise to customers. Our investigation highlighted how in social media analytics the status of visualizations is defined by the extent to which they communicate interesting and essential information to the customer in a simple and beautiful form. This status is constructed in the interplay between automated analysis processes and application of the analysts' visualization expertise.

Whether a visualization works, or is able to fulfil its business purpose, is hence not straightforwardly related to the perceived objectivity of analysis. Rather, the status of a visualization is evaluated on the basis of how well it manages to display interesting and useful results, which depend on customer needs, and the abilities of both the customer and the analyst. Despite the continuous contemplation of data quality and the acknowledgement of craftsmanship, the buyers' needs emerge as an important factor affecting the visualization practices. This makes the selection and interpretation of visualization models an example of an interested reading of reality (Rieder, 2017), where the actors are reading the data and visualizations in ways that show patterns or differences, following their business models' predications. Hence, the aim of generating interesting patterns that further business objectives can override the epistemic aim of realistic representation. Here, we witness an interplay between assessments of credibility and value in evaluating social media analytics visualizations. Visualizations emerge as devices which enable analysts to retain analysis credibility while constructing a valuable representation of patterns in the data.

Conclusion

Our investigation identified three intertwined practices that help analysts to achieve the aim of producing credible and valuable analytics: simple-boxing, flatter-boxing, and pretty-boxing. In addition to methodological norms guiding data analysis, social media analytics are shaped by the need to communicate intelligible and attractive results to customers. This criterion potentially stands in tension with the aim of realistic representation. When the two aims conflict, the analysts are faced with the challenge of conveying their results in a simple enough form to sell their product, while retaining the customers' interest and credibility of analysis. Using visualizations to simplify and beautify information provides a way to accomplish this goal, reconciling the pursuit of interesting results with credibility grounded in data-driven revelation.

Various features exhibited in our material influence how visualizations are constructed in social media analytics to serve the purposes of business. First, they are used to communicate exactness and objectivity of the analysis while hiding the complex processes of collecting, cleaning, and analysing the data. Visualizations are not necessarily shown in their original, valid form, but they are tweaked, simple-boxed, flatter-boxed, and pretty-boxed until they better communicate the desired message; for example, until they highlight the differences between measured units. As an end result, however, they still communicate objectivity based on numbers and impersonated analysis (cf. Porter, 1995). Second, after being tweaked, visualizations provide more clues to lay people to interpret the information in comparison to word lists or numerical representations, which again increases the perceived credibility of the analysis. Third, their credibility is intertwined with notions of usability, interestingness, and business value (cf. Frow, 2014). These are issues connected to the conventions accumulated in the business context over time.

Finally, the entire social media analytics business builds on the assumption that social media data can provide valuable insights to organizations. Visualization is what makes algorithmic output a form of human knowledge, but the interpretation and skills of the analysts are an essential part of the process. The algorithmic analysis process, human skills, and visualization all work together to build structure in and discern value from messy data. As Halpern (2014, p. 30) eloquently formulates, '[visualization] is a set of techniques by which to manage, calculate, and act on a world of incomplete information'. In business context, data visualizations act to increase the business value of data in concrete ways; they transform the data to an authoritative object, an artefact that yields for effects and outcomes, facilitating not only understanding but also business. By doing this, they play an important role in establishing the omnipotent nature of data analytics (cf. Beer, 2017). As more and more organizations are tempted by the possibility of taking advantage of automated analysis to understand their shareholders' discussions on social media, and to make better informed decisions, the practices of using visualizations within this field can have far-reaching effects in the society.

References

Amit-Danhi, E. R., & Shifman, L. (2018). Digital political infographics: A rhetorical palette of an emergent genre. *New Media & Society, 20*(10), 3540-3559. https://doi.org/10.1177/1461444817750565

Beer, D. (2017). Envisioning the power of data analytics. *Information, Communication and Society, 21*(3), 1-15. https://doi.org/10.1080/1369118X.2017.1289232

Beer, D., & Burrows, R. (2013). Popular culture, digital archives and the new social life of data. *Theory, Culture & Society, 30*(4), 47-71. https://doi.org/10.1177/0263276413476542

Coopmans, C. (2014). Visual analytics as artful representation. In: C. Coopmans, J. Vertesi, M. Lynch, & S. Woolgar (Eds.), *Representation in scientific practice revisited.* (pp. 37-60). Cambridge, MA: MIT Press.

Daston, L., & Galison, P. (1992). The image of objectivity. *Representations, 40*, 81-128. http://doi.org/10.2307/2928741

Elish, M. C., & boyd, d. (2018). Situating methods in the magic of big data and AI. *Communication Monographs, 85*(1), 57-80. https://doi.org/10.1080/03637751.2017.1375130

Frow, E. (2014). In images we trust? Representation and objectivity in the digital age. In: C. Coopmans, J. Vertesi, M. Lynch, & S. Woolgar (Eds.), *Representation in scientific practice revisited.* (pp. 249-268). Cambridge, MA: MIT Press.

Glaser, B., & Strauss, A. (1967). *The discovery of grounded theory: Strategies for qualitative research.* New Brunswick: Aldine de Gruyter.

Halpern, O. (2015). *Beautiful data: A history of vision and reason since 1945.* Durham & London: Duke University Press.

Kennedy, H. (2016). *Post, mine, repeat: Social media data becomes ordinary.* Basingstoke: Palgrave Macmillan.

Kennedy, H., & Hill, R. L. (2017). The feeling of numbers: Emotions in everyday engagements with data and their visualisation. *Sociology, 52*(4), 830-848. https://doi.org/10.1177/0038038516674675

Kennedy, H., Hill, R. L., Aiello, G., & Allen, W. (2016). The work that visualisation conventions do. *Information, Communication and Society, 19*(6), 715-735. https://doi.org/10.1080/1369118X.2016.1153126

Leonardi, P. (2011). When flexible routines meet flexible technologies: Affordance, constraint, and the imbrication of human and material agencies. *MIS Quarterly, 35*(1), 147-168.

Locke, J., & Lowe, A. (2012). Process flowcharts: Malleable visual mediators of ERP implementation. In: P. Quattrone, N. Thrift, C. Mclean, & F.-R. Puyou (Eds.), *Imagining organizations: Performative imagery in business and beyond.* (pp. 99-125). New York: Routledge.

Lynch, M., & Woolgar, S. (1988). Introduction: Sociological orientations to representational practice in science. *Human Studies, 11*(2-3), 99-116.

Porter, T. M. (1995). *Trust in numbers: The pursuit of objectivity in science and public life.* Chichester: Princeton University Press.

Quattrone, P., Thrift, N., Mclean, C., & Puyou, F.-R. (2012). Imagining organizations: An introduction. In: P. Quattrone, N. Thrift, C. Mclean, & F.-R. Puyou (Eds.), *Imagining organizations: Performative imagery in business and beyond.* (pp. 1-15). New York: Routledge.

Rieder, B. (2017). Scrutinizing an algorithmic technique: The Bayes classifier as interested reading of reality. *Information, Communication and Society, 20*(1), 100-117. https://doi.org/10.1080/1369118X.2016.1181195

Ruppert, E., Isin, E., & Bigo, D. (2017). Data politics. *Big Data & Society, 4*(2). https://doi.org/10.1177/2053951717717749

Scott, S., & Orlikowski, W. (2012). Imagining technology in organizational knowledge: Entities, webs, and mangles. In: P. Quattrone, N. Thrift, C. Mclean, & F.-R. Puyou (Eds.), *Imagining organizations: Performative imagery in business and beyond.* (pp. 83-98). New York: Routledge.

Shapin, S. (1984). Pump and circumstance: Robert Boyle's literary technology. *Social Studies of Science, 14*(4), 481-520. https://doi.org/10.1177/030631284014004001

Sørensen, B. (2014). The method of juxtaposition. Unfolding the visual turn in organization studies. In: E. Bell, S. Warren, & J. E. Schroeder (Eds.), *The Routledge companion to visual organization.* (pp. 46-63). New York: Routledge.

van Dijck, J. (2014). Datafication, dataism and dataveillance: Big data between scientific paradigm and ideology. *Surveillance & Society, 12*(2), 197-208.

Venturini, T., Bounegru, L., Jacomy, M., & Gray, J. (2017). How to tell stories with networks exploring the narrative affordances of graphs with the Iliad. In: M. T. Schäfer & K. van Es (Eds.), *The datafied society.* (p. 155-170). Amsterdam: Amsterdam University Press.

About the authors

Salla-Maaria Laaksonen (D.Soc.Sc.) is a postdoctoral researcher in the Centre for Consumer Society Research, University of Helsinki. Her research areas are technology, organizations, and new media, including organizational reputation in the hybrid media system, the organization of online social movements, and the use of data and algorithms in organizations.

Juho Pääkkönen (M.Soc.Sc, social and moral philosophy) is a PhD Candidate in Sociology at the University of Helsinki, Faculty of Social Sciences. His research focuses on epistemic conceptions and methodological issues pertaining to computational methods and big data across social sciences and data analytics.

7. Accessibility of data visualizations: An overview of European statistics institutes

Mikael Snaprud and Andrea Velazquez

Abstract

Access to public data is important for people to stay informed. Access to visualizations of national statistics can be essential in order to take part in political discussions and so to shape a democratic society. In this chapter we investigate accessibility for people with disabilities to data visualizations from a selection of European National Statistics Institutes (NSIs). We outline related practices and approaches to accessibility improvements and propose a way to evaluate and compare accessibility aspects of data visualizations. The findings indicate that in contrast to the recently harmonized European legal requirements, the degree to which the data visualizations meet the requirements, and the approaches to meet them, are very different among the NSIs across Europe.

Keywords: Data visualization; Accessibility; Web Accessibility Directive; National Statistics Institutes; NSI.

Introduction

Data visualizations can inform citizens about political topics, and access to them for all citizens, regardless of ability and related technology use, is essential for democratic processes. The United Nations Convention on the Rights of Persons with Disabilities requires that appropriate measures are taken to ensure access for persons with disabilities, on equal basis with others, to information and communication technologies, including the internet. The European Web Accessibility Directive (WAD), transposed into

Engebretsen, M. and H. Kennedy (eds.), *Data Visualization in Society*.
Taylor & Francis Group, 2020
DOI 10.5117/9789463722902_CH07

national law for all the EU member states from September 2018, makes web accessibility a legal obligation. It requires that public sector bodies provide accessibility statements, including a list of content that is not accessible, the reasons for the inaccessibility and accessible alternatives to it, and a feedback mechanism for users to report accessibility problems on all of their websites, including in relation to data visualizations. National Statistics Institutes (NSIs) are a key source of such visualizations. Therefore, in this chapter, we focus on the accessibility of data visualizations produced and provided by European NSIs. We present results from our evaluation of the accessibility of data visualizations on NSI websites and from research with NSIs regarding their preparations to conform with the Directive.

In order to be accessible, the data visualizations (DVs), like all other web content, need to be perceivable, operable, understandable, and robust. For this chapter we leave out understandable, as an evaluation of such would require many more resources than we had available for our study. However, we add findability, the ease by which a piece of information on a website can be found (Jacob & Loehrlein, 2009; Wikipedia, Findability, 2018), since it can have an impact on the ability of citizens to participate in democratic discourse.

Perceivability and operability of DVs are related to general website accessibility issues. For example, menus that cannot be used via keyboard navigation or input fields that are not properly labelled can cause problems for web users with disabilities. We used Tim Berners-Lee's 5-star scheme, described below, to assess the robustness of data formats (5-star data, 2015).

We used an automated accessibility checker tool, WTKollen, to test the accessibility of the websites of 44 out of 59 European NSIs (WTKollen, European NSI sites, 2018), the results from which are presented as scores.[1] The results presented below therefore refer only to the 44 websites we tested, and not to others which, for various reasons, it was not possible to test prior to publication of this chapter. We also carried out expert testing of data visualizations found on these websites, and conducted email surveys and semi-structured interviews with appropriate staff within the NSIs.

A limitation of our work is that we did not carry out user testing of the websites and DVs with web users with disabilities. This is widely deemed to be the most appropriate way of evaluating the accessibility of websites (e.g. Coyne & Nielsen, 2001), but it is resource-intensive, especially in the case of EU-wide research such as ours. Automated checker tools, like accessibility measures more generally, also tend to privilege the needs of people with certain disabilities, such as visual impairment, and ignore the needs of

1 The WTKollen project is suported by the Swedish Post and Telecom autorithy.

others, such as intellectual disabilities (Kennedy, Thomas, & Evans, 2010). It is also the case that some tests cannot be automated—for example, automatic image processing may not be able to distinguish a cat from a dog in a blurry picture to determine whether an alternative text for the picture is helpful or not—and this is another limitation of automated accessibility testing. Despite these limitations, we think that our research can provide valuable insights into the extent of the accessibility of DVs across European NSIs.

Data collection

Automated accessibility testing

The European Web Accessibility Directive is based on the Web Content Accessibility Guidelines (WCAG) from the World Wide Web Consortium (W3C). The guidelines are intended to cover any online content, including DVs, for people with disabilities, such as visual, auditory, physical, speech, cognitive, language, learning, and neurological disabilities. The WCAG 2.0 (W3C, 2008) was replaced by WCAG 2.1 in June 2018 (W3C, 2018).[2] Following these guidelines will often make web content more accessible as well as serving other purposes. Proper use of alternative text descriptions on images, for example, can enable search engines to provide accurate search results. To guide the testing process, the W3C published the WCAG Evaluation Method 'WCAG-EM' in 2014 (W3C, 2014). This methodology offers guidance on the expertise required to test web accessibility, how to select webpages from a website, and how to report the findings. WCAG2.0 and WCAG-EM are therefore the basis for a range of web accessibility testing methods, tools, and legislation in Europe.

We carried out website accessibility evaluations with the WTKollen checker tool (WTKollen Page checker, 2018), which is based on WCAG 2.1 and WCAG-EM 1.0. Whereas the WCAG-EM guidelines indicate what to look for to design the accessibility tests, they do not specify exactly how to implement tests. The applied tests are listed online (GitHub, 2018)—not all of them are equally relevant to all people with disabilities. For example, colour contrast can be important for a person with visual impairment while irrelevant for a blind user. Hence different user groups would assign different weights to the same test. For the score calculation, we needed to have one weight only for each test. Therefore, we decided to let all tests have the same impact on the score.

2 To better meet the needs for three major groups: users with cognitive or learning disabilities, users with low vision, and users with disabilities on mobile devices.

To select pages from websites, we used a crawler, to find up to 6,000 pages. A random sample of 600 of these pages was used to represent the site to be tested. The score for a webpage was computed as the ratio of passed tests to applicable tests for each success criteria where associated tests applied. Similarly, the score for a website was aggregated over the test results for all the success criteria.

Manual test procedure

We uses a DV grouping proposed by Kirk (2012, p. 76) for the analysis of the DVs:

Exploratory visualizations which aim to allow readers to discover features by interrogating the data themselves;

Explanatory visualizations which aim to convey specific information to readers, based on a predefined narrative;

Exhibitory visualizations which are also based on data, but contain an artistic element.

Further, we also grouped the DVs by ways of interacting with them:

Static visualizations, such as a PNG image;

Dynamic visualizations which move, but without users activating them;

Navigable visualizations, which change based on user interaction;

Configurable visualizations, which enable users to select graph types or numbers, to move levels, or to select variables.

Our evaluation proceeded according to the following steps:

1. Locate the selected DV through a search on Google and local search, record rank (automatic accessibility check of the website as indication for how easy it is to navigate is done earlier)
2. Examine data presentation and downloadable data formats (i.e. how the data are provided, discussed below)
3. Group DV (according to Kirk's groupings discussed above)
4. Group DV (according to mode of interaction commented above)
5. Carry out manual accessibility tests for keyboard navigation, zoom, and textual description of the image (discussed below)
6. Look for accessibility feedback option (i.e. verify if there is an accessibility feedback mechanism on the page)
7. Look for supplementary services that may help the user understand the DV:
   7.1. FAQ: Is there an FAQ available from the page? What kind of FAQ (general info, specific content)?
   7.2. Languages: Is there support for multiple languages?
   7.3. Glossary: Is there a glossary to explain terms used in the statistics on the page?

The outcomes of the tests were recorded in a spreadsheet together with screenshots and URLs so that a test can be repeated if needed and for accountability purposes.

An important accessibility feature for people with motor impairments is the ability to navigate by using the keyboard or other input device instead of a mouse. If the page with the DV is designed to enable this, then users can reach all elements on the page and navigate forward and backward through the elements in the browser window with the tab and shift keys. Keyboard navigability can also allow users to select data and configure DVs.

The zoom feature is essential to magnify both text and images for people with visual impairments. Text zoom should render the text so that there is no need to scroll sideways. To test zoom features, it is necessary to explore whether the page has an option to enlarge the content or not. We evaluated the ability of the page to present the screen content enlarged.

There are two ways to provide text alternatives to images on webpages, thus making them accessible to people with visual disabilities: a short alternative text (alt-text) and a longer description: longdesc. The purpose of the longdesc is to provide more elaborate information when a short alt-text does not adequately convey the function or information provided in a non-text element on a webpage (W3C, 2016). Our test recorded whether the longdesc was used for complex DVs.

In many cases the data behind a DV are provided for download from the NSI site. Data formats have a strong impact on whether users can access and reuse the data. If a person is not able to use the DV, then a reusable data format can be more accessible and thus enable users to understand the data. Reusable data formats are also in line with the intention of the European Public Service Information directive, 'PSI' (EUR-Lex, 2005). The 5-star scheme proposed by Tim Berners-Lee is a practical way of evaluating the extent to which a given dataset can be reused (5-star data, 2015).

1-Star: the data are open; however, they are locked-up in a document making it hard to get them out of the document, e.g. in a PDF or JPEG.
2-Star: the data are accessible on the web in a structured way; however, they are still locked-up in a document depending on proprietary software, such as Microsoft Excel.
3-Star: the data are available on the web and can be manipulated in any way, without the need to own any proprietary software package, e.g. in CSV format.
4-Star: as above, and the data items have a URL and can be shared on the web, for example via links.

5-Star: as above, and it is also possible to link data to other data to provide context, to discover more related data while consuming the data, thus benefiting from the network effect, e.g. through a link to a Wikipedia article.

More stars means more reusable and also to some extent more accessible. For example, a screen reader, used by blind users, will not be able to read data in a JPEG image. In addition, if access to data is only possible with proprietary software, then users without the software in question will be unable to access it. The context provided in the 4- and 5-star levels does not really matter for accessibility, but is helpful for automated assessment of what the data are about. For our test, we detected if the data could be downloaded, and recorded the data format mapped to this 5-star scheme.

Finally, for interactive DVs, we also recorded the following two properties.

Comparability: number of variables that could be represented in the same graph.

Number of representations available: different kinds of charts available for representing the data, such as bar charts, linecharts, maps, pie charts.

Web accessibility for the NSI websites

The automated evaluation of website accessibility was carried out in the period from October 25 to November 13, 2018. The NSIs with the 12 highest scores are listed in Table 7.1. The highest score is awarded to the Irish NSI, followed by a group of 8 NSIs with a score of 99. At the lower end of the list, we find the NSIs from Greenland (score 67), Cyprus (69), and Iceland (71). To view the full list and the details about the detected accessibility issues, visit the webpage (http://axe.checkers.wtkollen.se/en/benchmarking/testrunresults/d235468e-65a4-43b2-8428-5908f061fff9).

The above list was up-to-date at the time of publication. For the manual testing 14 NSIs were selected, based on high accessibility scores (in alphabetic order): Czech Republic (CZ), Denmark (DK), Germany (DE), Ireland (IE), Luxembourg (LU), The Netherlands (NL), Norway (NO), Poland (PL), Spain (ES), Sweden (SE), Switzerland (CH), and United Kingdom Visual ONS (UK-visual), and on a suggestion from Statistics Norway that they contain interesting DVs: Portugal (PT) and Slovenia (SI).

Table 7.1 Overview of NSI websites and accessibility score from the WTKollen checker tool

Rank	Score	Country	NSI short name	URL
1	100	Ireland	Nisra UK	https://www.nisra.gov.uk/
2	99	Spain	INE ES	http://ine.es/
3	99	Sweden	SCB SE	http://www.scb.se/
4	99	Denmark	DST	http://www.dst.dk/
5	99	Germany	Statistikportal DE	http://www.statistik-portal.de/
6	99	Switzerland	BFS CH	http://www.bfs.admin.ch/
7	99	Norway	SSB NO	http://www.ssb.no/
8	99	Luxembourg	Statistiques	http://www.statistiques.public.lu/
9	98	United Kingdom	ONS UK	https://www.ons.gov.uk/
10	98	Czech Republic	CZSO CZ	https://www.czso.cz/
11	98	Poland	Stat PL	http://stat.gov.pl/
12	97	The Netherlands	CBS NL	https://www.cbs.nl/

DV findability on the NSI websites

We searched for statistics about national population as a case to obtain an indicator of the DV findability on the NSIs websites. Population is well covered across all NSIs and it also seems to be a popular search topic. In a first attempt we used Google to search for 'population' and the name of the NSI. For Norway the search phrase was then 'population SSB'. For a corresponding search for each of the selected NSIs, all but one NSI appeared as the first item in the search results list. For thirteen out of fourteen NSI sites the local search returned the relevant page in rank one. Both the Google analytics data from Statistics Norway and the experience from Eurostat indicate that it is more common to search just for 'population' without any NSI portal name. For some searches the first result in Google is data from the World Bank. For Norway these data originate from Statistics Norway NSI.

A DV included in the Google search results list (Google public data, 2018), such as the one we found from SSB in Norway, can be convenient for the user, who may not need to look any further for the requested statistics. The DV we found had colour contrast issues, but otherwise it was quite accessible, using SVG graphics and offering the ability to present the data in several languages.

Our initial approach to test findability was to use similar English search phrases to find population DVs across all NSIs (e.g. 'population SSB' for Norway). In the course of the study we noted that the content on the NSI websites is mostly prepared for the national audience, to be searched in a national language. Therefore, a search in English across different NSI may produce results that are not relevant for the targeted national users. To refine this result we could do a new search for population in the local language. We also note that the search engine result can depend on who is doing the search and from where.

Data presentation analysis

The data presentation was assessed for the fourteen evaluated NSIs. The accessibility score is the result obtained in the period from September 2018 to November 2018. The accessibility scores of the population data presented are:

Score of 95 to 99, a few tests failed: DE
Score of 85 to 95, some tests failed: ES, LU
Score of 70 to 85, many tests failed: CZ, IE, NL, NO, PT, SE, UK-visual
Score below 70, most tests failed: CH, DK, PL, SI

The data presented of twelve NSIs are configurable in tables in which it is possible to select the variables to show; only one is navigable (DE) allowing movement or arrangement of the data presented; and one static (UK-visual). For nine out the fourteen NSIs the keyboard navigation is enabled. Only four NSIs (CZ, ES, NL, NO) have the zoom feature; and in all but one (DE), the option to download the data is possible: CZ is 1-star; 10 NSIs are 3-star (that is, non-proprietary open format), NL 4-star and LU 5-star.

Exploratory/Interactive visualization analysis

Twelve out of fourteen NSIs have interactive tools to graph the data which enable users to produce their own visual representations of the available data. CZ and UK-visual do not have interactive DVs. PT graphs need Adobe flash player which is not accessible for people with disabilities, and the DE tool is provided only in German and therefore not evaluated. Therefore, only ten DVs were manually tested.

Only four out of ten DVs could be checked automatically: DK, NL, NO, and ES, mostly because the generated graphics do not have an explicit link to enter into the checker tool. Single page applications will present the same URL independent of user configuration of the DV. The accessibility scores are:

ES scored 85 to 95, some tests failed
NO and DK scored 70 to 85, many tests failed
NL scored 65, most tests failed

We note that all of the exploratory DVs tested have data to download, but only in a format not necessarily accessible and not suited for machine processing, mainly JPG and PNG format. Zooming is only supported by the DK, ES, NL, and NO examples and keyboard navigation is only supported in four out of ten cases (CH, ES, NL, PL). None of the DVs has longdesc enabled.

The DVs can be presented in different graphs depending on the nature of the data selected. For example, if the data selected do not include territories the map visualization is not a valid option. Some of the options available include bars, pie, lines, points, pyramid, and map. The number of graphs available is variable among NSIs, with the maximum fifteen (CH) and the minimum two (ES and PL). All the tools tested can compare multiple variables. By their nature, all the DVs are configurable.

We can identify six different tools in use among the 10 NSIs assessed. The first one used in Norway, Ireland, Slovenia and Denmark. The second in Sweden and Switzerland. While Spain, the Netherlands, Luxembourg, and Poland all seem to use different tools.

Explanatory visualization analysis

A total of twelve NSIs were evaluated for explanatory DVs. For the Luxembourg and Slovenia NSIs we did not find any explanatory visualizations. The accessibility scores obtained are the following:

Score of 95 to 99, a few tests failed: CH, CZ, DE, DK, ES, NL, NO, PL, SE
Score of 85 to 95, some tests failed: IE, UK-visual
Score of 70 to 85, many tests failed: PT

Eleven of the explanatory DVs tested are static and one is dynamic (NL). Only five out of the twelve have the option to download data:

1-star: CZ, IE, NO, PL
3-star: PT

Only two support keyboard navigation (NO, PT), five have the zoom ability (DE, ES, NL, NO, PL) and two support the longdesc (PT, UK-visual). The accessibility scores are higher for this category than for other DVs. This is because the content is typically simpler, mainly consisting of text and numbers. Even though the explanatory visualizations are simple, the option to download the data is not common. This makes the data harder to reuse.

Exhibitory visualization analysis

Ten NSIs were assessed for their exhibitory visualizations; for the remaining four (DE, DK, LU, SE) we did not find any exhibitory DVs. Exhibitory DVs are produced like artistic posters, in different formats mainly in PDF formats and in PNG. The scores were also calculated with the PDF checker when necessary:

NO and PL: 95 to 99, a few tests failed
CZ, IE, NL, and UK-visual: 85 to 95, some tests failed
SI PDF checker score: pass 6, fail 3
CH, ES, and PT PDF checker score: pass 5, fail 2

Seven exhibitory DVs are static, two dynamic (CZ, NL), and one navigable (UK-visual). None has longdesc; and only NL support keyboard navigation. Five out of the nine exhibitory DVs have zoom ability, and seven have data to download, all are 1-star, making it hard to reuse the information.

Services to support users to understand the DVs

None of the fourteen evaluated NSIs has an accessibility feedback form, although most of them have a general feedback form to comment on the data or the page. One NSI page has a phone and email address for accessibility feedback and two NSIs have accessibility statements. By September 2020, all NSIs will need to have an accessibility feedback mechanism in place on their websites to conform to the Web Accessibility Directive.

FAQs are found on nine out of the fourteen NSIs. We found four FAQs containing general information about the page and five FAQs about specific content like consumer prices, wages, or summer prices. The option to select national language or English is provided by twelve out of fourteen NSIs. The UK-visual content is provided in English, and does not support any other language, possibly because this website is focused on DVs. The German NSI is only in German.

Glossary access to explain terms used on pages or in DVs is provided by nine out of fourteen NSIs. The glossary entries are found directly on the page, provided as references to an internal glossary or external ones like the ones from the OECD or from Eurostat's Statistics Explained (Eurostat, Statistics Explained, 2017).

NSI practices relating to DV accessibility

To supplement the analysis discussed above, we also used surveys and interviews. Statistics Norway helped us to distribute two surveys to the network of European NSIs. Both surveys had only three questions each, to keep them simple and to increase the response rate. The first survey was intended to get an idea about how the NSIs are preparing for the WAD, raise awareness about detected barriers with accessibility evaluation results, and to collect some input on how evaluation results can be shaped to enable the NSIs to understand them and to use them to repair the reported barriers. The second survey was designed to capture good DV examples and developments in the WAD preparations. The first survey, sent out by Statistics Norway in October 2017 to about 80 representatives from 39 NSIs in Europe, received a response from about 20 percent. The second survey, sent in April 2018 to the same group of respondents, had a lower response rate of 13 percent. One possible reason for the lower response rate may be a focus on the General Data Protection Regulation (GDPR) at this time. Interviews were carried out with Statistics Norway and with Eurostat.

The practical responsibility to make sure that DVs are accessible lies with software developers or with communications departments within the NSIs. Useful accessibility input has in several cases been obtained from colleagues with visual impairments. External consultants are sometimes contracted to audit overall websites. This is a costly operation and therefore not carried out regularly. Advanced and regular usability testing has been in place for a long time across a number of NSIs, to ensure that the statistics can be found and used. It seems that accessibility is an emerging topic to be included in regular testing activities for NSI online content.

Several different automatic tools are used by the NSIs to evaluate accessibility. Commonly used tools are Site Morse (see https://sitemorse.com/) and aXe Core (see http://deque.com/). One of the NSIs also reported that they intend to build a new tool. Such tools are helpful and cost-effective to operate, but not always straightforward to use. One important caveat is that these tools do not cover all conceivable tests.

From the first survey we found that respondents planned to pursue mainly two different approaches to improving their website and DV accessibility. They planned to invest in human resources which include staff training programmes and hiring accessibility expertise consultants, and facelifts or complete redesigns of their website.

Together with the survey we provided a benchmarking list similar to Table 7.1. The respondents were asked to comment on the results form. The checker tool (WTKollen, European NSI sites, 2017) was perceived as useful by respondents, and the findings verified that older or more complex webpages are more likely to have accessibility barriers than newer or simpler pages. For future tool development, respondents said that they would like to have a readability test and image evaluations. There was also a suggestion to group webpages according to complexity for better comparisons. To improve the presentation of the results in the checker report, the following suggestions were made:

- The results could have example images for better and faster understanding.
- The results could be sorted (e.g. by error status or importance)
- The page could be responsive
- It would be nice to be able to export all the errors to .csv/.pdf/.xlsx, as this would help the organization of corrections.

Some respondents raise concerns about tools since they do not always return the same result for the same content on a webpage. Such differences may even have prevented people from using checker tools at all. We also noted that some NSIs expected an official tool to be prescribed by the European Commission or their national ministry. However, the WAD is prepared in a way that is tool independent and there is no tool mentioned in the WAD implementation act.

In the second survey we requested users to provide examples of accessible DVs. Most respondents declined and indicated that they were working on this now. In our view, there is great potential in using DV templates from Eurostat to spread good accessible practices. We also asked about NSIs' preparations for the provision of accessibility statements and feedback mechanisms. Several NSIs have accessibility statements, sometimes linked from the page footer. However, in general, such statements do not list known deviations from accessibility requirements. Mostly they provide information about accessibility features of the sites. One NSI has had an accessibility statement since 2005 and regularly performs tests in cooperation with external experts. Many respondents aim to collect the information for the accessibility statement from the accessibility reports produced by officially adopted verification tools, or from user feedback.

In terms of preparations for the provision of feedback mechanisms, we recorded two approaches. One is to use a dedicated email to receive reports about accessibility problems. This approach can make it difficult for a user to remain anonymous, and it can also become hard to manage responses and task assignments for large volumes of feedback. The second approach is to use a general feedback mechanism already existing on the website. This may meet the formal requirements, but such feedback mechanisms are not designed to (automatically) collect data on accessibility problems, or to export reports about them to share good practices in terms of fixes or repair approaches.

Sharing good practices

There are at least three current approaches to reusing good DV examples. The Digicom project is an initiative to share good practices among the NSIs. As part of this project, Eurostat has developed templates to present DVs of statistics. These templates are designed so that they are easy to translate and connect with an API to the data from Eurostat.

There are also a range of DV libraries that can be used to reuse good DV examples, like D3, Google charts, or Highcharts. In more accessible solutions, DVs are scalable to allow for zooming, and have functionality to encourage or force the developer to describe the non-textual elements. Such encouragement could raise developers' awareness of inaccessibility impacts for users with disabilities. A simple export of the data can also be helpful to enable users to explore the data with a tool of their own preference.

Presentation through aggregators like Google can also be efficient. The Google Public Data Explorer (see https://support.google.com/publicdata) provides large, public-interest datasets from sources like Eurostat and the World Bank in a common presentation format. With this service, the user can find datasets and explore them with different chart types. Two important advantages of this approach are, first, that users will be familiar with the user interface and, second, that they will easily find it, since Google has over 90 percent of the European search market share (StatCounter GlobalStats, 2018). However, such intermediary access can also be used to track users and to prevent the user from finding the original data source with more updated data or further information about the dataset.

We have not been able to identify a reference library of DVs. The Internet Archive is a valuable resource for longtime references for a large portion of the online content. Unfortunately, this archive does not have all the relevant pages from the NSIs and it cannot store the dynamic features of most dynamic DVs. The Internet Archive also will not have direct access to the static databases often serving the 'live' data to the dynamic DVs.

Conclusions

There are good examples of DVs where we found few accessibility barriers. However, despite the Web Accessibility Directive, there is still lot of room for improvement. There are several different accessibility testing tools in use among NSIs to test the accessibility of their websites and their DVs. In our survey we were not able to find examples of NSIs who systematically apply user testing approaches to uncover accessibility issues. Several report that they occasionally ask a colleague with disabilities to test content. For the WAD preparations, we saw very limited work towards design of an accessibility statement or to organize a feedback mechanism.

In general, NSIs are aware of accessibility issues. Still, three factors seem to have hindered focused progress towards comprehensive accessibility provisions, and to prepare for the WAD. Several NSIs indicated that they would wait for the WAD implementation act to be finalized before they would take action. The General Data Privacy Regulation seemed to demand more attention, as there are high fines associated with a breach compared to accessibility problems which breach the WAD. The third reason is differences in accessibility checker tool reports for the same element on a webpage. Some NSIs expect that an official tool will be named. The draft implementation act for the WAD does not refer to any named accessibility tool, and there seems to be no intention to use a particular tool for the implementation from the regulators as far as we have been able to find out.

There are several different accessibility testing tools in use among NSIs to test the accessibility of their websites and their DVs. For exploratory, interactive visualization we have found six different tools in use. Given this relatively small number of tools, targeted improvements of them can have a large effect for many users. Whatever approach is used, the central role of the NSIs and their DVs in national democratic discourse calls for particular awareness of accessibility.

Acknowledgements

To the INDVIL team for all the fruitful discussions and inputs, to the WTKollen project for providing the accessibility checker tools to test websites, webpages with DVs, and PDF documents. And last but not least, many thanks to Helen Kennedy for her helpful comments and substantial support to shape this chapter.

References

5-star data. (2015). 5 star open data. Retrieved October 22, 2018 from https://5stardata.info/en/

Coyne, K., & Nielsen, J. (2001). *Beyond ALT text: Making the web easy to use for users with disabilities*. Fremont, CA: Nielsen Norman Group.

EUR-Lex. (2005). *Decision No 456/2005/EC of the European Parliament and of the Council of 9 March 2005 establishing a multiannual community programme to make digital content in Europe more accessible, usable and exploitable*. Retrieved from https://eur-lex.europa.eu/legal-content/EN/TXT/?qid=1535296253785&uri=CELEX:32005D0456

Eurostat—Statistics Explained. (2017). Welcome to statistics explained. Retrieved October 22, 2018 from http://ec.europa.eu/eurostat/statistics-explained/index.php/Main_Page

GitHub. (2018). Axe Core tests. Retrieved October 22, 2018 from https://github.com/dequelabs/axe-core/blob/master/doc/rule-descriptions.md

Google public data. (2018). Public data. Retrieved October 22, 2018 from https://www.google.com/publicdata/explore

Jacob, E. K., & Loehrlein, A. (2009). Information architecture. *Annual Review of Information Science and Technology, 43*(1). https://doi.org/10.1002/aris.2009.1440430110

Kennedy, H., Thomas, S., & Evans, S. (2010) Can the web be accessible for people with intellectual disabilities? *The Information Society*, 27(1), 29-39. https://doi.org/10.1080/01972243.2011.534365

Kirk, A. (2012) *Data visualization: A successful design process.* Birmingham: Packt Publishing Ltd.

StatCounter GlobalStats. (2018). Search engine market share Europe. Retrieved October 22, 2018 from http://gs.statcounter.com/search-engine-market-share/all/europe

W3C. (2008). Web content accessibility guidelines (WCAG) 2.0. Retrieved October 22, 2018 from http://www.w3.org/TR/WCAG20/

W3C. (2014). Website accessibility conformance evaluation methodology (WCAG-EM) 1.0. Retrieved October 22, 2018 from https://www.w3.org/TR/WCAG-EM/

W3C. (2016). Using longdesc. Retrieved October 22, 2018 from http://www.w3.org/TR/WCAG20-TECHS/H45.html

W3C. (2018). Web content accessibility guidelines (WCAG) 2.1. Retrieved October 22, 2018 from https://www.w3.org/TR/WCAG21/

WTKollen—European NSI sites. (2017). Checked sites from European NSI sites. Retrieved October 22, 2018 from http://checkers.wtkollen.se/eu-nsi

WTKollen—European NSI sites. (2018). Checked sites from European NSI sites. Retrieved October 22, 2018 from http://checkers.wtkollen.se/en/benchmarking/testrunresults/d235468e-65a4-43b2-8428-5908f061fff9?Sector&flags.fulltestrunresult

WTKollen—Page Checker. (2018). Check the accessibility of a web page. Retrieved October 22, 2018 from http://checkers.wtkollen.se/

Wikipedia, Findability. (2018). Findability. Retrieved October 22, 2018 from https://en.wikipedia.org/wiki/Findability

About the authors

Mikael Snaprud is the CEO of Tingtun AS. He has managed several projects in the area of e-Inclusion with European and national research grants. He has 20 years of experience from research and teaching, and has co-authored reports about ICT policies for the European Commission and the United Nations.

Andrea Velazquez works as scientific adviser at Tingtun AS. She has been part of the team of several e-Government projects funded by the European Commission and the Research Council of Norway. In the past she worked in processes benchmarking for quality engineering, and productivity programmes in the industry.

8. Evaluating data visualization: Broadening the measurements of success

Arran L. Ridley and Christopher Birchall

Abstract

This chapter investigates the evaluation of data visualizations using observational research in an award-winning design studio. It outlines some professional and commercial forces that are involved in the shaping of evaluative strategies and identifies differences in methods and forms of evaluation in projects with different aims and intended audiences. The research showed that alongside quantitative headline figures of consumption, such as audience reach and interaction, qualitative measures of audience experience—which consider the sociocultural context of consumption—were sometimes included in evaluation strategies, but this varied between projects depending on the level of access to, and knowledge about, the audience. This chapter highlights the importance of such measures, outlines attempts to develop them, and comments on the potential to do so.

Keywords: Evaluation; Data visualization; Sociocultural context; Audience; Design studio; Observation

Introduction

Data visualization plays an important role in the information environment, communicating complex messages through simplified yet powerful representations of otherwise opaque data. Data visualization is deployed in different contexts and in various settings, such as the representation of business information within or between companies, the delivery of

Engebretsen, M. and H. Kennedy (eds.), *Data Visualization in Society*.
Taylor & Francis Group, 2020
DOI 10.5117/9789463722902_CH08

products and services to personal consumers, and the communication of news and information in the public sphere. Within this broad landscape, data visualization is deployed to achieve different aims and objectives and can be tailored to do so for specific audiences. In a business setting, for example, the audience may be trained to expect, interpret, and use data visualizations, in formats with which they become familiar, to meet the requirements of the business. In contrast, consumers within the wider public sphere are a diverse and much more unpredictable audience within which the skills and experience, time, and environment required to successfully navigate complex data visualizations cannot be known. Evaluation of data visualizations can, therefore, be complex when these products are designed to meet goals which exist on a spectrum from cognitive, affective, behavioural, and even physiological human responses on the one hand (Zube, Sell, & Taylor, 1982), to the less personal modern commercial and communications imperatives, such as views, shares, click-throughs, and sales conversions, on the other. For this reason, strategies for evaluating data visualizations can include dimensions such as the metrics of reach, consumption, and audience interaction that are common in social media and web analytics (Aisch, 2017; Baur, 2017; Wattenberg, 2005) and the user testing and feedback of HCI (Human-Computer Interaction) and usability research (Freitas, Pimenta, & Scapin, 2014; Vogel, Kurti, Milrad, & Kerren, 2011). In some cases attempts may be made to measure understanding and impact generated by the visualization within audiences (Sheppard, 2005), and work such as the Seeing Data Project highlighted the potential for the collection of qualitative data from consumers to capture their opinions and feelings about a visualization (Kennedy & Hill, 2017).

This chapter describes how different production and consumption contexts can combine to create the need for different evaluative practices in different situations, where data visualizations operate under different conditions, with different audiences, and with different aims and goals. It also illustrates how professional production processes can support and prioritize some of these evaluative practices more than others. The existence of conventions within data visualization production—discussed in detail in the literature (Barnhurst, 1994; Coopmans, Vertesi, Lynch, & Woolgar, 2014; Few, 2012; Kennedy, Hill, Aiello, & Allen, 2016; Kosara, 2007a; Tufte, 2001)—is part of a network of influences that help to shape the processes of data visualization practice. Here, these processes are described also as an influence on the set of values used to evaluate data visualizations, which may limit the methods and forms of evaluation used. These values are derived from factors which exist largely on the production or supply side of data visualization, and while there have been studies of conditions on the demand side (Kennedy et al., 2016) the factors affecting consumption of data visualizations, such as the sociocultural context of consumption, are less well documented or established. Audience preferences, abilities, and experiences are often difficult to measure, particularly in real time, and so evaluation methods often include assumptions about audiences rather than involving them in evaluative processes.

Through observational fieldwork within a leading data visualization production studio, this chapter illustrates how the sociocultural context of consumption may be considered during the evaluation of visualizations in some production pathways, particularly where that context is most readily accessible, but not in others. From this we build a broader argument about the potential improvements to evaluative practices that could be enabled by the consideration of sociocultural contexts of consumption within evaluation, which might enable us to better understand how the social context of the consumer can impact the reception, and perhaps inform the design, of visualizations that perform important functions in the public sphere. Although the chapter is based on a small-scale exploratory study on only one agency, we propose that it nonetheless provides some useful food for thought.

Interrogating the sites of production

The decisions about evaluative practices on the supply side in which practitioners, clients, and other interested parties judge a product according to their preferred success criteria are made by various people in various roles. At its simplest definition, a practitioner can be considered as any person who produces a data visualization, such as the freelancers or academics who might be involved in collecting and sorting data, selecting tools and chart types, and producing and deploying finished data visualizations to suit their needs. However, data visualizations are often produced in professional settings such as design studios, within which multiple actors are present, each with varying degrees of influence on the production process. These actors extend beyond those directly involved in the design process (such as visual designers, coders, or UX/UI designers) to include project managers, PR staff, upper management, and clients (a catch-all term that itself includes a complex network of actors and stakeholders who may not all share the same goal). Although they may come from different disciplinary backgrounds, data visualization practitioners are sometimes expected to have a skill set that encompasses everything from the ability to collect, scrape, and sort data, through to the production of a data visualization itself (Kirk, 2016). According to Kirk (2016), skills such as strong numeracy and a familiarity with basic statistics, as well as some knowledge of common spreadsheet software are prerequisites, as are a sense of curiosity and an openness to creativity regardless of previous experience or understanding of design guidelines. Data visualization practitioners can also be categorized by their primary skill sets: 'designers' whose main experience is in design contrast with 'programmers' who 'create visualizations and visualization tools programmatically' (Bigelow et al., 2014). Bigelow et al. (2014) claim that both, however, are thought to share expertise in data analysis.

The diverse skill sets and interdisciplinarity of data visualization practitioners are important influences in the process of turning data, through the steps of analytical abstraction, into visual representations. These influences may be acknowledged by practitioners through efforts to provide transparency and authenticity to datasets, through the inclusion of data sources, annotations, or corrections in an attempt to create 'data provenance' (D'Ignazio, 2015; Hullman & Diakopoulos, 2011; Tufte, 2006). These practices—and the conventions that help to shape them—are part of the 'editorial layers' that shape visualizations (Hullman & Diakopoulos, 2011), but they can also impact decisions made about evaluation by influencing the values and priorities that are the focus of evaluation efforts. The production environment and processes are, therefore, an important factor in any investigation into the evaluation of data visualizations, and it is pertinent to include the site of data visualization production—the people, roles, relationships, rules and conventions, aims, goals, and pressures—in this analysis. As well as questioning the 'material economy behind the data' (D'Ignazio, 2015), investigating the site of data visualization production can make visible the role of different actors present and make it possible to ask questions about what influence each stakeholder has on evaluative practices and under what conditions these influences might be exerted.

The empirical evidence presented here was collected over four weeks on-site at an award-winning commercial design studio which specializes in data visualization. The studio produces data-driven products, such as business dashboards and static or interactive data visualizations for print and digital. They have around 30 employees, including permanent and temporary staff such as freelancers and interns, and serve domestic and international clients of varying size. Participant observation was undertaken to gain an understanding of production processes at the studio and to analyse the relationship of these with the design and implementation of product evaluation. Two live projects were followed throughout the period of observation. First, a large-scale production of a business dashboard, being designed to streamline the existing process of producing reports within a client's corporation. Second, a set of data visualizations intended to form the basis of a style guide to be utilized by a client. Alongside these live projects, data were collected about several completed projects of which one—a healthcare app designed to aid in the recovery of cancer patients—will be discussed in this chapter. During observation of the live projects, documentation of meetings and interviews with the project team members were undertaken, as well as analysis of the supporting documents produced. For the completed projects, interviews were conducted alongside analysis of the archived documentation of the project process.

Considering evaluation methods

The arguments of the previous sections have posited that design conventions and assumed best practices amongst communities of practitioners shape visualizations. Data visualizations are produced according to different aims and strategies and thus the mode of evaluation utilized in each case would be expected to vary accordingly to the goal of the producer. It is widely accepted that the general goal of data visualization is to visually communicate non-visual data (Kosara, 2007; Manovich, 2011; Munzner, 2014), but this oversimplifies the different goals of practitioners, who are working within different industries or disciplines and therefore have differing measures of success. A data visualization can be designed to be a clear communication of data for business purposes, a powerful political message, an emotive headline to attract eyeballs to a particular news item, or a beautiful artefact garnering attention in its own right (Rost, 2017).

Evaluating visualizations according to these different goals is challenging and some success criteria are easier to measure than others. For example, counting how many people have seen a visualization might be an appropriate and attainable evaluation strategy if the goal is to achieve a wide circulation, but measuring an emotional response to a data visualization, if such an outcome was the project goal, requires a much more complex evaluation method. Often the most accessible method available is selected and web and social media analytics provide convenient quantitative methods of evaluation. These popular approaches to evaluation are attractive to the producers of data visualization within a commercial design studio as they measure some dimensions of consumption that translate into commercial success. While mindful of the range of qualities required in the finished artefact, practitioners must consider how these artefacts can be utilized to generate more commissions. Popular measures such as shares or likes are a means of signalling the potential to reach prospective customers. One such analytics-based evaluation examining a visualization on the *New York Times* website found that only 10-15 percent of all visitors to the page clicked on the visualization, and concluded that this interactive graphic was therefore a 'waste of time and money' (Baur, 2017). The impact of the visualization on the consumer is not measured through this metric, however; any increase in knowledge or other impact on the consumer is not known. Herein lies the issue with employing numeric measures of consumption as a means of evaluation: it might capture the metrics of 'engagement', the amount of times something has been interacted with, shared, or viewed, but it doesn't capture *why* this is happening. Engagement with the process of consumption can help to answer these more difficult questions.

Evaluation is not only done after distribution of a final product. It can also take place at various places along the development process, at the predesign, design, prototype, deployment, or redesign stage (Lam et al., 2012). User testing is a common form of evaluation within design processes, which can be utilized in different ways. As part of academic research, it often takes place within laboratory settings and deals with specific elements such as memorability (Borkin et al., 2013), speed of task completion or recall (Chin et al., 2009), or the effectiveness of particular visual elements (Skau & Kosara, 2016). When combined with measurement of participant satisfaction, such studies sometimes aim to judge whether certain visualization techniques are more or less effective than others for representing and communicating data (Chin et al., 2009; Haroz & Whitney, 2012). The focus of such approaches to user testing is often on the data visualization itself, the effectiveness of the visual elements, and presentation styles. These are important factors in the design of visualizations, but so is their consumption. Within the controlled conditions of the lab, such tests can evaluate certain elements of data visualizations, but they do not take into account the conditions of consumption in 'the wild'. Outside of the lab, where consumers encounter media products in diverse, often unpredictable situations, the impact of visualizations may vary as consumers experiment with or appropriate products, devoting different amounts of attention, or utilizing different emotional or cognitive processes (Oudshoorn & Pinch, 2003). As Kennedy, Hill, Allen, and Kirk (2016) argue 'who users are, contexts of visualization use and other factors outside of the visualization text are also important in determining visualization effectiveness'. Some such studies, situated in the field of Human Computer Interaction and UX/UI design, do not provide much information about the participants themselves, even in studies with low numbers of participants, as Kennedy, Hill, Allen, and Kirk (2016) note. Participants in these studies are rarely representative of the general public or even indeed the intended audience, reflecting the emphasis on putting new features or chart types to the test.

Observations of evaluative practices within a design studio

Three cases examined during fieldwork, including the two live projects and one completed project, were designed with different goals and with different audiences in mind. These differences led to different processes being utilized during production and these differences were mirrored in the evaluation strategies used for each project.

The first live project studied—the business dashboards—involved a product used by the client to analyse data and to inform decisions. This was produced for a specific business context, defined by business processes and needs, which determined specific, measurable success criteria. As the product was used by its employees, the client brought an understanding of the environment and conditions in which the product was consumed. Design studio staff working on the project made considerable efforts to understand the end users themselves through interviews and workshops, to extend the knowledge brought by the client. Participants at these workshops were asked to detail their thoughts and feelings in relation to their current workflow, which helped the practitioners to understand the key moments and actors in the decision-making process and where within this process the tool would sit. Consistent contact with the client was maintained through an iterative development and evaluation process, in which versions of the tool were released to the client who in turn tested them and fed back to the practitioners, who then amended the product and initiated further cycles of this process until the tool was considered complete. In addition to this testing carried out with the client, further user testing was undertaken internally with other practitioners not attached to the project, in which novel or new functionality or features were tested or 'validated' before being revealed to the client. This process allowed for a great deal of information to be gathered about the sociocultural context of consumption of the visualization within the client organization, within an evaluation method embedded within the production process.

In one archived project that was observed, which involved the design of visualizations within a mobile app to be used by survivors of cancer, the sociocultural context of consumption was also considered during design. This time, however, this inclusion was more limited, reliant upon the expertise of the client rather than user testing and iterative design. The app was designed to provide personalized information to help users to respond to the demands of the illness and its treatment and to aid recovery by relaying custom instructions for the user to follow. The context of consumption was therefore quite specific, and the client was able to relay clear expectations of the desired outcome to the practitioners based on detailed knowledge of the consumer. However, unlike the controlled setting of the corporate office in which the business dashboards would perform, these visualizations were likely to be consumed in a wide variety of contexts by consumers who might share the common experience of being cancer survivors but have different preferences, capabilities, contexts, and experiences of consumption. The potential users of an app like this are harder to recruit for user testing, due to the very specific nature of the audience, and their condition being sensitive in nature. In this case, knowledge of the sociocultural context of consumption was provided by the expert client. Where this project differs from the business dashboard project is in the greater potential scope of consumption contexts and the limited access to the consumer. In these situations, practitioners made use of other available resources for user testing (studio staff members, for example); a practice not unusual in the design community (Dickey-Kurdziolek, 2018).

The second live project observed during the research period was the design of a style guide for a client and this project involved far less engagement with the sociocultural context of consumption. Rather than producing public-facing visualizations directly, this product was designed to enable their production by the client themselves, who could utilize different datasets to produce visualizations for different audiences, in accordance to the design guidelines laid out in the style guide. These guidelines were produced using 'fake' or 'dummy' data as placeholders for future content, in contrast to the business dashboard which involved a lengthy investigation into the properties of the available data before design. The design and evaluation process of this project relied upon assumptions about data visualization best practice rather than on considerations relating to the end users of the client's visualizations.

The three projects discussed here show how different production scenarios can influence evaluation strategies and specifically how the sociocultural context of consumption may or may not be included within them. The business dashboards project involved a consumption context that was known and defined, through business logics, expertise, and testing carried out by the client and the practitioners. This context could therefore be incorporated into comprehensive evaluative strategies within an iterative development process. The app for cancer sufferers involved descriptions of potential sociocultural context of consumption, provided by the client. However, while the client was able to detail some specific information about the potential end user, the consumers themselves—and their personal consumption environments and situations—remained out of reach for the practitioner. Finally, the style guide was designed to direct future visualization production by the client and as such was produced without knowledge of either the datasets to be included or the audiences to be reached. In this case, evaluation incorporating the client's consumers and their context of consumption was unlikely, with assurances of quality instead provided by the assumed best practices embedded in the style guide.

Evaluating the sociocultural context of consumption

Visualizations are often produced and consumed in diverse environments, and both the processes of encoding, during production, and decoding, during consumption, are subject to the 'sociocultural milieu' of the producer and consumer (Hall, 1973). These diverse environments could possibly lead to unpredictable consumption effects, and may result in the data visualization being decoded by the consumer in ways that may be unexpected by the producer. The consumer undertakes a complex process of decoding through cultural, perceptual, cognitive, and psychological lenses to extract meaning (Hullman & Diakopoulos, 2011). Educational and class background can be one influence on this (Bourdieu, 2010; Hall, 1973); the role of emotions in engagement with media artefacts, particularly when content matter is emotive, may be another (Hakone et al., 2017). The examples discussed above illustrate how this decoding process is evaluated to different extents, and that the sociocultural context of consumption can be given greater or lesser attention during evaluation in different projects.

As discussed earlier, consumption of online visualizations can be measured in digital spaces in terms of audience reach and interaction through popular web and social media analytics, but metrics relating to the impact of visualizations on individuals are harder to come by. How does one measure the amount of knowledge gained from a visualization, or if the knowledge taken away is what the producer intended? How can the emotional impact of visualizations be determined? Moreover, other factors relating to consumption, such as whether the consumer is alone or with others, concentrating on a subject or viewing casually, technological devices used, or viewing environments (such as at home, on the bus or in a pub), are rarely considered and difficult to capture. These factors—the sociocultural context of consumption—provide a significant challenge to efforts to build comprehensive evaluation methods for data visualizations, and therefore to our ability to build a complete picture of their consumption and impact.

The Seeing Data project shed light on this issue, identifying six factors which could impact engagement with data visualizations: subject matter; source/media; beliefs and opinions; time; emotions; and confidence and skills (Kennedy et al., 2016). Echoing aspects from media effects theory, these factors can have an impact on various groups in society and on individual media users (Potter, 2012; Valkenburg, Peter, & Walther, 2016). Amongst the findings of the Seeing Data study were clear examples of how engagement with visualizations varied according to factors other than the design decisions made during production. Consumers engaged more closely with visualizations which focused on one of their pre-existing interests; visualizations from particular sources were judged to be more or less trustworthy; enjoyment of visualizations varied depending upon whether preformed opinions were challenged (Kennedy et al., 2016). The sociocultural context of consumption is therefore an active factor influencing impact and must be taken into consideration within the evaluative process. Beyond the best practice principles of professional designers is the messy social world where consumption takes place and a myriad of personal and cultural influences impact engagement, enjoyment, and comprehension.

While rare in practice, opportunities for sociocultural evaluation do exist. The Seeing Data project experimented with a 'widget' (2017) within data visualizations that allowed consumers to submit emotional or cognitive feedback about their experiences. Contemporary digital communications technologies offer potential for more evaluation like this, with possibilities for increasing the scale of such studies offered by technical solutions such as browser plug-ins and mobile apps that allow feedback to be gathered on visualizations as they are encountered in everyday digital consumption. Like the 'widget' of the Seeing Data project, these technologies provide the potential for consumers to participate actively in evaluation, reporting on their experiences and reactions, understanding, or other dimensions of impact, offering the possibility of richer, deeper evaluation which considers the personalized environments of consumption and the effects of visualization upon individuals, rather than simply the rate of consumption in general. Of course, within any attempt to produce more context-specific methods of evaluation, the communicative strategies and aims of production projects need to remain in focus, be considered, and matched to evaluation methods, so that the most suitable evaluative method, which meets both the aims of producers and the context of consumers, can be implemented.

Conclusion

Different project goals can be associated with different success criteria, and these can lead to different evaluative methods being employed. These goals and success criteria are likely to be focused on supply-side demands, such as audience size or interactions, and less often engage with the demand side—the sociocultural context of consumption. Observational fieldwork in a data visualization studio suggested that the most quantitative and simplest measurements of consumption through popular analytics methods were employed where client or project goals gave them value, and the sociocultural context was only embedded within evaluative methods where resources and expertise allowed, and where commercial imperatives made it appropriate and gave it value, too. Where intended audiences were known and definable, and where the sociocultural context of consumption was relatively well understood, efforts to incorporate this context into both production and evaluation could be comprehensive. However, where data visualizations were aimed at audiences that were more diverse and less well understood, the sociocultural context of consumption was much less likely to feature in evaluation. The measurement of the diverse consumption contexts present in the public sphere is of course difficult. It is within these public sphere environments, though, that sociocultural context is at its most variable and likely to assert its greatest effect. Data visualizations are produced to serve many purposes, but by making information accessible and available in the public sphere they can have implications for democratic functions such as decision-making and preference formation by citizens. Of course, in a commercial production environment the needs of the practitioner and clients must be acknowledged, and context-specific evaluation needs to be considered in relation to, rather than in place of, existing evaluative methods. If some of the affordances of digital media—in apps, websites, and browsers—can be utilized to incorporate sociocultural evaluation into existing evaluative practices, however, a richer and more rounded form of evaluation could be developed that includes measurements of impact and experience alongside the quantitative metrics that exist already.

References

Aisch, G. (2017). In defense of interactive graphics. *Vis4*. Retrieved November 2, 2017 from https://www.vis4.net/blog/2017/03/in-defense-of-interactive-graphics/

Barnhurst, K. G. (1994). *Seeing the newspaper.* New York: St. Martin's Press.

Baur, D. (2017, March 13). The death of interactive infographics? *Medium*. Retrieved from https://medium.com/@dominikus/the-end-of-interactive-visualizations-52c585dcafcb

Bigelow, A., Drucker, S., Fisher, D., & Meyer, M. (2014). Reflections on how designers design with data. *Proceedings of the 2014 International Working Conference on Advanced Visual Interfaces—AVI '14*. http://doi.org/10.1145/2598153.2598175

Borkin, M. A., Vo, A. A., Bylinskii, Z., Isola, P., Sunkavalli, S., Oliva, A., & Pfister, H. (2013). What makes a visualization memorable? *IEEE Transactions on Visualization and Computer Graphics, 19*(12), 2306-2315. http://doi.org/10.1109/TVCG.2013.234

Bourdieu, P. (2010). *Distinction: A social critique of the judgement of taste.* (R. Nice Trans.). London: Routledge.

Chin, G., Singhal, M., Nakamura, G., Gurumoorthi, V., & Freeman-Cadoret, N. (2009). Visual analysis of dynamic data streams. *Information Visualization, 8*(3), 212-229. http://doi.org/10.1057/ivs.2009.18

Coopmans, C., Vertesi, J., Lynch, M., & Woolgar, S. (2014). Introduction: Representation in scientific practice revisited. In: C. Coopmans, J. Vertesi, M. Lynch, & S. Woolgar (Eds.), *Representation in scientific practice revisited.* (pp. 15-36). Cambridge, MA: MIT Press.

D'Ignazio, C. (2015). What would feminist data visualization look like? Retrieved September 5, 2018 from https://civic.mit.edu/2015/12/01/feminist-data-visualization/

Dickey-Kurdziolek, M. (2018, January 4). Discovery on a Budget: Part I. Retrieved from https://alistapart.com/article/discovery-on-a-budget-part-i

Few, S. (2012). *Show me the numbers: Designing tables and graphs to enlighten* (2nd ed.). Burlingame, CA: Analytics Press.

Freitas, C. M. D. S., Pimenta, M. S., & Scapin, D. L. (2014). User-centered evaluation of information visualization techniques: Making the HCI-InfoVis connection explicit. In: W. Huang (Ed.), *Handbook of human centric visualization.* (pp. 315-336). New York, NY: Springer. http://doi.org/10.1007/978-1-4614-7485-2_12

Hakone, A., Harrison, L., Ottley, A., Winters, N., Gutheil, C., Han, P. K. J., & Chang, R. (2017). PROACT: Iterative design of a patient-centered visualization for effective prostate cancer health risk communication. *IEEE Transactions on Visualization and Computer Graphics, 23*(1), 601-610. http://doi.org/10.1109/TVCG.2016.2598588

Hall, S. (1973). *Encoding and decoding in the television discourse.* Birmingham: Centre for Cultural Studies, University of Birmingham.

Haroz, S., & Whitney, D. (2012). How capacity limits of attention influence information visualization effectiveness. *IEEE Transactions on Visualization and Computer Graphics, 18*(12), 2402-2410. http://doi.org/10.1109/TVCG.2012.233

Hullman, J., & Diakopoulos, N. (2011). Visualization rhetoric: Framing effects in narrative visualization. *IEEE Transactions on Visualization and Computer Graphics, 17*(12), 2231-2240. http://doi.org/10.1109/TVCG.2011.255

Kennedy, H., & Hill, R. L. (2017). The feeling of numbers: Emotions in everyday engagements with data and their visualisation. *Sociology, 52*(4), 830-848. https://doi.org/10.1177/0038038516674675

Kennedy, H., Hill, R. L., Aiello, G., & Allen, W. (2016). The work that visualisation conventions do. *Information, Communication and Society, 19*(6), 715-735. https://doi.org/10.1080/1369118X.2016.1153126

Kennedy, H., Hill, R. L., Allen, W., & Kirk, A. (2016). Engaging with (big) data visualizations: Factors that affect engagement and resulting new definitions of effectiveness. *First Monday, 21*(11). https://doi.org/10.5210/fm.v21i11.6389

Kirk, A. (2016). *Data visualisation: A handbook for data driven design*. London: Sage.

Kosara, R. (2007a). A Tale of Two Types of Visualization and Much Confusion. Retrieved June 22, 2017 from https://eagereyes.org/criticism/tale-of-two-types

Kosara, R. (2007b). A Tale of Two Types of Visualization and Much Confusion. Retrieved June 22, 2017 from https://eagereyes.org/criticism/tale-of-two-types

Lam, H., Bertini, E., Isenberg, P., Plaisant, C., & Carpendale, S. (2012). Empirical studies in information visualization: Seven scenarios. *IEEE Transactions on Visualization and Computer Graphics, 18*(9), 1520-1536. http://doi.org/10.1109/TVCG.2011.279

Manovich, L. (2011). What is visualisation? *Visual Studies, 26*(1), 36-49. https://doi.org/10.1080/1472586X.2011.548488

Munzner, T. (2014). *Visualization analysis and design*. Hoboken: CRC Press.

Oudshoorn, N., & Pinch, T. (2003). *How users matter: The co-construction of users and technologies*. Cambridge, MA: MIT Press.

Potter, W. J. (2012). *Media effects*. Thousand Oaks, CA: Sage Publications.

Rost, L. C. (2017, March 10). Why Do We Visualise Data? [Blog post] Retrieved from https://lisacharlotterost.github.io/2017/03/10/why-do-we-visualize-data/

Seeing Data. (2017). Rate these visualisations! *Seeing Data*. Retrieved November 17, 2017 from http://seeingdata.org/developing-visualisation-literacy/rate-these-visualisations/

Sheppard, S. R. J. (2005). Landscape visualisation and climate change: The potential for influencing perceptions and behaviour. *Environmental Science & Policy, 8*(6), 637-654. http://doi.org/10.1016/J.ENVSCI.2005.08.002

Skau, D., & Kosara, R. (2016). Arcs, angles, or areas: Individual data encodings in pie and donut Charts. *Computer Graphics Forum, 35*(3), 121-130.

Tufte, E. (2001). *The visual display of quantitative information* (2nd ed.). Cheschire, CT: Graphics Press.

Valkenburg, P. M., Peter, J., & Walther, J. B. (2016). Media effects: Theory and research. *Annual Review of Psychology, 67*(1), 315-338. http://doi.org/10.1146/annurev-psych-122414-033608

Vogel, B., Kurti, A., Milrad, M., & Kerren, A. (2011). An interactive web-based visualization tool in action: User testing and usability aspects. In: *2011 IEEE 11th International Conference on Computer and Information Technology.* (pp. 403-408). IEEE. http://doi.org/10.1109/CIT.2011.68

Wattenberg, M. (2005). Baby names, visualization, and social data analysis. *Proceedings of the 2005 IEEE Symposium on Information Visualization* (INFOVIS'05), 1-7 http://doi.org/10.1109/INFVIS.2005.1532122

Zube, E. H., Sell, J. L., & Taylor, J. G. (1982). Landscape percetion: Research, application and theory. *Landscape Planning, 9*(1), 1-33. https://doi.org/10.1016/0304-3924(82)90009-0

About the authors

Arran L. Ridley is currently working on 'Seeing Data', a funded ESRC and White Rose DTC project that is investigating the possibilities of how the effectiveness of data visualization can be defined and measured and the potential for capturing and understanding people's engagement with data visualizations.

Chris Birchall is a Lecturer in Digital Media at the School of Media and Communication, University of Leeds. His research interests include digital methods and their application to the study and practice of digital citizenship and political communication online, and mobile and digital technologies, social change, and the human experience.

9. Approaching data visualizations as interfaces: An empirical demonstration of how data are imag(in)ed

Daniela van Geenen and Maranke Wieringa

Abstract

This chapter points out data visualization's double role as explorative and communicative means in humanities research. We draw from science and technology studies looking at the mediation process at stake: the interaction between visualization tool and researcher. To emphasize this mediation process and expose the various decisions at its heart we introduce the term 'data interface'. We highlight *how* visualizations function as data interfaces and visualization practices allow for *interfacing with* data biographing a network graph's 'life'. Using the lens of the 'data interface' underscores that a particular (network) visualization provides just one perspective on the data. Moreover, we examine *if* and *how* the used data interfaces encourage scholars to critically position their investigative work, during research processes and communication.

Keywords: Data interface; Critical positioning; Mediation; STS; Visual network analysis

Introduction

In the introduction of *Science in Action*, science and technology studies scholar Bruno Latour (1987) illustrates how particular scientific findings and technological developments led to the three-dimensional model of DNA with which we are familiar today. This introduction is the prelude to Latour's call to study the production of knowledge 'in the making', instead of merely

Engebretsen, M. and H. Kennedy (eds.), *Data Visualization in Society.*
Taylor & Francis Group, 2020
DOI 10.5117/9789463722902_CH09

focusing on outcomes (1987, p. 4). In this chapter, we respond to Latour's call to bring the epistemic process, and particularly, the ways in which data are imag(in)ed, to the foreground of the research communication. Our approach 'stages' the 'cultural life' of a data visualization in scholarly research. That is to say, we write a biographical account that depicts the visualization's making and distribution. This approach alludes to the double role of data visualization: first, visualization as an activity employed during the research process to get a perspective on data by the application of 'exploratory data analysis' (EDA) (Tukey, 1977). Second, data visualizations as images used as representations and research results, which are publicly communicated (e.g. Lynch & Woolgar, 1990; Coopmans et al., 2014).

We point out data visualization's double role by looking at the interaction and negotiation between the software tool used for visualizing the data and the researcher using this tool. The notion of the 'data interface' is introduced to emphasize this mediation process. By biographing the 'life' of a single network visualization, we pinpoint its role as a data interface. The network graph was created in a humanities research project, which investigated the Dutch-speaking Twittersphere as communication infrastructure (van Geenen et al., 2016). Using the case of this specific visualization, created with the network visualization tool Gephi (Bastian et al., 2009), we underscore how the various steps and decisions in the construction and the subsequent circulation of a graph play a role in defining it.

Defining 'data interfaces'

We look at data visualizations by approaching them as interfaces. Interfaces 'are the point of juncture between different bodies, hardware, software, users, and what they connect to or are part of' (Cramer & Fuller, 2008, p. 150). As such, interfaces function as mediators between different entities in situations in which 'users are not simply the audience, but also the actors' (Chun, 2011, p. 65). Like the graphical user interfaces (GUIs) which are familiar to us, (the process of) data visualization mediates between the data and its beholder. This mediation is especially emphasized when graphical representations of data are offered as, and have been derived from an interaction with, a GUI. Software scholar Wendy Chun (2011) aptly noted that GUIs, which offer tangible entry points to engage with abstract information, should be understood as 'programmed visions'. This notion pinpoints the non-neutral, preprogrammed quality of the graphical (re) presentations interfaces present to the user.

Moreover, 'programmed vision' implies that certain visions of the developers, implemented into the software programs, are reified by the use of these tools. In the mediation processes featured by these tools, the actual execution of the underlying code stays invisible, thus obscuring the developers' choices (Chun, 2011). The exploration presented in this chapter highlights the mediation process, in which the preprogrammed and, therefore, inscriptive quality of software plays an important role. We show that this 'situatedness' of the research methods, in combination with the scholars' academic, cultural, and social background (Haraway, 1988), is vital to knowledge production. The notion of the 'data interface', then, emphasizes the sense-making process of visualization performed by both the researchers and the future 'readers', who are faced with the data visualization as research outcome.

We are not the first to consider the interfacing aspect of data visualization: practitioners like Citraro and Rees (2015) have made a good case for a complementary line of argument. Stephen Few (2014) and Gray et al. (2016) likewise note the mediative character of data visualizations. In contrast and addition to these previous approaches, our aim is to demonstrate *how* data visualization figures as data interface in diverse situations, in scholarly practice and (public) research communication.

Taking account of the 'life' of data visualizations

Gephi, the software tool we used for our exploration, is a popular open-source software program for mapping, manipulating, and analysing all kinds of network data (Bastian et al., 2009). The software tool was designed to enable social and cultural scholars with little technical expertise to encounter complex relational data at the level of the GUI (Heymann, 2010). Consequently, it is often used in humanities and social science research. Its designers present Gephi as a tool for 'Visual Network Analysis' (Heymann, 2010; Venturini et al., 2015), thereby placing the emphasis on interfacing with and 'reading' network visualization. In this contribution we wish to sketch two reading positions with regard to network visualizations: that of the researcher engaging in exploratory data analysis (EDA), and that of an audience member to whom research results are offered via the (scientific) image of a network diagram.

EDA was coined by statistician John Tukey (1977), and is often one of the initial stages of a research project. In this stage, researchers are striving to grasp the examined phenomenon and to comprehend the data on which they are working (O'Neil & Schutt, 2014, pp. 34-36). EDA uses plots, summary statistics, and most applicable to our contribution, graphs (O'Neil & Schutt,

2014, p. 35). EDA using Gephi features Social Network Analysis (SNA) (Bastian et al., 2009; Jacomy et al., 2014), which is a branch of the social sciences that builds on mathematical principles of graph theory to chart interpersonal relations and examine social structures (Marin & Wellman, 2011). Insights in SNA are derived from the graph, as the positions of nodes—for example, individuals—are dependent on their connectedness and thereby the positions of all other nodes (Marin & Wellman, 2011).

Thus, EDA mobilizes particular forms of knowledge and methodological principles. The rise of the application of software tools in humanities and social science research has prompted some scholars to reflect on their effect on the research process and outcomes through the ways in which 'our digital helpers are full of "theory" and "judgement" already' (Rieder & Röhle, 2012, p. 70). We account for the way in which our 'digital helper' frames the research process by outlining the interaction with Gephi and stressing the relevant steps we take to make sense of the data *and* the tool.

In relation to visualizations' communicative capacity, it is not so much the exploratory process of gaining insights into the data that is relevant. Rather, the network graph provides a very specific kind of 'interface' to the data, displaying one of many possible perspectives on this research material. In other words, it is here that the visualization's—unnoticed—rhetoric power and, therefore, the question of understandability come to the fore (e.g. Haraway, 1988; Kennedy, Hill, Aiello, & Allen, 2016; Latour, 1986). In this contribution we discuss *how* a visualization functions differently at various stages of its life by focusing on specific aspects of its rhetoric power and pointing out how it requires particular forms of 'reader' engagement.

The life of a network visualization

Before a network visualization—or any information visualization—is 'born', data need to be selected, extracted, cleaned of irregularities in data formatting, and filtered on specified parameters. It is only after the data have been prepared for analysis that we usually start to visualize them. This, however, does not mean that the visualization stage is the final stage of the research. We will discuss how visualizations can also feed back into one's analysis, and thereby, become a particular kind of interface for working with the underlying data. To do so, we will biograph a network visualization with which both authors are familiar: a network visualization displaying day-to-day communication practices (@replies) in the Dutch-speaking Twittersphere (van Geenen et al., 2016).

Staging the mediation process

In the beginning of our network graph's life, we departed from tabular information extracted from Twitter's application programming interfaces (APIs) (see van Geenen et al., 2016 for detailed information on the corpus collection). The collected data sample contains more than 3.5 million Dutch tweets sent between 4 and 12 September 2016 (van Geenen et al., 2016). After cleaning the data (i.e. fixing formatting errors and dealing with missing information due to the partially black-boxed data extraction from Twitter), we started preparing them for an exploration in Gephi. As we were interested in communication between users, we filtered out solely replies (i.e. tweets that start with @username). Simultaneously, we added the usernames of accounts these replies addressed as an additional column to the spreadsheet. In doing so, we were accommodating the use of Gephi in our analysis, since the tool requires two types of data points in order to visualize relations as the foundation for the network graph: a source and a target. The following sections will concentrate on the mediation process in Gephi, exploring the data visually, on the one hand, and preparing the network graph as communicable visualization, on the other.

Gephi's focus on sociality

Gephi's analytical strength resides in its layout algorithms (Bastian et al., 2009). The 'ForceAtlas 2' layout algorithm was specifically developed for use in the Gephi application software and is optimized for handling large sets of relational data (Jacomy et al., 2014, pp. 5-11). The application of ForceAtlas 2 is stimulated by Gephi's design and stimulated by the Gephi core team (see e.g. van Geenen, 2018, for a more elaborate discussion of this matter). This technical specification makes it suitable for the processing and exploration of our dataset, which consists of 224,305 nodes (accounts), connected by 499,485 edges (809,871 sent replies; in case of double connections these were merged to weighted and thus thickened edges). Put in motion in 'Overview', one of Gephi's three tabs, ForceAtlas 2 causes a gradually perceivable spatialization of the graph. Next to 'Overview', Gephi features 'Data Laboratory' (i.e. allowing for inspection of the tabular data) and 'Preview' (i.e. allowing preparation and export of the static network graph). 'Overview' plays a vital part in knowledge production in Gephi, in the data processing and graph spatialization. The ForceAtlas 2 spatialization is force-directed. Thus, connections (replies, in our case) attract nodes (accounts) whereas nodes themselves repulse each other (van Geenen, 2018, pp. 2-3).

This simulation clusters the graph based on the number of connections nodes possess (degree), a clustering principle termed 'modularity' (van Geenen, 2018, p. 2). The application of modularity can be understood as a 'distant reading' strategy that features the lens of sociality. This strategy helps structuring the EDA approach to a large dataset that comprises social interactions. The research project at stake studied Twitter as everyday communication infrastructure (van Geenen et al., 2016). In order to identify this infrastructure, we used Gephi's 'Modularity Class' community detection algorithm (set to resolution 0.5 to identify also smaller communities), which classifies nodes based on shared connections (Blondel et al., 2008). Starting from a single node, the calculation process 'snowballs' through the entire graph and measures with which cluster each node has the most connections, and based on this, generates node metadata. Subsequently, we coloured and 'partitioned' the nodes based on the communities inferred by the algorithm.

While we are aware of the flaws of representing modularity in such a fashion, we used this strategy with a quantitative orientation as an initial exploration to follow up with a 'close reading' of these clusters. Modularity does not express, for instance, whether a particular node is strongly or loosely affiliated with a particular cluster, which erases the nuances we touched upon. We used modularity to initiate the qualitative encoding of the encountered clusters, and simultaneously, question the validity of these 'inferred data publics' (de Lange, 2017) based on the nodes' connectedness. In that we performed a close reading of both the research material *and* the data interface.

A close reading of visual network analysis in Gephi

According to Gephi developers, the ForceAtlas 2 algorithm provides 'transparency' in offering a continuous, manipulable simulation of the graph spatialization process (e.g. Jacomy et al., 2014, p. 2). There is a feeling of directness when working with the program, especially when the algorithm is 'running', set into operation by the push of one button. As with large graphs such as our communication network, it takes time to render the spatialization and reach a point at which the node positioning is moderately stable. Yet, when algorithm properties are tweaked during this process, through the settings panel, one sees the network visualization's instantaneous response (see Figures 9.2 and 9.3). While Figure 9.1 shows the 'raw' graph, Figure 9.2 displays the graph after running the algorithm, adapting and experimenting with the 'Scaling' (i.e. the adaptable graph size that takes the node positioning into account) and the 'Gravity' (i.e. the simulated forces

Figure 9.1. 'Raw' version of the network graph in the 'Overview' after the data import into Gephi. Created by D. van Geenen using Gephi.

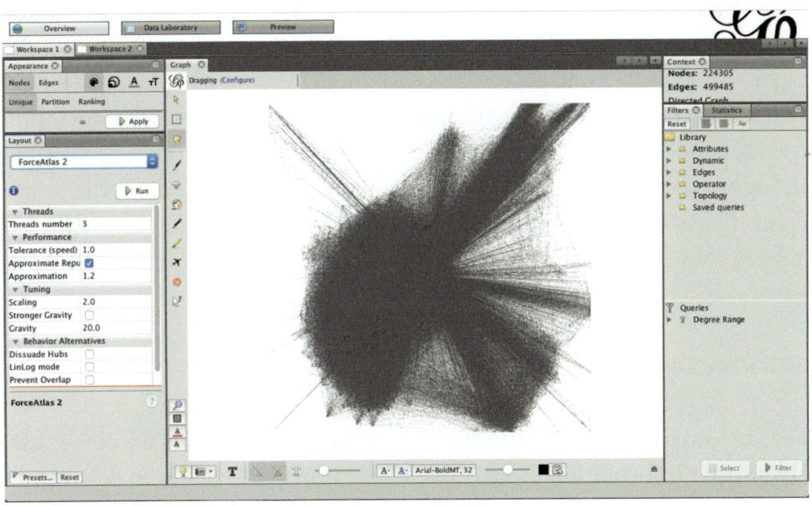

Figure 9.2. Spatialized graph after the application of ForceAtlas 2 (Scaling: 2.0, Gravity: 20.0; node size based on degree). Created by D. van Geenen using Gephi.

attracting nodes to the centre) of the graph. Thus, through Gephi's software affordances, the designed action possibilities the tool offers (e.g. Gaver, 1991), the visualization itself becomes an interface to the data, offering (the promise of) 'direct manipulation' (Shneiderman, 1982; for detailed analysis of Gephi's affordances, see van Geenen, 2018).

Direct manipulation can be described as the 'representation of the object of interest, rapid incremental reversible actions and physical action instead of complex syntax' (Shneiderman, 1982, p. 237). It can be understood as an immediate, visual feedback on a user's given action. For network visualization in Gephi, most of these characteristics of 'direct manipulation' apply. An exception are conveniently reversible actions, as Gephi does not offer an 'undo button' (cf. van Geenen, 2018). In tweaking the algorithm, though, settings can be 'reversed' by means of changing properties such as scaling back to the previous value, which results in a similar node positioning. (Since the software program presents a graph *simulation*, exact node positions can slightly differ.)

Apart from tweaking the running layout algorithm, applied filters are another example that have a kind of 'live' effect on the appearance of the network graph (e.g. Bastian et al., 2009). Based on the detected modularity clusters, we started filtering out all extremely small communities, which appeared to be unconnected from the main graph. Moreover, as a strategic focus in the preparation of the close reading of the graph, we decided to delete all nodes with less than ten connections (degree). In other words, we chose to concentrate on the most active accounts, which had sent or received more than ten replies (Figure 9.3). To sum up, we initially approached the algorithmic processing of the data with an EDA strategy: 'playing around' with the layout algorithm's settings and filters through direct manipulation in order to come to a first legible spatialization. Whereas the visualization in Figure 9.1 is not helpful in providing insights into the data, the graph spatialization assists in 'reading' the data (see Figure 9.2). This spurred subsequent tweaking, research questions, and explorations, resulting in Figure 9.3 as the 'final' graph.

Situating Twitter publics beyond modularity

For the purpose of situating the identified communities we close read the profile information of a sample of accounts per cluster to define and classify these communities. We combined these observations with the knowledge that we had gained through our long-term engagement with the Dutch Twittersphere. The graph presented us with the 'usual suspects' in communication research on Twitter: a dense cluster of highly connected, politically interested professionals such as politicians, media organizations, and journalists. However, modularity also opened the way for a new perspective on the data: we saw users coming together around particular occupations (e.g. foresters, or people involved in teaching and education), topics (e.g. public debate on sustainability), or interests (e.g. the Dutch theme park *De Efteling*). Some of these communities were beyond our expectations, for instance the gaming and vlogging communities. The visualization,

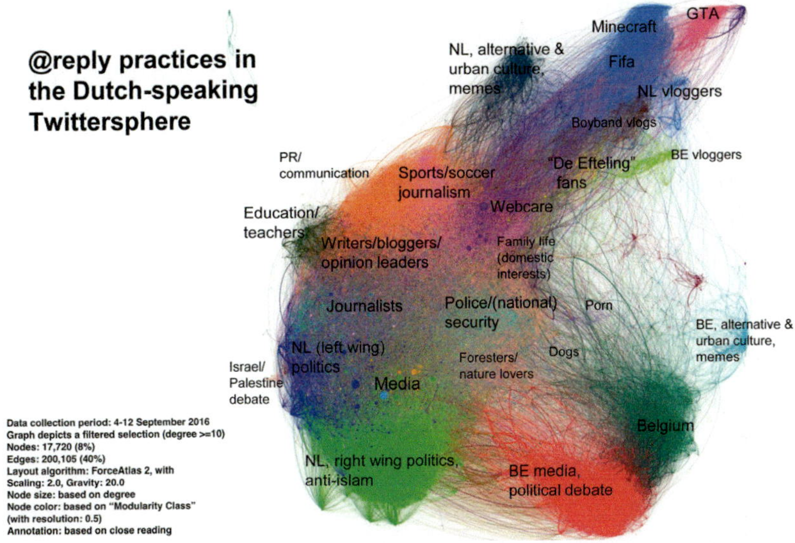

Figure 9.3. Exported, spatialized, filtered, and annotated graph. Created by D. van Geenen and M. Wieringa using Gephi and Photoshop in preparation of conference presentations.

then, became an interface to the data: it added to the perception of the research material by transforming tabular data into a palpable structure and manipulable form.

The network graph above provides more clues about the data such as different kinds of detectable media practices, that is the diverse ways in which users made use of Twitter's specifications such as the @reply functionality. As we studied reply practices, webcare accounts 'skewed' our sample. These are accounts of diverse (commercial) organizations that provide customers with the opportunity to address their concerns about products and services. The webcare account of the Dutch Railways (@ns_online) was in fact the most connected account in our graph (with 4,968 edges compared to an average degree of 3.6 edges for the whole graph), followed by @postnl and @kpnwebcare. We discussed whether we should exclude such webcare accounts, which due to their interaction with a diversity of other accounts function as central connectors in the reply network. Since the spatialization (solely) builds on the 'social hierarchy' of degree, it does not discriminate between the different media practices. Eventually, we decided to include all the different media practices, using the network visualization as a point of departure for further research.

Other research sparked by this graph included an investigation of the media practices of Dutch-speaking politically interested communities and the 'locality' of Twitter publics (see van Geenen et al., 2016). The first study originated from the observation that particular politics groups are well-represented in the communication infrastructure, such as accounts that interacted with the official account of right-wing politician Geert Wilders, the first politician in the list of highly connected accounts. Here we also observed that many of the tweets sent in this cluster surrounding @geertwilderspvv were concerned with a local incident in the Dutch city of Almere. It deepened our interest in the dynamics between the national and local spheres of public debate, and the role of local engagement in everyday communication on Twitter.

Between exploration and communication

We understand the function of mapping the data in the shape of an evolving network graph, and in this interfacing with the data, as 'augmenting human intellect', to borrow from interface design pioneer Douglas Engelbart (1962). In the context of EDA, then, network visualization helps scholars to come to a 'degree of comprehension in a situation that was previously too complex' to fathom (Engelbart, 1962, p. 1). The practice of visualization can be said to further scholarly thought, by making sensible what otherwise remains an overload of tabular information (e.g. Gray et al., 2016). As we have demonstrated, network graphs can figure as useful data interfaces.

However, we also need to be aware of what they do and do not show, and for what reasons. That is, we need to take the time to reflect *in which way* network visualization imposes shape on our research. Donna Haraway (1988) argues for the need to account for the positioning of the researcher, when she considers 'situated knowledge'. It has been argued elsewhere that, since network analysis and visualization thrive on software, the tools researchers use and become familiar with should be considered part of their positioning (van Geenen, 2018). In light of Haraway's observation, the biography of the network visualization we are sketching accounts for our research practice, since it depicts our engagement with relational data, based on our backgrounds and, thus, the (preliminary) knowledge we draw upon. With this account, then, we reveal our own 'critical positioning' and the 'partial perspective' (Haraway, 1988, pp. 585-586) we have on the research material as scholars. By doing so, we show that graphs do not exist in a void, but come into being through a complex series of interactions, based on certain preset conditions, of which the decisive moments should be made available to the public.

The cultural circulation of a network visualization

At some point in the research process we have to start formalizing our findings and translate them into something communicable to an audience. We used Gephi's 'Preview' tab to tweak the readability of the visualization. Furthermore, the tab's interface affords exporting the graph as static image (i.e. as PNG, PDF, or SVG file): a 'screenshot' of the sense-making process. Here, the network visualization—literally—moves from a state of mutability to a form of 'immutability'. While EDA draws on the graph's mutability, for example, by means of tweaking the layout algorithm's settings, the preparation of the findings for communication purposes results in a single static representation of the graph, which can be used in all kinds of (scholarly) publications. It becomes an 'immutable mobile' (Latour, 1986, pp. 7-13), which, as we discuss below, affects the public's possibility to engage critically with the presented research findings.

After annotating the exported network diagram in Photoshop, this image has led a particularly 'eventful' life: the visualization was presented at academic conferences (van Geenen et al., 2016) and has been featured on television for a broader audience (Boeschoten, 2017). Furthermore, we touched upon its complexities at the Impakt Festival, an annual, popular scientific festival around new media (van Geenen & Wieringa, 2017). Below, we will discuss how this 'screenshot' functions in two of these contexts: the academic conferences and the Impakt Festival.

Academic conferences: Scientific mediation versus public research communication

In presenting our research at several conferences, we were faced with a familiar dilemma articulated by numerous STS scholars: in which ways should we make use of graphical representations, which are expected to be the 'objective product' of a systematic knowledge production process (e.g. Haraway, 1988; Latour, 1986)? Moreover, such static images can easily be shared due to scholars' access to media platforms that invite sharing visual information such as Twitter. During and after our presentations we found the prepared network diagram circulating on Twitter, some versions with more comprehensive annotations on its making process than others. Due to their compressed nature, such 'screenshots', we argue, do not live up to the dynamic data interfaces that helped to imagine the data. This observation encouraged us to think of forms of 'methodological reflexivity' (Rieder & Röhle, 2012, p. 80) in the research communication that could do justice to the complex mediation process from which the network visualization originated. For instance, we developed an internal distribution policy: a code of conduct for the contextual information, which should be featured on slides and included in papers or non-academic articles (e.g. Figure 9.3).

Furthermore, for the purpose of catering to the traceability of, and stimulating reflection on, the meaning-making process of researchers interfacing with the data, we built a 'fieldnotes' plug-in for Gephi (Wieringa et al., forthcoming). It provides a comprehensive time-stamped version of both a text file of the applied settings in Gephi and a network graph file.

Impakt: Tackling data interfaces in interaction with the public

In our contribution to the Impakt Festival we demonstrated how single static images do not do justice to the complexities of the network graph (van Geenen & Wieringa, 2017), if only because graphs are nearly illegible if they are comprised of a vast amount of nodes. During our Impakt presentation, we elaborated and reflected on the network graph's making: rendering the research process visible, allowing the public critical engagement with the visualization.

Our talk addressed the diversity of reply practices the visualization represents and argued that forms of procedural mapping and interactive engagement should be on the agenda for (critical) data studies approaches. We exemplified the mediation process in Gephi to the public in a video that showed how ForceAtlas 2, in interaction with the researcher, handles the data. In this we made an effort to confront the idea that vision, especially that of an expert viewer—such as ourselves—working with algorithmic, standardized visualization tools, will automatically lead to absolute objectivity (e.g. contributions to Coopmans et al., 2014). This problem was aptly expressed in Haraway's notion of the 'god trick' (1988, p. 589), a phenomenon that is amplified through the 'programmed visions' our data interfaces present us with. Showing and explaining the video to the public, we *mobilized*—literally and figuratively speaking—the network graph to demonstrate the importance of reflection on such visualization practices. Concluding, we made an effort to demonstrate how data can be imagined and imaged, and how the resulting visualization is streaked with particular norms, conventions, and rhetoric (Haraway, 1988; Kennedy et al., 2016).

To summarize, we used the network graph as an illustration stating that data interfaces feature the potentiality to 'augment human intellect'. However, this augmentative capacity depends on their affordances to facilitate a continuous assessment of the principles applied to make sense of the data. As such, we vouch for 'account-ability' (Garfinkel, 1967 as cited in Eriksén, 2002) implemented into tool design processes. We ended our talk on data interfaces at Impakt Festival advocating for the need to think of design strategies that provide the opportunity to access data interfaces in (actual) interaction. During our talk we positioned our work critically. Our objective in writing this article is to further stimulate research in which data interfaces are approached critically, and that questions how modes of 'tool criticism' could be built into our data interfaces.

Conclusion

In conceptualizing the process of data visualization as a form of interfacing, we have added new perspectives on data visualization. The focus on data interfaces stimulates an understanding of the process of data visualization as mediation, and the resulting image as one of many possible interfaces to the data. Using the vehicle of the biography of a data visualization, we highlighted how visualization practices allow for *interfacing with* data, and exposed the choices and selections at the heart of this process. We emphasized data visualizations' constructedness, and thus their role as results of a specific knowledge production process. We examined in which ways the data interfaces we use are capable of encouraging the scholars to critically position their work during the research process. By doing so, we argue that researchers should practise tool criticism (cf. van Es, Wieringa, & Schäfer, 2018; van Geenen, 2018). Moreover, we strive to stimulate them to provide the audience of their research outcomes with the possibility of a critical engagement with their network graphs.

References

Bastian, M., Heymann, S., & Jacomy, M. (2009). Gephi: An open source software for exploring and manipulating networks. *Proceedings of the Third International ICWSM Conference,* 361-2. Retrieved from https://aaai.org/ocs/index.php/ICWSM/09/paper/view/154

Blondel, V. D., Guillaume, J.-L., Lambiotte, R., & Lefebvre, E. (2008). Fast unfolding of communities in large networks. *Journal of Statistical Mechanics: Theory and Experiment, 2008*(10), P10008. https://doi.org/10.1088/1742-5468/2008/10/P10008

Boeschoten, T. (2017, December 4). Hoe schrijven boze witte mannen onze geschiedenis? [Video file]. Retrieved from https://www.universiteitvannederland.nl/college/hoe-schrijven-boze-witte-mannen-onze-geschiedenis/

Chun, W. H. K. (2011). *Programmed visions: Software and memory.* Cambridge/London: MIT Press.

Citraro, D., & Rees, K. (2015). Epiphanies through interactions with data. In: D. Bihanic (Ed.), *New challenges for data design.* (pp. 175-185). London: Springer-Verlag.

Coopmans, C., Vertesi, J., Lynch, M., & Woolgar, S. (2014). *Representation in scientific representation revisited.* Cambridge/London: MIT Press.

Cramer, F., & Fuller, M. (2008). Interface. In: M. Fuller (Ed.), *Software studies: A lexicon.* (pp. 149-152). Cambridge/London: MIT Press.

de Lange, M. (2017, March & April). *Data publics in the smart city.* Paper presented at the Data Publics conference, Lancaster University, UK. Retrieved from https://themobilecity.nl/2017/11/24/my-keynote-talk-at-data-publics-conference-lancaster-university-march-31-april-1-2-2017/

Engelbart, D. (1962). *Augmenting human intellect: A conceptual framework.* Summary Report AFOSR-3233. Menlo Park, CA: Stanford Research Institute.

Eriksén, S. (2002). Designing for accountability. *Proceedings of the Second Nordic Conference on Human-Computer Interaction*, 177-186. http://doi.org/10.1145/572020.572041

Few, S. (2014). Data visualization for human perception. In: Interaction Design Foundation (Ed.), *The encyclopedia of human-computer interaction* (2nd ed.). Aarhus: Interaction Design Foundation.

Garfinkel, H. (1967). *Studies in ethnomethodology.* Englewood Cliffs: Prentice-Hall.

Gaver, W. W. (1991). Technology affordances. *Proceedings of the SIGCHI Conference on Human Factors in Computing Systems*, 79-84. https://doi.org/10.1145/108844.108856

Gray, J., Bounegru, L., Milan, S., & Ciuccarelli, P. (2016). Ways of seeing data: Toward a critical literacy for data visualizations as research objects and research devices. In: S. Kubitschko & A. Kaun (Eds.), *Innovative methods in media and communication research.* (pp. 227-251). Cham: Palgrave Macmillan.

Haraway, D. J. (1988). Situated knowledges: The science question in feminism and the privilege of partial perspective. *Feminist Studies, 14*(3), 575-599.

Heymann, S. (2010, February 1). Gephi initiator interview: How 'semiotics matter' [Blog post]. Retrieved from https://gephi.wordpress.com/2010/02/01/gephi-initiator-interview-how-semiotics-matter/

Jacomy, M., Venturini, T., Heymann, S., & Bastian, M. (2014). ForceAtlas2, a continuous graph layout algorithm for handy network visualization designed for the Gephi software. *PLoS ONE, 9*(6), 1-12. https://doi.org/10.1371/journal.pone.0098679

Kennedy, H., Hill, R. L., Aiello, G., & Allen, W. (2016). The work that visualisation conventions do. *Information, Communication and Society, 19*(6), 715-735. https://doi.org/10.1080/1369118X.2016.1153126

Latour, B. (1986). Visualization and cognition: Drawing things together. *Knowledge and Society, 6,* 1-40.

Latour, B. (1987). *Science in action: How to follow scientists and engineers through society.* Cambridge, MA: Harvard University Press.

Lynch, M., & Woolgar, S. (1990). *Representation in scientific practice.* Cambridge/London: MIT Press.

Marin, A., & Wellman, B. (2011). Social network analysis: An introduction. In: P. Carrington and J. Scott (Eds.), *The sage handbook of social network analysis.* (pp. 11-25). London: Sage.

O'Neil, C., & Schutt, R. (2014). *Doing data science: Straight talk from the frontline.* Sebastopol: O'Reilly.

Rieder, B., & Röhle, T. (2012). Digital methods: Five challenges. In: D. M. Berry (Ed.), *Understanding digital humanities.* (pp. 67-84). London: Palgrave Macmillan.

Shneiderman, B. (1982). The future of interactive systems and the emergence of direct manipulation. *Behaviour and Information Technology 1*(3), 237-56. https://doi.org/10.1080/01449298208914450

Tukey, J. W. (1977). *Exploratory data analysis.* Reading: Addison Wesley.

van Es, K., Wieringa, M., & Schäfer, M. T. (2018). Tool criticism: From digital methods to digital methodology. *Proceedings of the 2nd International Conference on Web Studies,* 24-27. https://doi.org/10.1145/3240431.3240436

van Geenen, D. (2018). *The role of software tools in critical data studies practices: Studying the affordances of Gephi as a sociotechnical system* (Master's thesis, Utrecht University, Netherlands). Retrieved from https://dspace.library.uu.nl/handle/1874/367489

van Geenen, D., Schäfer, M. T., Boeschoten, T., Hekman, E., Bakker, P., & Moons, J. (2016, October). *Mining one week of Twitter: Mapping networked publics in the Dutch Twittersphere.* Paper presented at The 17th Annual Conference of the Association of Internet Researchers, Berlin, Germany.

van Geenen, D., & Wieringa, M. (2017, October 26). *Data visualization as interface* [Video file]. Retrieved from https://youtu.be/QPGcJkHqpxc?t=42m17s

Venturini, T., Jacomy, M., & Pereira, D. (2015). *Visual Network Analysis.* Working Paper. Retrieved August 15, 2016 from https://www.tommasoventurini.it/wp/wp-content/uploads/2014/08/Venturini-Jacomy_Visual-Network-Analysis_WorkingPaper.pdf

Wieringa, M., van Geenen, D., van Es, K., & van Nuss, J. (forthcoming). The Field Notes Plugin: Making network visualization in Gephi accountable. In: A. Daly, K. Devitt and M. Mann (Eds.), *Good Data.* Amsterdam: Institute of Network Cultures.

About the authors

Daniela van Geenen is a lecturer and researcher in data journalistic practices and data visualization at the University of Applied Sciences Utrecht and an affiliated researcher at the Datafied Society research platform. She investigates the scholarly ethos that working with digital methods calls for and studies the design of accountable software tools.

Maranke Wieringa is a PhD candidate at Utrecht University, where she is part of the Datafied Society research platform and teaches various courses on (scholarly) data analysis. Her academic interests lie at the intersection of software and data. For her dissertation, Maranke investigates algorithmic accountability in Dutch municipalities.

10. Visualizing data: A lived experience

Jill Simpson

Abstract

Researching data visualization as a lived experience provides a perspective from which to explore its social life. Borrowing elements from feminist autobiographical research and critical making, this chapter uses the personal story of the design and circulation of a hand-drawn, *small-data* visualization depicting the author's experience of Obsessive Compulsive Disorder. By critically reflecting on the visualization's design and circulation, this chapter engages with wider academic debates about data visualizations' subjectivities. Furthermore, by interrogating notions of authenticity and honesty associated with hand drawing, it introduces the idea of a politics of hand-drawn visual representations of data.

Keywords: Critical data studies; Hand-drawn; Situated knowledge; Feelings; Lived experience

Introduction

Adopting a critical data studies perspective, this chapter explores the processes at work in the design and circulation of one particular data visualization. It draws on a case study documenting the experience of manually gathering and visualizing my own data, through the analogue medium of hand drawing. By borrowing from elements of *critical making*, a participatory method which combines the process of making with critical thinking (Ratto, 2011), it was possible to engage with similar decision-making processes to those that designers perform in the production of data-driven visualizations.

Exploring data visualization through this very personal and lived experience of its design and circulation has provided an opportunity to critically reflect on three important issues. Firstly, how slow, manual data

Engebretsen, M. and H. Kennedy (eds.), *Data Visualization in Society*.
Taylor & Francis Group, 2020
DOI 10.5117/9789463722902_CH10

gathering reveals the subjectivities inherent in all datasets (Crawford, 2013) and their visual representations. Secondly, the ways that images circulate online through social media and how images can be operationalized and repurposed. And thirdly, the politics of hand drawing visual representations of data in a field dominated by digital design.

These seemingly separate issues are connected through discussions of how feelings and emotions are integral to the design of data visualizations. The significance of feelings in people's engagement with complex charts and graphs has already been recognized within the field of critical data studies (see, for example, D'Ignazio & Klein, 2016; Kennedy & Hill, 2018). However, this chapter will also reflect on the ways that data visualizations can be *felt* by those who produce them and the people about whom they tell a story.

Using personal experience as a starting point from which to explore broader social and theoretical issues is part of the feminist tradition of autobiographical research (see, for example, Miller, 1991; Stanley, 1992). Stanley and Wise (1983) argue that drawing on personal experiences (of the researcher or the researched) makes it possible to put theory into context and in doing so, researchers can explain not only what is known, but how it has come to be known (Stanley & Wise, 1983). Within a broad tradition of autobiographical feminist research I draw more specifically on Miller's definition of personal criticism (1991, p. 1). This, Miller describes, 'entails an explicitly autobiographical performance within the act of criticism' (p. 1). Here, Miller is talking in the context of cultural criticism. However, drawing on Stanley and Wise (1983), I believe that the same approach can be used to provide a contextual critique of data and its visualization.

I have suffered with an anxiety disorder called Obsessive Compulsive Disorder (OCD) for most of my life. The data visualization which is the focus of this case study represents my compulsions to check and recheck the same things over and over again. The motivation to quantify and visualize my experience of OCD was to help people, who might never have suffered from a mental illness, to understand the way it impacts people's everyday lives. However, as I quantified, analysed, and visualized the data I also began to see it as a form of critical making (Ratto, 2011), in which I was re-enacting the same conscious and unconscious decisions made by data visualization designers (D'Ignazio & Klein, 2016), albeit on a much smaller scale.

Critical making combines critical thinking with the act of making something and involves participatory design and prototyping (Ratto, 2011). The method I used was not participatory in a group sense, nor was it limited to prototyping, having set out with the intention of producing a finished data visualization which could be published. In these ways it differs from

Ratto's definition of critical making (2011, p. 253). However, the design work I undertook was informed by critical literature on data and its visualization, and it was an iterative and reflective process, both important elements of critical making (Ratto, 2011). Furthermore, in line with one particular aim of this methodological approach, through making I was able to connect process and theory in order to produce academic critique (Ratto, Wylie, & Jalbert, 2014). Combining personal criticism and elements of critical making has provided an original approach from which to explore the lived experience of data visualizations.

Visualizing mental illness

Mental ill health tends not to produce visible physical symptoms on the body, making it largely invisible to others. This can make it hard for people to imagine the different ways a mental illness might negatively impact a person's life in the same way they might empathize with a physical disability. This was reflected in the words of a senior aide to the UK's prime minister, when the aide suggested that state-funded benefits should go to people who were really disabled, and not to those at home on medication for anxiety (BBC News, 2017). The aide was rightly criticized, yet while reading about what the aide had said I began to wonder whether data visualization might offer a medium through which to make the *affects* and *effects* of mental illness more visible to others.

In order to visualize my experience of OCD it was necessary first to quantify it. To do this I drew on *Dear Data,* a small-data, analogue art project by Giorgia Lupi and Stefanie Posavec (Dear Data, n.d.). By *small-data*, I mean data that are collected about a very small number of participants and include only a few variables. In the case of *Dear Data* each dataset comprised of only one person's data collected about a single topic. This is in contrast to big datasets which, for example, might be made up of tweets published by millions of people (boyd & Crawford, 2012). *Dear Data* was a year-long project in which the artists manually gathered self-tracked data about themselves, and visualized them. Each week they decided on a topic together on which to collect their data. Examples include: a week of laughter, a week of doors, and a week of complaints (Dear Data, n.d.). At the end of every week they would hand draw a visualization of their data onto the front of a post card before posting it to one another (Dear Data, n.d.). Through the process of gathering and visualizing their personal data about everyday topics, they were able to learn about one another's lives. I was drawn to this project because of its warmth and contextual detail, qualities which are not commonly associated with quantitative data.

Influenced by *Dear Data,* I began to manually self-track my illness over the course of one day. The most obvious way to quantify my experience was to record my compulsions to check and recheck the same things over and over again. Every time I checked a door handle, the floor area around where I had been sitting or standing, or even the URL of my email login page I would write it down. Using pen and paper I recorded what it was that I was checking, how many times I checked it and if the compulsion was repeated a specific number of times. If the checking incident was particularly frustrating or distressing I would also record some contextual detail about how I was feeling at that moment, or what had prompted the incident, to help me remember it in more detail when I came to analyse the data.

With the data recorded I was able to read through it and begin to identify patterns and trends in my checking behaviour. Whilst reviewing the data I also began to think about what I wanted to communicate to people through the visualization. In order to increase awareness and understanding of OCD it was important to draw people's attention to the time and effort given over to my compulsions and how this impacts my day-to-day life. To achieve this, the visualization needed to convey the number and repetitive nature of the compulsions. The story I wanted to tell not only shaped the way I organized the data, but also the design of the visualization. Metaphors are an important design tool through which to make the message or concept which is being communicated visually clearer to the recipient (Ursyn, 2008). Utilizing this design strategy I designed the chart so that it resembled the shape of the human brain (see Figure 10.1). With this metaphor I aimed to communicate the idea of physicality, so as to represent mental ill health as an embodied and lived experience. The notion of physicality was also intended to encourage people to think about how mental illnesses are often invisible to others, yet can still be debilitating.

The lines which form the shape of the visualization do not carry meaning about the data. Instead they are an aesthetic design choice intended to mimic the lines visible on the surface of a brain, thus strengthening the visual metaphor. At the end of each line sits a bubble, with each bubble representing a separate incident of checking. Their positioning is purely aesthetic and as such it is not possible to read the visualization as a timeline. The colours of the bubbles relate to categories of checking, for example, checking door and drawer handles, or checking the floor area around where I had been

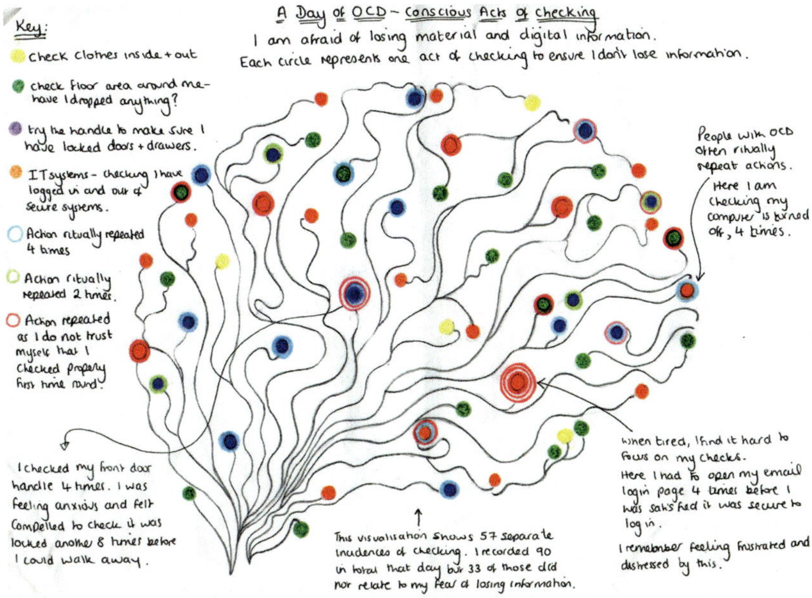

Figure 10.1. Visualizing mental illness: A day of OCD. Copyright 2017 by J. Simpson. Reprinted with permission.

standing or sitting down. Some of the bubbles have either a blue or green halo which indicates that the checking was ritually repeated a specific number of times, while a red halo represents when I felt compelled to repeat the action/s again. I chose bright colours and a recognizable shape as I wanted the visualization to be inviting and accessible. My aim to engage people who might never have experienced a mental illness led me to choose a pretty and whimsical design to encourage curiosity and to counter what I perceive to be negative cultural stereotypes about mental ill health.

Big data are critiqued for their lack of contextual detail (boyd & Crawford, 2012; Kitchin, 2014). To help people connect with the individual behind the data, the visualization I designed included handwritten notes highlighting particular points in the data and explaining them. These describe the way I was feeling or explain the reasoning behind an incident to help people better understand the lived experience behind the numbers. The motivation to design a data visualization had been to raise public awareness and understanding of OCD. The visualization, alongside an article about the potential for the medium of data visualization to make personal experiences of anxiety disorders more visible, was published in openDemocracy in May 2017. It was titled 'Visualising Mental Illness' (Simpson, 2017).

Data subjectivities

By manually gathering, analysing, and visualizing my own data I was faced with some of the same conscious and unconscious decisions that data visualization designers encounter in their work (D'Ignazio & Klein, 2016). The process was particularly revealing about the subjectivities that are at the heart of all datasets and their visual representation.

There is a labour involved in making data meaningful (Nafus, 2014). Yet, the people working behind the scenes to collect data, to store them, to analyse them, and to make them visible remain hidden to the audiences of visual representations of data (D'Ignazio & Klein, 2016). Through the process of slow, manual, small-data gathering the labour of sense-making became visible to me. In collecting and analysing such personal data I connected with the emotive and embodied experience of conducting research (Law, 2017). At times I felt emotional while recording the data and, when reviewing them, I felt shocked at the number of incidences I had recorded. It seems to me that the emotive and embodied experiences of research undoubtedly shape data collection and the way they are analysed. Yet these experiences are almost always left out of data's visualization.

Kennedy & Hill (2018) describe the importance of feelings and emotions when people who are not experts in data engage with and make sense of data visualizations. The process I undertook suggests that emotions can also play a significant role for the people who are working with data. I found collecting, analysing, and visualizing my data, so that they made sense to other people, required me to connect emotionally with the data and design process. Of course the extent to which design labour is emotional is dependent on the subject matter of the data and their relationship to it. Not only was the topic of my data visualization sensitive, it was extremely personal, and this does not reflect the majority of data visualization work.

There are, however, other examples of data visualizations which represent sensitive topics and whose design prompts emotional engagement. Digital interactive examples include Valentina D'Efilippo's (n.d.) *Poppy Field* (a beautiful and poignant representation of those who have lost their lives in wars during the nineteenth century) and Periscopic's (n.d.) hard-hitting visualization *Gun Killings in the U.S.* (a visualization of the lives lost and years stolen by gun violence). While the data illustration work of Mona Chalabi (n.d.) and Dear Data (n.d.) demonstrate how data can be represented with warmth and a sense of connection to the person who designed the visualization. These examples add weight to the argument that the embodied and emotional labour required to make sense of data should be considered as part of the lived experience of data visualization. Being open about these aspects of data visualization design makes it possible to appreciate the situated nature of data visualizations. In recognizing this, the idea that there are multiple alternative truths within data might become more apparent to those viewing a data visualization (D'Ignazio & Klein, 2016).

In order for the visualization to tell a coherent story about my personal experience of OCD I was faced with choices. These involved what data to include, what data to leave out, which characteristics of the data to conceal, and which to make visible, so that the visualization makes sense to another person. Data visualizations are always partial, they can never represent all that a dataset might have to tell us (Boehnert, 2015). It is generally accepted that in order to present data graphically, it is necessary to discard much of what characterizes the data through processes of reduction (Manovich, 2011). Looking through my own data I made the decision to focus on one main theme in order to tell a coherent narrative; my fear of losing information. Every time I repeatedly checked the gas hob was turned off, for example, or the iron unplugged, was excluded from the visualization. In order to be transparent about the missing data I included a handwritten annotation acknowledging the missing data points. This kind of contextual detail helps to make visible both the situated nature of data visualization and the subjectivities of data visualization designers, by hinting at the ways in which their work might be influenced by their own decision-making (D'Ignazio & Klein, 2016).

The circulation of data visualization

When 'Visualising Mental Illness' (Simpson, 2017) was first published it featured as the lead story on the home page of openDemocracy. The front page feature led with the title, a one-line summary and a large image of the visualization with a click-through link to the full article. The hand-drawn visualization stood out among the digital photographs which accompanied many of the other articles. As with all the stories they publish, openDemocracy tweeted about the article to their followers who number over 60,000. My colleagues and friends retweeted openDemocracy's tweets, and composed their own to share the article, often embedding an image of the visualization into their tweets. It was a surreal experience watching this deeply personal visualization, designed in my flat in the North of England, spread across the world via social media.

As I watched my data visualization circulate on Twitter I became very aware that I had no control over who shared the image or the message they attached to it. This made me feel uncomfortable; this was my data, my story, and yet it was being repurposed and circulated to support other people's agendas. John Berger (1972) argued that once an artwork has been reproduced it becomes a form of information. People can use a reproduced artwork to support their own arguments, which may differ from how the artist intended their work to be interpreted (Berger, 1972). Although Berger is talking about the art world, it is interesting to draw on him to think about how the meanings of data visualizations are also not fixed. Through the digital reproduction and circulation of this hand-drawn data visualization, it lost some of its original intended meaning and gained new meanings. This has implications when thinking about the politics of data visualizations as they circulate through social networks. Data and their visualization are always bound up in politics and power relations, which are played out in the kinds of data collected and what, from that data, is made visible (Boehnert, 2015; Kennedy & Hill, 2017). The way in which people might repurpose the information to fit their own political agendas adds another layer of complexity in unpacking the ideological work that data visualizations do (Kennedy & Hill, 2017, p. 773).

The politics of hand drawing

Thanks to its circulation on Twitter, the article was picked up by *Scientific American* in the US, which featured the data visualization on its visual blog. It also caught the attention of a national newspaper in Australia which featured the visualization in an online article on mental ill health. Significantly, the publication of the article in openDemocracy coincided with Mental Health Awareness Week in the UK and Mental Health Month in the US. Its timely publication may explain some of the interest in the visualization on social media and by the aforementioned publications. Nevertheless, in a field dominated by digital design it was surprising to see this static, hand-drawn, small-data visualization capture people's attention. This has led me to consider the significance of the medium of hand drawing in producing alternative visual representations of data.

Digital data visualizations are presented as complete and neutral reflections of data. The organizational conventions that data visualization designers draw on play an important role in making them appear objective (Kennedy, Hill, Aiello, & Allen, 2016). This is in sharp contrast to the way in which hand-drawn images are more often presented as imperfect and incomplete representations of a concept (Dexter, 2005). Rather than being associated with technical neutrality, drawing appears subjective and linked to qualities like 'intimacy, informality, and authenticity' (p. 5). The ways in which drawing is perceived to be subjective hints at the work digitalization does in portraying data visualizations as objective, technical, and accurate. Therefore, at first, drawing may appear to be incompatible with data visualization. However, drawings are particularly good at expressing 'emotion, experience, and feeling', all important elements in developing a narrative (p. 8). Using affect in design is an important strategy in engaging people in an issue and one that is utilized by some data visualization designers (D'Ignazio & Klein, 2016). Indeed, many designers recognize that good data visualization design provokes *feeling* in their audiences (Kennedy & Hill, 2018).

In hand drawing my visualization, its imperfections were made visible. For example, the wobbly circles representing individual data points, or the rubbed out pencil lines still faintly visible in the background. Dexter (2005) argues that it is the visibility of such mistakes that gives drawing an air of authenticity and honesty. These imperfections, combined with the personal nature of the data and the handwritten annotations, made for a powerful and *affective* data visualization, which captured people's attention. Yet, the notion of authenticity troubles me. Although the subjectivities of my small-data visualization might appear to be more visible, it is not necessarily more honest than a digital graphical expression of data.

Using hand-drawn images to communicate serious topics is nothing new. In the field of comic journalism, graphic representations in a comic's style are used to depict 'hard news' (Weber & Rall, 2017, p. 382). In their research, Weber and Rall (2017) identified several strategies that comic journalists use to give their work authenticity, including the use of 'visual stylistic devices' (p. 386). Citing the use of colour, mark making and handwritten text, they explain how visual styles help to 'remind the reader' that the comic is a subjective representation, *made* by the author (p. 386). Indeed, Weber and Rall (2017) say how comic journalism's obvious subjectivity makes it appear transparent and honest. They note how hand drawings produce a sense of honesty as they make readers aware that they represent the author's interpretation of an event.

In the hand-drawn visual style of my own data visualization I subconsciously adopted many of the authenticity strategies which Weber and Rall (2017) identified in their research. These are in contrast to the visual strategies and conventions used by digital data visualization designers, which often work to portray a sense of objectivity (Kennedy et al., 2016). However, I find it hard to argue that either a digital or hand-drawn approach is inherently more or less *honest,* even though honesty is a quality associated with the authenticity of the hand-drawn. Although one visual style might appear more authentic, does this mean it is more honest? This raises questions around claims to authenticity and honesty which are communicated through the visual style of data visualizations. These questions are significant in the context of the ideological power of data visualizations and the politics embedded within them (Boehnert, 2015; Kennedy & Hill, 2017). As David Beer (2013) argues, 'There is something convincing about visuals, however it is that they have been created' (p. 118). Although the hand-drawn nature of the data visualization I designed suggests the subjectivities within the data collection and design process, the decisions I made which have shaped my representation of the data remain hidden. Perhaps then, we need to consider extending our conversation about the politics of data visualizations to explicitly include analogue, hand-drawn, designs.

Conclusion

The subjectivities of data and their visual representation tend to be obscured in digital data visualization design. Although research suggests that designers are aware that the design process involves decisions which will prioritize certain viewpoints of the data, the conventions they work within play a role in communicating an 'aura of objectivity' (Kennedy et al., 2016, p. 723). Making the subjectivities of visual representations visible to audiences through the medium of hand drawing can work to imply authenticity (Weber & Rall, 2017). However, just because data subjectivities appear more visible it should not be assumed that the visual representation is any more *honest.* The situated nature of data and their visualization always shapes their design in ways that are invisible to the audience. The design process I embarked upon required methods of selection and reduction which produced a very particular view of the data. This complemented the story I wanted to tell, while alternative possible narratives within the data remained hidden. Developing ways of making these design decisions more visible to audiences might work to unpack some of the ideological power data visualizations possess (Kennedy & Hill, 2017), by introducing the idea that multiple alternative truths might exist within the data (D'Ignazio & Klein, 2016).

By borrowing from elements of personal criticism and critical making, this chapter has brought theoretical and conceptual ideas into a personal context. In doing so it has revealed the different ways in which data can be embodied, emotive, and felt (Kennedy & Hill, 2018). This supports Kennedy and Hill's (2018) argument that we must look beyond technological structures, to consider data visualization as experienced as part of everyday life. Critically engaging with my own experience of visualizing mental illness has demonstrated how existing conceptual ideas can be interrogated, and new ones emerge, when data visualizations are explored as a lived experience.

References

BBC (2017, February 26). Disability benefits: PIPs should be for 'really disabled'. *BBC.* Retrieved from https://www.bbc.co.uk/news/uk-39097019

Beer, D. (2013). *Popular culture and new media: The politics of circulation.* Basingstoke: Palgrave Macmillan.

Berger, J. (1972). *Ways of seeing.* London: Penguin Books Ltd.

Boehnert, J. (2015). *Viewpoint: The politics of data visualization.* Retrieved August 3, 2016 from http://discoversociety.org/2015/08/03/viewpoint-the-politics-of-data-visualization/

boyd, d., Crawford, K. (2012). Critical questions for big data: Provocations for a cultural, technological and scholarly phenomenon. *Information, Communication and Society,* 15(5), 662-679.

Chalabi, M. (n.d.) *Mona Chalabi.* Retrieved October 30, 2018 from https://monachalabi.com/illustrations/

Crawford, K. (2013). The Hidden Biases in Big Data. Retrieved June 9, 2016 from https://hbr.org/2013/04/the-hidden-biases-in-big-data

Dear Data (n.d.). *Dear data.* Retrieved September 4, 2016 from http://www.dear-data.com/theproject

D'Efilippo, V. (n.d.) *Poppy field.* Retrieved October 30, 2018 from http://www.valentinadefilippo.co.uk/projects/poppy-field/

Dexter, E. (2005). *Vitamin D: New perspectives in drawing.* London: Phaidon Press Limited.

D'Ignazio, C., & Klein, L. (2016, October). *Feminist Data Visualization.* Paper presented at and published in the workshop proceedings from the Workshop on Visualization for the Digital Humanities at IEEE VIS Conference, Baltimore, Maryland.

Kennedy, H., Hill, R. L., Aiello, G., & Allen, D. (2016). The work that visualisation conventions do. *Information, Communication and Society, 19*(6), 715-735. https://doi.org/10.1080/1369118X.2016.1153126

Kennedy, H., Hill, R. L. (2017). The pleasure and pain of visualising data in times of data power. *Television and New Media, 18*(8), 769-782.

Kennedy, H., Hill, R. L. (2018). The feelings of numbers: Emotions in everyday engagements with data and their visualisation. *Sociology*, 52(4), 830-848. https://doi.org/10.1177/0038038516674675

Kitchin, R. (2014). *The data revolution: Big data, open data, data infrastructures and their consequences.* London: Sage Publications Ltd.

Law, J. (2016). Modes of knowing: Resources from the Baroque. In: J. Law & E. Ruppert (Eds.), *Modes of knowing: Resources from the Baroque.* (pp. 17-56). Manchester: Mattering Press.

Manovich, L. (2011). What is visualization? *Visual Studies, 26*(1), 36-49.

Miller, N. K. (1991). *Getting personal: Feminist occasions and other autobiographical acts*. London: Routledge.

Nafus, D. (2014). Stuck data, dead data, and disloyal data: The stops and starts in making numbers into social practices. *Distinktion: Journal of Social Theory, 15*(02), 208-222.

Periscopic (n.d.) *U.S. gun deaths in 2013*. Retrieved October 30, 2018 from https://guns.periscopic.com/?year=2013

Ratto, M. (2011). Critical making: Conceptual and material studies in technology and social life. *The Information Society, 27*(4), 252-260.

Ratto, M., Wylie, S. A., Jalbert, K. (2014). Introduction to the special forum on critical making as research program. *The Information Society, 30*(2), 85-95.

Simpson, J. (2017, May 10). Visualising mental illness. *openDemocracy*. Retrieved from https://www.opendemocracy.net/transformation/jill-simpson/visualising-mental-illness

Stanley, L. (1992). *The auto/biographical I.* Manchester: Manchester University Press.

Stanley, L., Wise, S. (1983). *Breaking out: Feminist consciousness and feminist research.* London: Routledge & Kegan Paul plc.

Ursyn, A. (2008). Digital drawing, graphic storytelling and visual journalism. In: S. Garner (Ed.), *Writing on drawing: Essays on drawing practice and research.* (pp. 169-177). Bristol: Intellect Books.

Weber, W., Rall, H.-M. (2017). Authenticity in comics journalism: Visual strategies for reporting facts. *Journal of Graphic Novels and Comics, 8*(4), 376-397.

About the author

Jill Simpson is a doctoral researcher in the Sociology Department at the University of York. Her research explores the meaning making processes at work in the design and interpretation of data visualizations. More broadly Jill's research interests combine critical data studies, interdisciplinary social research, and public engagement through creative practice.

11. Data visualization and transparency in the news

Helen Kennedy, Wibke Weber, and Martin Engebretsen

Abstract

This chapter explores the role of data visualization in relation to transparency in the news, a field in which a decline in trust and a subsequent need to reassert credibility is an ongoing challenge. Being transparent about how the news is produced is seen as one way of generating trust, yet there has been very little empirical research into transparency practices in newsrooms. Our chapter fills this gap, focusing on transparency and data visualization. We argue that working with data visualization involves particular enactments of transparency, many of which are surprisingly not visual.

Keywords: Transparency; Uncertainty; Objectivity; The news; Journalism.

Introduction: Data visualization in the news

Visual representations of data play a central role in the recent expansion of data-driven news. From simple bar charts and line charts to more sophisticated chart types, data visualizations (or dataviz) are assumed to have the capacity to engage audiences, a view that extends beyond the news. At the same time, the news is experiencing other changes and challenges. At the time of writing, the global political climate is characterized by claims that we are living in a 'post-truth' world, in which people have had enough of objective facts and data. In this context, transparency, seen for some time as a trust-generating mechanism appropriate to the networked age, is believed to make it possible for audiences to see how the news is produced and therefore to establish trust (Singer, 2010).

However, there has been very little empirical research exploring how transparency gets done in newsrooms (Coddington, 2015 is one exception)

Engebretsen, M. and H. Kennedy (eds.), *Data Visualization in Society.*
Taylor & Francis Group, 2020
DOI 10.5117/9789463722902_CH11

and none that focuses specifically on data visualization. This is surprising, because scholarship on data visualization frequently addresses similar debates and concerns to those outlined above. For example, commentators note that data visualizations are associated with characteristics such as truthfulness and objectivity (e.g. Masson & van Es, 2017), and this can make them seem trustworthy.

In this paper, we address the empirical gap in the literature by exploring the role of data visualization in relation to transparency and trust in the news. Drawing on empirical research into the uses, roles, and forms of data visualizations in newsrooms in six European countries, we argue that for respondents in our research, working with data visualization involved particular enactments of transparency, many of which are surprisingly not visual. We suggest that dataviz transparency is an increasingly important journalistic norm, understood as a 'moral prescription for social behavior' (Schudson, 2001, p. 151), but that how to 'do' transparency remains in a state of 'interpretative flexibility', undetermined and still under negotiation (Wyatt, 1998). We proceed to situate our research in the context of relevant debates, after which we present our methods and findings.

Transparency in the news

Transparency, or revealing 'as much as possible about sources and methods' (Kovach & Rosenstiel, 2007, p. 92), is increasingly important in the news. This can be seen in the fact that, in 2014, the Society of Professional Journalists revised its ethical code to include transparency as a key ethical principle (Vos & Craft, 2017). Karlsson (2010) argues that transparency represents a form of openness in news practices, which makes it possible for audiences to see how the news is produced, and so makes news producers more accountable to their audiences. Kovach and Rosenstiel (2007) understand transparency as journalists being honest about what they know and how they came to know it. Similarly, Allen (2008, p. 328) defines it as 'making public the traditionally private factors that influence the creation of news'.

While some writers believe that transparency affects how audiences trust the news, others disagree. Karlsson, Clerwall, and Nord (2017) propose that efforts to promote transparency may be limited in their ability to restore trust. Others are cautious about its implementation: Karlsson (2010) argues that transparency can become routinized and separated from its normative intent, and Singer (2010) argues that some journalists see the requirement for transparency as an intrusion to their automony.

A number of commentators pit transparency against the enduring and contested journalistic norm of objectivity. Some argue that transparency enables a superior form of truth-telling to objectivity. Weinberger (2009) proposes that whereas objectivity was suited to a paper age, transparency is a more appropriate trust-generating mechanism in a networked age, in which links direct readers to the sources that have been consulted and the choices that journalists have made, persuading readers to accept ideas as credible the way that objectivity used to. Belief in the need for transparency predates the current, so-called crisis of trust in the news, but the need becomes more pressing in this context.

Conceiving of transparency and objectivity as distinct is not inevitable, as the same practices which are seen by some commentators to enhance transparency are seen by others as relating to the objectivity norm. McNair (2017) notes that objectivity has historically been achieved through mechanisms such as using credible sources and corroboration of information, precisely the things that transparency practices aim to reveal and enable. McNair claims that practices like making storytelling choices explicit and providing audiences with tools to look behind the scenes and interact with news stories represent journalists' engagement with the objectivity norm, albeit in the form of an acknowledgement of its limitations.

While many writers evoke the objectivity norm when discussing transparency, Anderson (2018), also concerned with re-establishing trust in the news, focuses on uncertainty. He proposes that for journalism to be trusted and to be seen to be pursuing honesty and sincerity, it needs to be more embracing of its uncertainties. Tracing its recent history, Anderson argues that journalism has come to professional maturity by honing its drive for factual certainty. As a result, it ends up proclaiming to be 'more scientific than science itself' (p. 181), given that science more readily acknowledges the uncertainties in which it deals. News journalism's increasing confidence in its ability to 'convey reality with a type of a scientific certainty' (p. 178) has led to the distrust in journalistic truth claims that we are currently witnessing, in Anderson's view.

McNair's call for 'the reassertion of objectivity as an aspirational quality standard' (2017, p. 1328) and Anderson's proposal that news journalism needs to acknowledge its uncertainties are both motivated by a concern to re-establish trust, and both point towards the need for greater transparency. What McNair sees as objectivity work and Anderson sees as uncertainty work both require transparency practices. As the use of data visualization in the news proliferates, it is important to investigate empirically whether news professionals see working with dataviz as enabling them to 'do transparency', navigate uncertainties, and re-establish trust, especially given related debates in data visualization research.

Transparency in data visualization and its relationship with objectivity and uncertainty

Debate about data visualization focuses on similar issues to debate about transparency in the news, especially in its relationship to objectivity. Kennedy, Hill, Aiello, and Allen (2016) argue that data visualizations are imbued with 'the quality of objectivity', which is in turn associated with characteristics such as trustworthiness. Data visualizations' appearance of objectivity has a number of origins: they report numbers, historically trusted because they appear universal, impersonal, and neutral (Porter 1995); and they are associated with science, meaning they are sometimes seen to be objective and trustworthy (Tal & Wansink, 2016).

Despite data visualizations' appearance of objectivity, critics and data visualization experts argue that dataviz do not provide us with neutral windows onto data. Rather, they are the result of numerous choices, it is claimed (Ambrosio, 2015). To engender trust, professional data visualizers may therefore need to be open about the choices they have made in the visualization production process. As news journalists increasingly include data visualizations in their professional toolkit, and because objectivity is an enduring and contested journalistic norm, it is important to examine how journalists perceive the dataviz that they produce in relation to objectivity, and whether and how their perceptions inform transparency practices.

Uncertainty is also central to debate about data visualization. Dasgupta, Chen and Kosara argue that uncertainty is an 'intrinsic part of any visual representation in visualization' (2012, p. 105). They note that multiple aspects of visualization design introduce uncertainty, and the data on which visualizations are based may also contain uncertainties. Thus they distinguish between *data* uncertainty, which relates to the numeric stratum and is what concerns Anderson, and *visual* uncertainty, which is specific to data *visualization* and relates to the visual stratum of dataviz production. We also understand uncertainty in this broad sense, as relating to data, the visual production process, and contexts of consumption, as these also introduce uncertainties. For example, some writers have identified that limited graphicacy (Balchin, 1972), or data visualization literacy, amongst audiences (Archer, this volume, and Tønnessen, this volume) produces uncertainty. Goodchild (2009) argues that a further uncertainty relating to consumption results from the ability to share visual information at speed across digital networks, a process in which images are often extracted from their original locations and from related contextualizing information. These consumption-related uncertainties suggest the need for transparency about how data visualizations are produced, and they raise the question of whether and how dataviz practitioners' thinking about audience graphicacy and contexts of consumption informs their transparency practices in dataviz production.

Mechanisms to make uncertainty transparent are widely debated in dataviz literature, such as fuzziness, the location of visual objects, the use of boxplots or related variations (e.g. MacEachren et al., 2012). However, Boukhelifa and Duke argued in 2009 that there was a gap between *rhetoric* about the importance of visualizing uncertainty and dataviz *practice*, in which uncertainty is rarely represented outside of laboratory experiments. This raises the question of whether visual techniques are used to make uncertainties transparent in dataviz in the news.

Synthesizing these debates, the overarching question that this chapter addresses is: what is the relationship between dataviz and transparency for news professionals? To answer this primary question, we ask: how do journalists perceive the dataviz that they produce, and to what extent do their perceptions inform transparency practices? To what extent does practitioners' thinking about audience graphicacy and contexts of consumption inform their transparency practices in the dataviz production process? What techniques are used when journalists working with dataviz make uncertainties transparent? We provide some answers to these questions below, after a discussion of our methods.

Methods

Our chapter draws on interviews with 60 editorial and newsroom leaders, data journalists, visualization designers, and developers in 26 major news organizations in Norway, Sweden, Denmark, Germany, Switzerland, and the United Kingdom. We used a purposive sampling technique, recruiting a balance of newsroom types, from international news providers, national broadcasters, national broadsheet and tabloid newspapers to regional broadcasters and newspapers, all of which had an online presence. Interviewees had many, varied job titles, drawn from journalism, design, IT, data science, and elsewhere. Interviews were conducted face-to-face or via video-conferencing, according to a semi-structured interview guide. Each interview lasted about one hour, was audio recorded and then transcribed and anonymized. To aid analysis, the main aspects of the Scandinavian and German-speaking interviews were translated into English. The data were coded and analysed in part deductively according to pre-defined themes and codes, in part inductively as new themes emerged.

Interviewing newsroom practitioners gives access to self-reports and perceptions: our respondents talked about their perceptions of dataviz and they self-reported on their transparency practices. Interview methods do not allow access to actual practices, which would need a different method, and the discussion that follows should be read with this in mind. Below, we discuss how respondents perceive the visual representations of data that they produce in relation to objectivity. Then we focus on mechanisms for 'doing' dataviz transparency in order to build trust amongst audiences, highlighting how reported techniques were surprisingly not visual. Finally, we reflect on transparency strategies for addressing uncertainties relating to audience graphicacy and contexts of consumption.

Perceptions of dataviz and how they inform transparency practices

To explore how newsroom professionals perceive the data visualizations that they produce and commission in relation to objectivity, and the extent to which their perceptions inform transparency practices, we asked respondents what they saw as the primary function of dataviz in the news, and whether they saw dataviz as offering neutral windows onto data or as shaping the data they represent in certain ways. A small number of respondents said they saw dataviz as a form of truth-telling. Data visualizations add empirical evidence to claims made in news texts, and as such they support the norm of objectivity, these respondents observed:

> I think that diagrams may corroborate facts and support credibility. (Data journalist)

However, most respondents felt that data visualizations serve to emphasize the angle of the story in which they are embedded. In this sense, dataviz are shaped by the perspective of the news story. Indeed, one respondent (Developer) described it as lazy *not* to provide an angle onto data, because doing so is the essence of journalistic work. Other respondents concurred:

> If you use a data visualization as a central element in a news story, that data visualization also has to carry the angle of the story. (Data journalist)

On the whole, respondents appeared to see *data* as objective, factual, or truthful, whereas data *visualization* was more readily seen as a process involving interpretation, and so less objective. In this way, respondents were more likely to acknowledge the uncertainties that can be introduced in visual production and that result from presentational choices than data uncertainties which relate to the numeric stratum that provides the basis for the visualization. One respondent claimed that the visual character of dataviz gives them a false 'quality of objectivity', as Kennedy et al. argue (2016). He said:

> The allure of dataviz is it has this visual sense of being objective. There's no adjectives. It looks more neutral than writing a paragraph that says something, which will contain trigger words that make people feel like they're being guided. (Data visualization editor)

This respondent was the only one who questioned the objectivity of the data on which dataviz are based, noting that 'the existence of some data means someone has made a decision to collect it or to compile it, and that decision will usually be made with some ultimate goal in mind'. It was more common for respondents to question the objectivity of the visual representation process, by commenting that producing data visualizations means selection, interpretation, and transformation, as seen in the following quote:

> The moment I choose a colour, I have added extra information. Unemployment figures, typhus have no colours. I have to choose one. That is the beginning of interpretation. (Art director)

Many respondents acknowledged that they shape data through the visualization choices that they make. Thus although a small number of respondents indicated that they see dataviz as objective, most did not share this view, instead seeing the process of visualizing data as involving interpretation. This interpretation needs to be made transparent, these respondents noted, and they described a number of practices through which they seek to achieve transparency.

Most respondents felt that being transparent about the dataviz production process was important, regardless of whether they saw data visualizations as forms of truth-telling or as involving interpretation. For those respondents who saw dataviz production as a process of selection and interpretation, their views informed their practices, because they believed that these very processes should be made transparent. Many explicitly linked their transparency practices to trust-building:

> We have as a principle here to be very transparent. If we have a story that is controversial because we have hit a few choices, so we will tell it, be open about the choices that we have made. (Digital editor)

Respondents described widespread uses of transparency practices which aim to build trust and establish credibility, which we discuss below according to the categories we identified above: data uncertainty, visual uncertainty, and consumption uncertainty.

Data transparency: Linking to sources, sharing datasets

Crediting sources and linking to sources were seen by respondents as ways of making the process of producing a data visualization transparent. According to respondents, these techniques are taken seriously by the organizations in which they work, although their implementation varies. All respondents said that they credit sources, and some organizations also link to them. Some do this consistently, others do it some of the time. Others have different linking practices for different sources. For example, when using data from its national statistics organization, one newsroom links to the organization's homepage, not to the specific dataset, but this is not how they link to other sources.

There are also differences within newsrooms and across types of stories. One editor said that whether and how they link to data 'depends on what kind of data it is'. A small number of respondents put all the data they have used in a story into a publicly available document, though those who do so believe that these are not widely read. Two respondents noted that their newsrooms have changed their approach to transparency. Previously they provided links to sources, but the 'mobile first' principle of contemporary journalism makes this increasingly unfeasible, so now they give thorough accounts of their methods, which we discuss further below.

Some respondents acknowledged that their organizations' transparency practices could be improved, an indication that they felt such practices were important. One reason for limited transparency is that linking to or sharing data is time-consuming. Some respondents reflected on the social role of journalism when talking about crediting, linking to, and sharing datasets. One respondent contemplated how far a news provider should go in the provision of full datasets:

> A lot of news media are now offering datasets that the readers can explore more or less freely. I think it's fun, because I work with data. But I don't believe it's journalism, offering no particular angle to the matter. I really don't. (Digital editor)

Another, whose organization no longer links to datasets, explained that he and his colleagues 'build a narrative into the story, rather than giving the data like that, dumping it' (Visual journalist). Another respondent concurred, stating that 'You do not want to publish a 136-page PDF to people and say: here you go. No, we need to break it all down, it is our responsibility to understand what the data say' (Developer). Thus we saw some differences amongst our respondents. A minority saw the sharing of *full* datasets as a transparency mechanism, but others felt that doing this without telling a story or providing explanation would constitute lazy journalism, because it is the role of journalism to interpret available data.

Many of these practices aim to show that sources are credible and make it possible for audiences to corroborate information. They are intended to show trustworthiness and generate trust. But practices are diverse and not adopted consistently across or within newsrooms. We see this diversity as resulting from the 'interpretative flexibility' of data visualization in the news, a term used within science and technology studies (STS) to characterize socio-technical assemblages for which a range of meanings exist, definitions are as yet undetermined, and uses are still under negotiation (Wyatt, 1998). Regardless of journalists' views on the objectivity or otherwise of data visualizations, using dataviz in the news involves doing transparency in some way. For those who see data visualizations as objective, transparency practices provide evidence that they are so. For those who see them as the result of interpretation and selection, transparency practices make visible these processes. This is especially the case in relation to the visual stratum of data visualization, as we explain below.

Visual transparency: Accounting for methods

Another way of 'doing transparency' when working with dataviz in the news is to account for methods. This was seen by many respondents as a way of making the interpretative work of visualizing data visible. Most respondents stressed that they give thorough accounts of their methods. Some said that being transparent about visual representation process is the right thing to do, suggesting an implicit moral dimension to the practices they described. For others, the moral dimension is more explicit:

> First of all it's ethically correct to provide it. Then some people will feel reassured probably, but also it's promoting some sort of culture of using data, reading data. (Data visualization designer)

Ethical standards increase the credibility of a profession, yet in the case of dataviz in the news, such standards are not yet stable, another element of its interpretative flexibility. This, combined with limited audience graphicacy, makes it hard for audiences to evaluate whether ethical standards have been met. Transparency practices provide evidence that ethical standards have been followed, according to this respondent. Thus there is a moral dimension to the emerging dataviz transparency norm. As Schudson noted, journalistic norms are not simply customs, they are also 'moral prescriptions for social behavior' (2001, p. 151).

For a small number of respondents, transparency practices like accounting for methods play a role in the negotiation of objectivity. For these respondents, as for some scholarly commentators, there is a relationship between transparency and objectivity. One respondent suggested that acknowledging the presence of subjective decision-making by making methodology transparent is a way of achieving maximum objectivity, or of convincing 'users that your work is as objective as possible' (Data journalist).

Most respondents said that in their newsrooms, they combine both transparency practices discussed thus far: explaining methods and crediting or linking to sources. A minority goes even further, answering questions about methods on Twitter, even though this is time-consuming, or sharing background work on Pinterest or GitHub. But as with data transparency, some respondents acknowledged that they could 'do transparency' better. The digital editor at a Danish national broadsheet noted that although his organization was good at crediting sources, it was not consistent 'when it comes to accounting for our methods'. Another respondent who worked for a national broadcaster noted that while he and his colleagues provided a lot of methodological detail for online visualizations, similar information in relation to broadcast output could be improved.

Techniques not reported: Visual strategies

Amongst the transparency practices that our respondents described, one thing that was striking was that none of them involved deploying *visual* strategies for communicating uncertainty, even though the same uncertainties discussed in the dataviz literature concern our respondents. Fuzziness, boxplots, and other visual strategies for communicating uncertainty were not discussed, except by one respondent (Consultant data journalist) who said he was interested in exploring ways of communicating uncertainty in the future.

Instead, respondents reported widespread uses of *textual* practices through which they aim to be open about their methods and processes and related limitations, as can be seen in the two sections above. Visual design techniques for visualizing uncertainty might exist, but our research suggests that they are not yet established as conventions in the European newsrooms in which we carried out our research. In the absence of established visual conventions through which visualizations can show 'perhaps' or 'probably', language is used—a fact box, a caption, a link to a dataset, to the source of a dataset, or an explanation of methods. These textual strategies were the main mechanisms for 'doing transparency' that our respondents used.

How thinking about audiences and contexts of consumption informs transparency practices

Respondents' views about audiences, their graphicacy and the contexts within which dataviz circulate also informed their transparency practices, in a small number of cases. Some respondents felt that audiences naively assume that dataviz represent truths about the world—one said, 'At first sight, maps and graphs appear more objective' (Head of data journalism) and another concluded, 'That's why infographics bear such a big responsibility' (Art director). For some, this is problematic. A data journalist said, 'It is problem that people regard numbers and graphics—or to exaggerate, everything that is produced by a machine—as objective truth'. Another respondent stated that people are 'too naive about the truthfulness of dataviz'. He continued:

> It's like: 'Look here, such is the world!' If there is a map, or a graph, or a chart saying so and so. They will say, 'I have found the evidence of how the world is!' (Data journalist)

In complete contrast, some respondents felt that audiences are too sceptical about the dataviz that they see in the news. Some believed that audience scepticism combined with the proliferation of misinformation to make audiences perceive data visualizations as biased or fake, even when they are not. Two data journalists at the same broadsheet newspaper discussed this problem, expressing concern that despite implementing transparency practices, audiences respond with 'Fake news!', 'This is [your newspaper's] data', or 'This is not true' (Data journalists 1 & 2).

Kennedy et al. (2016) note that data visualizers understand graphicacy to include the ability to critically assess the trustworthiness of dataviz. Perceived audience naivety (or believing that dataviz represent the truth) and perceived audience scepticism (or the belief that dataviz are biased or fake) can therefore both be understood as limited graphicacy. When data visualizations are shared online, stripped of context and combined with limited audience graphicacy, they introduce uncertainties relating to consumption, as a minority of our respondents noted.

Espeland and Sauder (2007) argue that 'numbers are easy to dislodge from local contexts and reinsert in more remote contexts. Because numbers decontextualise so thoroughly, they invite recontextualisation' (p. 18). In other words, once 'in the wild', data can become separated from the transparency practices discussed in the previous section, which are designed to inform audiences about what the numbers can be taken to represent. This is even more of a problem for visual representations of numbers, as images have even greater 'shareability' than numbers and text (Bruns & Hanusch, 2017).

One of our respondents talked at length about his organization's attempts to address this problem, noting that 'data visualizations can take a life of their own' because 'it's very easy for a graph that you've done to be robbed of context and taken out' (Editorial developer). This respondent had produced a visualization which explored whether the UK would still have voted to leave the EU if constituency boundaries for this vote were the same as for general elections, and found that it would indeed have done so. This prompted the respondent and his colleagues to reflect on the possibility of the visualization being used in political propaganda, asking themselves, 'What can go wrong here, what can go right here? How do we write this up, how do we explain all these things in a way that heads it off?'. To counter the potential decontextualization of the dataviz and their use in misinformation contexts, one strategy he and his colleagues adopted was to embed explanatory text into the graphic file, so that when the image is circulated, explanatory text circulates with it.

Although this concern was not articulated by many respondents, these comments nonetheless indicate that the issue of how to anticipate uncertainties on the consumption, or decoding, side of data visualization are on the agenda in some newsrooms. Encoding transparency into visualization production in ways that acknowledge that consumption contexts are marked simultaneously by audience naivety and scepticism, by debates about truth and post-truth, is an emerging practice. The context of misinformation and the technological assemblages of social media platforms combine to produce a new challenge for journalists, which is heightened by data visualization's visual character, numeric foundations, and contexts of circulation.

Conclusion: Data visualization as enabling transparency and re-establishing trust?

While Anderson (2018) and McNair (2017) conclude their historical analyses by arguing that there is a need for more transparency and openness about uncertainty in *future* journalism, in our empirical study of current practice, respondents suggested that these things are well underway. Our empirical research thus fills a gap in the literature, advancing understanding of uses of data visualization and enactments of transparency in contemporary newsrooms.

On the whole, news professionals see working with dataviz as contributing to journalistic transparency in particular ways. Our respondents attempted to be transparent in relation to both data and visual process, regardless of their views about the objectivity or otherwise of the visual and numeric strata of dataviz. The data visualization process demands a series of visual choices which are distinct from the choices made in text-based journalism and which are not yet established as conventions, and so distinct enactments of transparency result from the particular characteristics of data visualization.

Our findings suggest that the networked circulation of news visuals and the context of misinformation both present new possibilities and make new demands with regard to transparency in data visualization, and in journalism too. Kennedy et al. (2016) argue that the transparency practices like those that our respondents described not only serve the practical purpose of being transparent; they also serve the rhetorical purpose of *performing* transparency. They quote Latour, who argues that traceability in the creation of visuals is a key component of their ability to 'transport truth' (Latour, 1995, p. 180). This is another reason for 'doing' transparency.

Our respondents described dataviz transparency practices which are primarily textual accompaniments to visual information, the diversity of which suggests that conventions have not been established. We introduced the concept of interpretative flexibility to explain this indeterminacy. This is something that may change, and the extent to which conventions become established, and whether practices become more visual, should be the subject of future research.

Studies using content analysis (e.g. Engebretsen, 2017; Zamith, 2019) have found transparency work to be less widespread than the picture that our respondents painted. Our respondents' descriptions of their practices suggest that this may be changing, or that there may be a difference between what people say and what they do. Follow-up research using quantitative content analysis could seek to verify what our respondents reported. Our research provides some explanation of why transparency practices are sometimes not undertaken, for example because of limited resources or the view that news professionals should do the work of interpreting data and not leave this to audiences.

Our findings were relatively consistent across the countries in which we carried out our research—the quotes included in this paper come from respondents working in all six of them. Newsroom data visualizers and data journalists belong to a global community which is connected via social media and face-to-face conferences, as a number of our respondents acknowledged. As such, our respondents could be seen as belonging to a global epistemic community which shares similar challenges and experiments with similar solutions. Both the news and dataviz are fields with super-national forms and norms, the development of which crosses borders at digital speed.

Acknowledgements

The research reported in this chapter was supported by grants from the Norwegian Research Council (NFR) and the Norwegian Media Authorities (RAM).

References

Allen, D. (2008). The trouble with transparency: The challenge of doing journalism ethics in a surveillance society. *Journalism Studies, 9*(3), 323-340. https://doi.org/10.1080/14616700801997224

Ambrosio, C. (2015). Objectivity and representative practices across artistic and scientific visualization. In: A. Carusi, A. S. Hoel, T. Webmoor, & S. Woolgar (Eds.), *Visualization in the age of computerization*. (pp. 118-144). London: Routledge.

Anderson, C. W. (2018). *Apostles of certainty: Data journalism and the politics of doubt*. Oxford: Oxford University Press.

Balchin, W. G. V. (1972). Graphicacy. *Geography, 57*(3), 185-195.

Boukhelifa, N., & Duke, D. J. (2009). Uncertainty visualization—Why might it fail? In: *CHI Extended Abstracts'09*. (pp. 4051-4056). Boston, MA: ACM.

Bruns, A., & Hanusch, F. (2017). Conflict imagery in a connective environment: Audio-visual content on Twitter following the 2015/2016 terror attacks in Paris and Brussels. *Media, Culture and Society, 39*(8), 1122-1141. https://doi.org/10.1177/0163443717725574

Coddington, M. (2015). Clarifying journalism's quantitative turn. *Digital Journalism, 3*(3), 331-348. https://doi.org/10.1080/21670811.2014.976400

Dasgaptu, A., Chen, M., & Kosara, R. (2012). Conceptualizing visual uncertainty in parallel coordinates. *Computer Graphics Forum, 31*(3), 1015-1024. https://doi.org/10.1111/j.1467-8659.2012.03094.x

Engebretsen, M. (2017). Levende diagrammer og zoombare kart: Datavisualisering som nyskapende fortellerform i journalistikken. *Norsk Medietidsskrift, 24*. http://doi.org/10.18261/issn.0805-9535-2017-02-02

Espeland, W., & Sauder, M. (2007). Rankings and reactivity: How public measures recreate social worlds. *American Journal of Sociology, 113*(1), 1-40. http://doi.org/10.1086/517897

Goodchild, M. (2009). Agendas on display [Review of the book *PIcturing an uncertain world: How to understand, communicate, and control uncertainty through graphical display,* by H. Wainer]. *American Scientist, 97*(5), 422-424.

Kennedy, H., Hill, R. L., Aiello, G., & Allen, W. (2016). The work that visualisation conventions do. *Information, Communication and Society, 19*(6), 715-735. https://doi.org/10.1080/1369118X.2016.1153126

Karlsson, M. (2010). Rituals of transparency: Evaluating online news outlets' uses of transparency rituals in the United States, United Kingdom and Sweden. *Journalism Studies, 11*(4), 535-545. https://doi.org/10.1080/14616701003638400

Karlsson, M. Clerwall, C. & Nord, L. (2017). Do not stand corrected: transparency and users' attitudes to inaccurate news and corrections in online journalism. *Journalism & Mass Communication Quartlery, 94*(1): 148-167.

Kovach, B., & Rosenstiel, T. (2007). *The elements of journalism: What news people should know and the public should expect.* New York: Three Rivers Press.

Latour, B. (1986). Visualization and cognition: Drawing things together. *Knowledge and Society, 6*, 1-40.

MacEachren, A. M., Roth, R. E., O'Brien, J., Li, B., Swingley, D., & Gahegan, M. (2012). Visual semiotics and uncertainty visualization: An empirical study. *IEEE Transactions on Visualization and Computer Graphics, 18*(12), 2496-2505.

Masson, E., & van Es, K. (2017). Visualizing connectivity: Data as evidence in *The Architecture of*Radio. *First Monday, 22*(10). http://dx.doi.org/10.5210/fm.v22i110.8039

McNair, B. (2017). After objectivity? Schudson's sociology of journalism in the era of post-factuality. *Journalism Studies, 18*(10), 1318-1333. https://doi.org/10.1080/1461670X.2017.1347893

Porter, T. M. (1996). *Trust in numbers: The pursuit of objectivity in science and public life.* Princeton: Princeton University Press.

Schudson, M. (2001). The objectivity norm in American Journalism. *Journalism* 2(2), 149-170. http://doi.org/10.1177/146488490100200201

Singer, J. (2010). Norms and the network: Journalist ethics in a shared media space. In: C. Meyers (Ed.), *Journalism Ethics: A philosophical approach.* (pp. 227-239). Oxford: Oxford University Press.

Tal, A., & Wansink, B. (2016). Blinded with science: Trivial graphs and formulas increase ad persuasiveness and belief in product efficacy. *Public Understanding of Science, 25*(1), 117-125. https://doi.org/10.1177/0963662514549688

Vos, T. P., & Craft, S. (2017). The discursive construction of journalistic transparency. *Journalism Studies, 18*(12), 1505-1522. https://doi.org/10.1080/1461670X.2015.1135754

Weinberger, D. (2009). Transparency: The new objectivity. *Knowledge Management World.* Retreieved January 23, 2019 from http://www.kmworld.com/Articles/Column/David-Weinberger/Transparency-the-new-objectivity-55785.aspx

Wyatt, S. (1998). *Technology's arrow: Developing information networks for public administration in Britain and the United States.* Maastricht: Universitaire Pers Maastricht.

About the authors

Helen Kennedy is Professor of Digital Society at the University of Sheffield. Her research traverses digital landscapes and is currently focused on datafication in everyday life and its visual dimensions.

Martin Engebretsen is Professor of Language and Communication at the University of Agder, Norway. His research areas include text and discourse studies, multimodality, digital journalism and visual communication.

Wibke Weber is Professor of Media Linguistics at ZHAW Zurich University of Applied Sciences. Her areas of research are data visualization, information graphics, visual storytelling, visual semiotics, multimodality, and augmented and virtual reality.

Section III

Data visualization, learning, and literacy

12. What is visual-numeric literacy, and how does it work?

Elise Seip Tønnessen

Abstract

This article explores the concept of literacy related to the use of data visualizations. Literacy is here understood as the ability to make sense from semiotic resources in an educational context. Theoretically the discussion is based in social semiotic theory on multimodality in the tradition of New Literacy Studies. Empirical examples are taken from observations in two Social Science classrooms in upper secondary school in Norway, where the students work with publicly available data visualizations to answer tasks designed by their teacher. The discussion sums up factors that affect reading and learning from such complex resources: taking time to explore axis system, variables, and digitally available options; questioning data; and contextualizing results.

Keywords: Literacy; Numeracy; Multiliteracies; Reading for learning

Introduction

The development of innovative data visualizations creates new demands on the ability to make meaningful use of such resources. This ability may be seen as a kind of literacy, which requires certain skills that may be related to the meaning-making resources applied, to the digital technology, and to the understanding of specialized conventions in statistics. This article will explore this concept of literacy theoretically and discuss it in light of empirical examples. The examples were observed in Social Science classrooms in upper secondary schools in Norway. The students were asked to use digital data visualizations to answer specific questions and complete tasks designed by the teacher. The empirical cases will be used to discuss these literacy

Engebretsen, M. and H. Kennedy (eds.), *Data Visualization in Society.*
Taylor & Francis Group, 2020
DOI 10.5117/9789463722902_CH12

practices, and how they appear as strong, weaker, or even failing in relation to the planned learning outcomes. My aim is to explore the relationships between the understanding of semiotic modes, of how digital media work, and more specifically of how familiar the students are with conventions for statistics and visual graphs.

Theoretical perspectives

Literacies

The New London Group (1996) calls for a plural concept of *multiliteracies* to meet the challenges posed by new media and globalization in society. Gunther Kress (2003, p. 23), one of the participants in the New London Group, points to the complexity of literacies, claiming the concept needs to take into account the relevant semiotic modes as well as the ability to use media for production and distribution in multimodal communication. In this chapter I will explore literacy as cultural practices that are shaped by and adjusted to a certain context (Barton, 2007). My interest is in *literacy practices*, but empirically these can be studied through situated *literacy events*. The events studied in this chapter are situated in a school context, but the learning resources come from a research context, and are read in a digital medium. This complicated context is in line with the learning outcomes related to developing the students as 'budding researchers':

> Students should be able to use a variety of digital search strategies to find and compare information that describes problems from different points of view and evaluate the objectives and relevance of one's sources. (KP, 2013, Social Science curriculum)

The use of the term *literacy* extended from verbal language to other semiotic systems and media has been criticized for its lack of precision (e.g. Kress, 2003, p. 23; Hasan, 1996). In this article I will take the idea of literacy as the ability to make sense from semiotic resources in an educational context (Hasan, 1996) as my starting point and discuss how this may be more accurately described based on my empirical examples. A preliminary label for this form of literacy may be *visual-numeric literacy*, which draws on an understanding that reading such graphs requires the mastery of certain modes, mainly visual and numeric. In previous research there is a tendency to focus either on the visual (Chevalier et al., 2018; Allen, 2018) or on the numeric dimension (Prince & Archer, 2014) of related literacies.

The *visual* modes relevant to reading graphs are organized in a composition (van Leeuwen, 2005), where elements in a defined space make meaning in terms of size, direction, and relative distance to the axes defining the space. This connects to the *numeric* dimension, where specialized conventions have been developed within mathematics and statistics. Some of these are spatial conventions about how systems of axes or columns and rows work. They are connected to methodological conventions about relations between variables (independent and dependent) and how they are placed spatially, combined with more general conventions about what directions mean in our culture (such as developments in time moving from left to right, positive values moving upwards and to the right).

Such complexities can be comprehended on different levels. Ruqaiya Hasan (1996) distinguished between three aspects of literacy. Firstly, *recognition literacy* is necessary to understand the relevant meaning resources. For writing the central resource is the alphabet; in visual-numeric literacy relevant semiotic resources are e.g. lines, bars, bubbles, colours, and labels. In digital media it may also include knowing how to find the graphs and the options for changing them.

Hasan argues that recognition literacy, if taught in isolation, is not sufficient. Literacy also requires discursive abilities, connected to 'enabling the pupils to do something with their language' (1996, p. 399). This entails the ability to produce and interpret connected texts within the genre suited for the context in question. This is important to enable users to achieve their goals, whatever they may be, and be active participants in society. In visual-numeric literacy this may involve posing relevant questions and understanding how variables can be combined and how to choose displays that best visualize a point. Action literacy is developed through practice; reading several data visualizations gives the experience necessary to make meaningful choices.

The third aspect, *reflection literacy*, involves the ability to reflect, enquire, and analyse (p. 408). This includes reflection over reading practices in society, questioning the values they carry and the perspectives they entail. Reflection literacy is what it takes to question choices and readings, critique sources, and contextualize findings. These three aspects of literacy are interrelated. Hasan claims that 'reflection literacy includes a well-informed variety of action literacy [...] and the latter includes recognition literacy; the reverse is, however, not true' (p. 417).

Case study

The empirical examples for my discussion were observed in two classrooms in secondary schools in Norway. In both cases the subject was Social Studies, with a basic course for students aged 17 in case A and an advanced course in Human Geography for students aged 18 in case B. In both cases we visited the classroom to observe the students working with tasks that were a normal part of teaching, planned by their teacher. In case A the class used *Google Public Data* (2018) in their work on unemployment, as part of the topic 'Working life and business'. The tasks were integrated in a lecture where the teacher introduced the topic before the students went online, and afterwards he summed up the findings in a classroom dialogue. In case B the topic was 'Demography', and the tools used were from *Gapminder* (2018). The class had spent one lesson getting acquainted with the tool previously and handed in their findings in writing after the double lesson spent on the tasks.

The data visualizations used in these two cases are both available to the general public. The unemployment graph in *Google Public Data* is based on big datasets from Eurostat, displaying data on unemployment rates as a line chart, showing time on the x-axis and percentage of unemployed on the y-axis (Figure 12.1). The tool includes three other options for visualization: bar charts, bubble charts, or maps, but none of these were used in the classroom we observed.

The *Gapminder* tools used in case B are developed to visualize publicly available data in order to promote 'a fact-based worldview' (Gapminder, 2018) among the general public and include instructions for teachers. Links to *Gapminder* are included among the external resources suggested for Social Studies by the *National Digital Learning Arena*, the official portal for digital learning materials in Norwegian upper secondary schools (NDLA, 2018). The default settings display the relations between income (x-axis) and life expectancy (y-axis) as a bubble chart (Figure 12.2). In addition to the relations between the two axes, it offers coding of the bubbles in size and colour (the default settings of size indicating population size and colour indicating world region may be changed in the search fields to the right, Figure 12.2). In addition, a time dimension is shown as an animation. The tool includes options for various ways of displaying the data: trends (line charts), ranks (bar charts), maps, population pyramids, and stacked area graphs, but none of these were activated in the classroom we observed.

In each class students volunteered to take part in the study, 8 students in case A and 10 students in case B. The study followed the standards of Norwegian ethical guidelines and was approved by the Data Protection

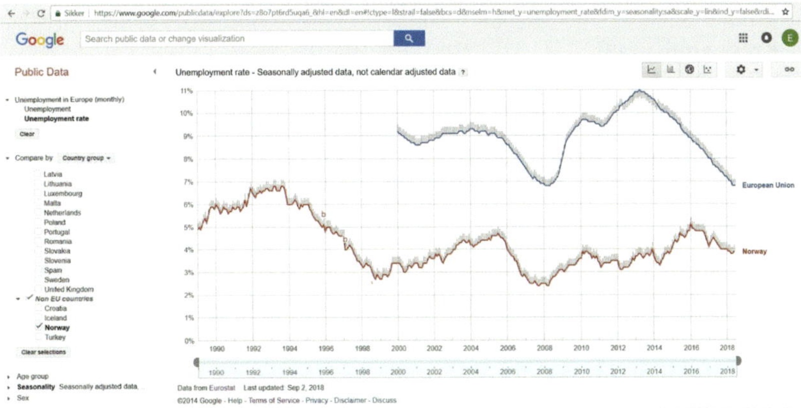

Figure 12.1. Screenshot of Google Public Data. Based on free material from Google Public Data. Source of data: Eurostat, CC-BY licence.

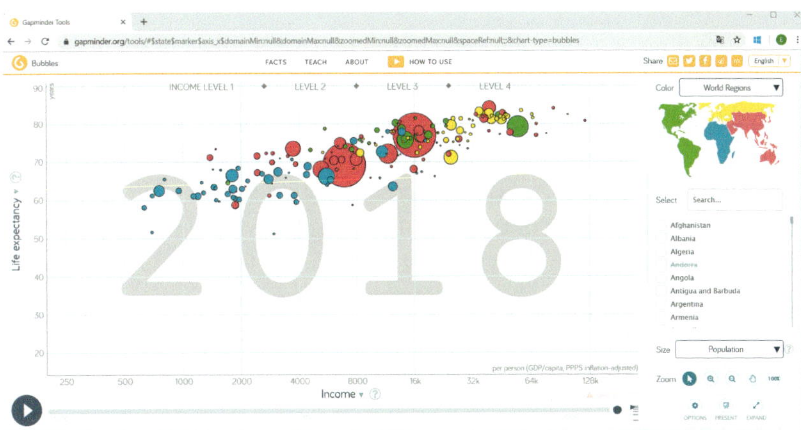

Figure 12.2. Screenshot of the starting image on Gapminder tools. Based on free material from gapminder.org, CC-BY license.

Authorities in Norway. The students worked in pairs with the tasks given by the teacher, and we made screen recordings and recordings of the pairs and their discussions while working.

In the analysis my focus is on how working with these data visualizations may contribute to learning, and on identifying factors that enhance learning, or represent obstacles to learning. My understanding of learning is inspired by Bezemer and Kress (2016) who emphasize that learning requires engagement and rests on interpretation:

> Instead of measuring the *transmission* of knowledge, our interest is in uncovering and describing the transformative *principles* that learners

> bring to bear as they engage with the world around them. (Bezemer & Kress, 2016, p. 38)

Hence, when learners engage with the world through textual and mediated means, learning is closely connected to literacy, to handling the semiotic resources, and in our cases the digital media involved. As a basis for the analysis below, I looked systematically through the screen recordings, noting which semiotic and digital resources were used, how they were interpreted, and how they contributed to completing the tasks given by the teacher. This allowed me to point out factors that lead to more or less meaningful literacy practices.

Analysis of literacy practices

Out of the nine pairs we observed in the two cases, most of them worked steadily through the lesson to answer the questions designed by the teacher. Analysing the literacy practices, I assessed them as situated in an *educational* setting, where success is seen in relation to learning, understood as active engagement in transformative processes in line with Bezemer & Kress (2016). I found instances of successful readings as well as weaker readings or direct misreadings in each literacy event. In the following I will explore the factors leading to good, weak, or failed reading events across the groups.

What characterizes successful literacy practices?

The best practices I observed were characterized by the students taking the time to understand how the graphs worked before they started exploring them and answering the specific tasks designed by the teacher. In case B the two girls in group 2 started by asking what the colours of the bubbles in Gapminder stood for and agreed that it indicated on which continent the country was located. Group 5 used this knowledge to ascertain the location of countries with which they were not familiar. Group 3 took time to check that they understood the labels in the axis system, translating from English to Norwegian.

In case A I also found instances where the students posed questions relevant to reading the graphs. Group 7 asked the teacher how they could access data from before 2000, which was the starting point for EU data, and he helped them discover that some countries were represented with a longer time span. They also asked about the difference between 'unemployment' (total number) and 'unemployment rate' (percentage of work force unemployed), but in this case they were just told to choose the latter, not to investigate the help information available by clicking the question mark besides the label. These examples show that a basic factor in visual-numeric literacy is getting an overview of the composition of graph, variables, and options included.

In case B one of the tasks was to reflect upon why some of the bubbles in *Gapminder* were not moving during the first part of the time series from 1800 to 2015. Two of the groups passed quickly over this question by saying that it meant no change. But three groups questioned whether there were data available for all countries back in the 1800s. Group 2 ran the relevant time series a couple of times to determine which countries this applied to, and found that it was mostly African countries, where public statistics may not go that far back. However, none of the groups consulted the label 'data doubts' (bottom, right-hand side, Figure 12.2). If they had, they would have found the information that 'countries on a lower income level have lower data quality in general, as less resources are available for compiling statistics. Historic estimates before 1950 are generally also more rough' (Gapminder, 2018). These examples show the need for critical assessment of the numbers and statistics behind the graphs, which may be supported by information not immediately visible on the screen. Hence it also points to the need to understand the relationship between what is available at the (screen) surface of digital texts and what may be accessed through links and clicking.

As can be seen from these examples, meaningful readings depend on background information. While exploring the graphs, the students leaned on their previous knowledge about society and history. In general, these were not very sophisticated, which is not surprising given their young age. The students in case B related what they saw to well-known historic events such as the World Wars, or the Wall Street Crash of 1929. The students exploring the unemployment rates in case A were aware of the financial crisis, and how it affected Greece in particular, but their background knowledge was more approximate when it came to what caused the crisis, and how this connected to unemployment rates.

How the tasks were designed carried consequences for how the readers engaged with the graphs, both in terms of personal engagement and background knowledge. In case A the first question was about comparing the unemployment rate now to when the teacher was young in the 1990s. When group 9 compared the 6.7 percent unemployment in 1994 to the recent rise from 3.2 percent to 4.9 percent (2014–2016) they reflected: 'We think there is a crisis now, but it was so much higher then!' In this case the personal contextualization provided a longer time span for assessing the numeric information.

Whereas the students in general used the data visualizations mainly to confirm and—at best—expand the knowledge they already possessed, one example illustrates how the teacher designed a task that encouraged the students to learn something *new* from data visualizations. They were asked to focus on China in the time span from 1957 to 1962 and were specifically challenged to search for information about the great famine following from Mao's agriculture and industry reform policy. For most of the groups this led to reasoning about how natural conditions in combination with politics may affect ordinary people. For group 5 this led to emotional responses as they realized the suffering involved. Going back to the graph after updating their background knowledge, they followed the big pink bubble as it bounced downwards to indicate the fall in life expectancy, along with a left move to indicate a parallel fall in income. They were touched by the facts:

- Wow, that was a lot [1958]
- Yes. [moving forward to 1960]
- Oops!
- Yeah, there was a famine!
- But a life expectancy of 30 years. How is that possible?
- It is quite sick!

Concluding from these examples, I find literacy practices that enable the students to expand and reflect on their knowledge when they establish an understanding of how the variables and values on the axes define the graphic space. Furthermore, these readings were characterized by an active engagement in the topics studied, where the students formulated their expectations in advance, based on prior knowledge, while they were still open to include new information and reflect about reasons and consequences. A distinctive feature was that the students posed questions and spent the time and effort it took to find the answers, combining the displayed data with supplementary information.

Why do some readings appear weaker?

The most prominent trend in our observations seemed to be literacy practices that did not harvest the full learning potential from the data visualizations. This has to do with the specific skills required in visual-numeric literacy, but also with the readers' degree of engagement with the graphs and the

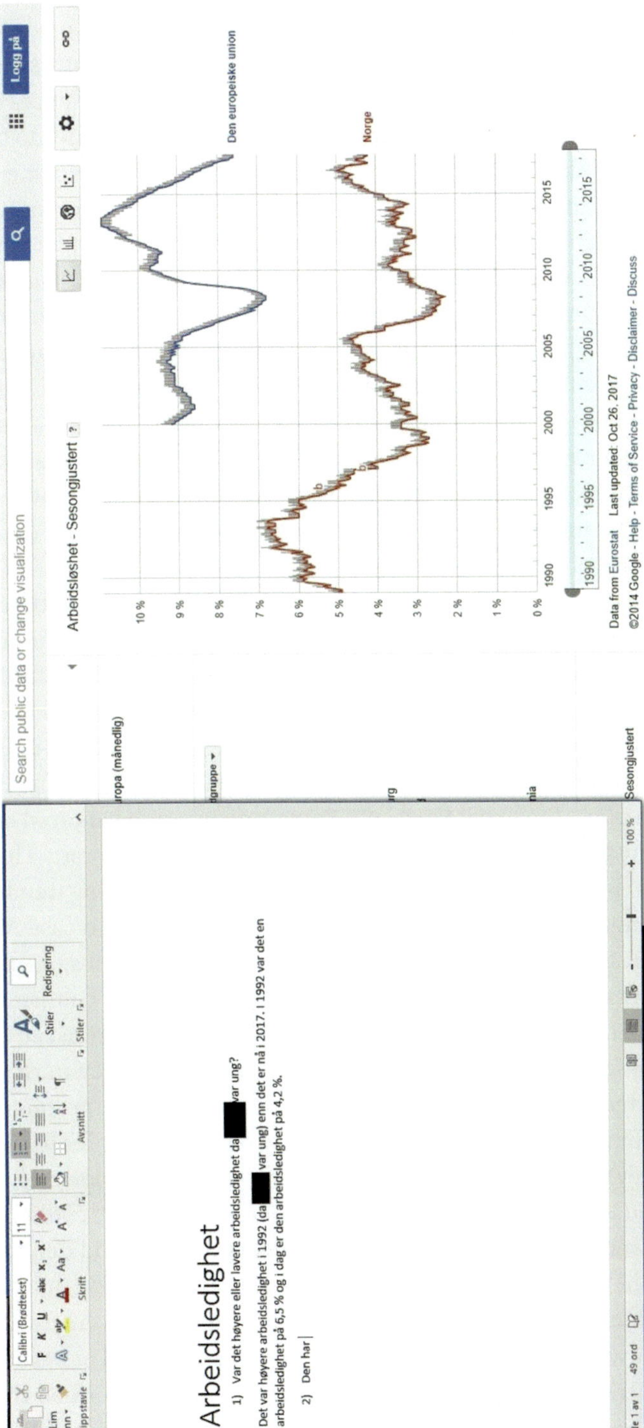

Figure 12.3. Screenshot of group 7's screen after placing the word document side by side with the same graph as the one displayed in Figure 12.1. Based on free material from Google Public Data. Source of data: Eurostat, CC-BY licence.

tasks given by the teachers. When the students in case A first opened the graph on unemployment, they were typically looking for sudden turns and dramatic changes. Their engagement increased when they detected crises in Greece, Spain, Iceland, or Estonia around 2008. However, they seemed to be more interested in the changes as such than in the level of unemployment over time. Group 7 at first estimated the unemployment rate in Norway to be quite stable. Then they decided to place the Word document where they were typing the answers side by side with the graph on the screen. This led to a compressed x-axis that made the slope of the rising and falling curves steeper (Figure 12.3). Not taking this relative change into consideration, the girl who was typing exclaimed: 'Why did I say it was stable?' and they adjusted their answers accordingly. If they had compared the variation observed with unemployment rates from earlier years, or in other countries, they might have modified their assessment, as another group did when asserting that the unemployment rate in Norway was overall lower than in other countries or regions.

The tendency to extract the most visible facts from the graph without seeing them in relation to other available information was even more striking for the groups working on *Gapminder*, since this tool contains more information and more options for display. When the students first approached the bubble chart with the default settings of income (x-axis) and life expectancy (y-axis), they focused mainly on the extreme cases; the lowest or highest life expectancy or income, and when they moved on to the following tasks, the highest child mortality and fertility rate. This led to readings that picked out single facts, rather than discovering trends.

Several examples of such isolated readings were observed. In case A I found that the students described the development in countries one by one, apparently not noticing the option to compare groups of countries (upper left corner in Figure 12.1). When the students in case B were asked to comment on the connections between income and life expectancy, most of them just asserted that the better the income the longer the life expectancy. Only a few formulated reasons, e.g. how better economy allows for better healthcare. They were also asked to find the four countries with lowest life expectancy today and reflect on which parts of the world they could be found in. Answering that three of them (Lesotho, Swaziland, and Central African Republic) were in Africa and one (Afghanistan) was in Asia did not really do credit to the level of detail included in the tool they were using. And when they were asked to compare the development of child mortality for three countries (USA, Norway, and Mali), they mostly described the countries one by one, rarely commenting on the relations between them.

Furthermore, I found few examples where the students reflect on the meaning of the values on the axes. Even though the teacher in case B specifically told them to note that the values for income on the x-axis were logarithmic (each interval doubles the value), they did not question what this meant and how it affected the shape of the graph. When they changed the axes to child mortality and time, they did not notice that now the y-axis had a logarithmic scale. When discussing child mortality, they did not seem to take in the realities of the measurement: '0-5 year olds dying per 1,000 born' (explanation along the y-axis). In the case of Mali this meant that every second child died before the age of 5 throughout the nineteenth century, and the situation did not improve until well into the 1960s (Figure 12.4a).

One reason why the students rarely exhausted the full potential of the graph may be that they did not take the time to get fully acquainted with it. Several functions were never activated, such as the background information marked with a question mark where there are options for choice, or the information videos placed right underneath the graph.

I did, however, find a few examples of students discussing the meaning of the labels. This occurred when the wording on their task sheet was not exactly the same as on the screen. The students in group 5 discussed whether there was a difference between 'Children per woman', which was the label used in task 3, and 'Babies per woman', which was the label they found in *Gapminder*. One of the boys claimed that the term babies was limited to the first year in life, while children would be used for those past age one. This was knowledge from the textbook, and it would have been relevant for the variable 'child mortality', which was used in task 2.

The main factor weakening these reading events was the lack of time and effort invested in reading and interpreting the visualizations and the data they were based on. These data visualizations are packed with information and require careful and thorough interpretation. The combination of several variables in one display is its specific strength, but this strength was not exploited to its full extent in the cases I observed. From my observations it seems relevant to ask how many dimensions the students are able to take in at once. In the cases I observed, none of the students finished all the tasks given by the teacher. This meant that they never got to the stage where they were allowed to pose their own questions to the data, which is the learning outcome envisioned in the national curriculum. Hence the time available compared to the workload would seem to suggest that quick reading and short answers are more realistic outcomes.

A specific case of misreadings

There were not many direct misunderstandings in our examples. But one specific task in case B led to a row of very different choices that it is illuminating to study in depth. The misreadings happened when the students were asked to change the variables on the axes. The task formulated by the teacher said:

2. Choose the indicator *Time* on the first axis and *Child Mortality* on the second axis.

a) Describe how child mortality has developed in the USA, Norway, and Mali.

The problem appeared when the students had difficulties finding the small triangle next to the labels that allowed them to choose other variables. The resulting graphs can be seen in Figure 12.4a-d. Group 2 and 3 established the graph with the intended axes variables on their first try (Figure 12.4a) and had no specific difficulties reading the graph. Displaying time on the x-axis made it easy to see development over time. They commented on the general trend that child mortality had been lower and decreased more rapidly in Norway and the USA than in Mali, and questioned why the curve for Norway had so many ups and downs throughout the 1800s. Group 2 also questioned the sudden rise in child mortality in the US in 1918 and found the explanation through a search that led them to information about the Spanish flu.

Group 1 searched for 'life expectancy' in the search field for coding the size of the bubbles (bottom, right), and ended up changing this to 'number of child deaths', and not changing the axes variables (Figure 12.4b). In the resulting graph child mortality was indicated in total numbers by the size of the bubbles. At the time the screenshot was taken, they approached the teacher to ask why Mali was not moving at all. She directed them to the right axis variables, resulting in Figure 12.4a. The next time they needed to change the axes for task 3, they had no trouble applying this literacy skill to a new task.

Group 4 got the axes mixed up; they changed the y-axis to 'Time' and the x-axis to 'Child mortality'. One of them suggested that it would be more natural to have Child mortality on the vertical axis, but after some changes back and forth they ended up with the graph in Figure 12.4c. Displaying time on the y-axis is counter-intuitive to established conventions of reading time development from left to right (Kress & van Leeuwen, 2006). In addition, the value on the x-axis was negative, which meant that the movement over time in this graph went from bottom right to top left. Within Western reading

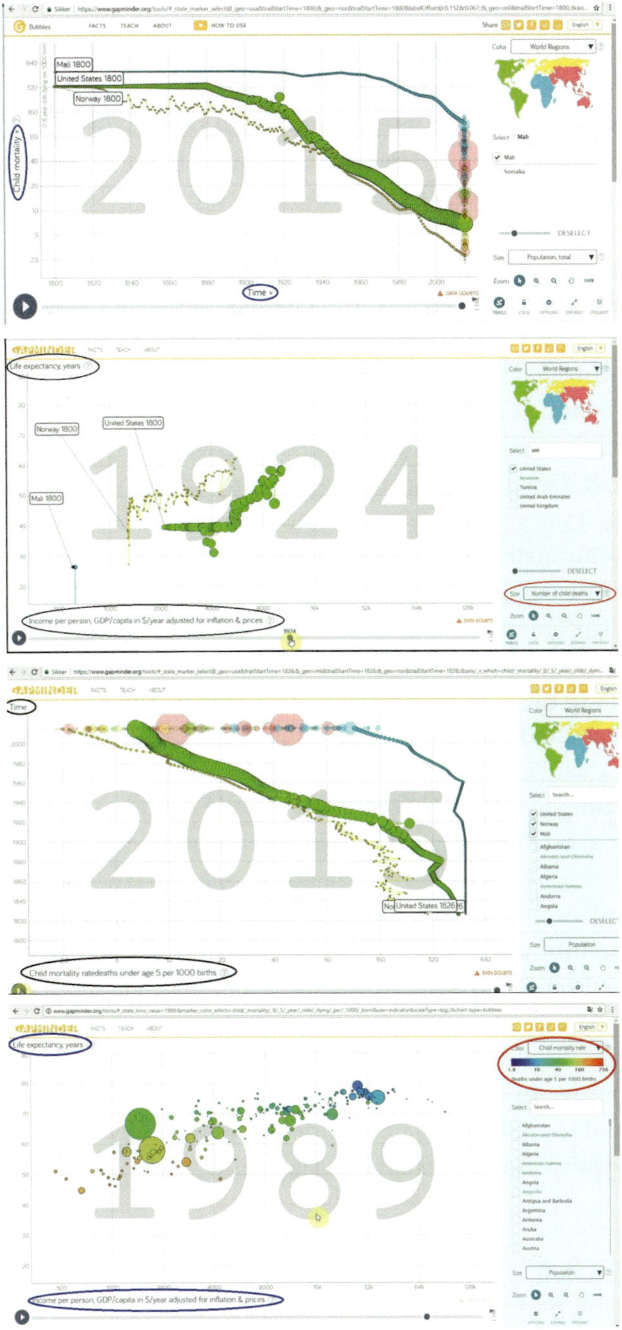

Figure 12.4 Versions of graph to answer task 2 on child mortality in three countries. a) Group 2 and 3 with intended axis variables, b) Group 1, c) Group 4, d) Group 5. Based on free material from gapminder.org, CC-BY licence.

conventions this is hard to interpret. Due to this confusion, and time limits, the group ended up not answering question 2.

Group 5 also had problems changing the axes variables, and although one of the students questioned the result, they did not proceed to finding out what the problem was. They searched for child mortality in the search field for colour-coding of the bubbles, resulting in Figure 12.4d. Here child mortality was visualized in colour, indicating high mortality with warm colours and low mortality with cold colours. Keeping income and life expectancy as axes values resulted in a rising pattern of bubbles. In their discussion the boys talked about Norway and the USA 'peaking upwards', and in writing they first formulated the rise as an improvement: 'In the USA and Norway child mortality has developed steadily upwards', but then they corrected the last two words to 'in a positive direction'. Hence their answer appeared correct, but it was taken from their general knowledge rather than from their reading of the graph. The teacher would probably never know that they needed some instruction on the rather simple task of finding out how to change axis variables in this specific tool.

These misreadings are interesting since the problem is media-related rather than semiotic in nature. The options for choosing variables and coding are inherent in the dynamics of digital media that afford exploratory work with data visualizations. The problems in our case B would not occur in a textbook where the display of graphs is stable and designed by experts for explanatory use. The more options given to the reader, the more demanding it gets to establish a graph that can be meaningfully read. In the classroom, misreadings are mostly avoided because the students are led by hand through the tasks designed by the teacher, but the independent and actively researching student envisioned in the national curriculum needs to understand which variables can be meaningfully combined and what forms of display will give a clear visualization. More experience with data visualizations is needed to foresee the results of chosen values, and consequently to be able to discover and correct mistakes. The ability to notice mistakes, and to analyse and correct them, and generally question readings, is vital to any kind of literacy (Roe, 2008, p. 96).

Concluding discussion

Our observations include groups on different levels, one working in a basic course, the other on a more advanced level. The students in case B demonstrate a higher level of literacy in their ability to activate their pre-understanding and contextualize their reading of the graphs. Still, this does not prevent them from encountering problems when they are asked to change the axis system and explore new datasets. One might argue that what I have termed 'misreadings' in this article is mainly due to students' problems in handling the many choices given by the digital *Gapminder* tool. This finding means that the literacy discussed in this article is not merely visual and numeric, it is also about how digital media work, and how they allow the user to interact with preprogrammed affordances in data visualizations. Consequently, the literacy I gave the preliminary label 'visual-numeric' may be more complex than this term suggests.

This complexity involves connections of statistic, technological, and semiotic resources that work on different levels. On a fundamental level, the axis system defines a space that is semiotically charged, and hence functions as an overall framework for reading the graph. Within this framework the lines and bubbles require the reader to take notice of slope and area respectively, and also codings of colour (Cairo, 2016, p. 128). Interpreting or producing a meaningful space between the axes requires specialized statistical knowledge of variables, values, and other conventions. The digital medium is the means to systematize, save, and reshape data, often too big to handle in any other medium, but also to display and interact with these data. This requires both general and more specialized digital literacy.

As pointed out by Hasan (1996), literacy works on different levels. The students *recognized* several semiotic resources and digital functions from their general experience with digital media, e.g. using search functions, pressing the play button. They may have recognized the triangle opening the menu of variables (see arrow in Figure 12.2) if it had been shown to them when needed. But this simple act of recognition is related to a more general insight in how digital media facilitate access to layers of information behind the screen surface.

The *action* aspects of literacy seem to need guidance and teaching in our example. The teachers designed a progression of tasks to build experience for the tasks to come, e.g. in case B asking why some bubbles were not moving in the early years, before the students approached the task of comparing Mali to two Western countries. Our observations reveal a need to find *teachable moments* in school literacy practices. One appeared when the students first were asked to change the axis variables. Those students who had the teacher's attention at that moment avoided 'misreadings' and carried this understanding on to the following tasks.

Reflection literacy involves the ability to critically question the ways data are presented, how they are used, and what for, and also to question one's own reading practices. In the misreadings I observed, some of the students did pose questions, but they rarely went back to correct their mistakes. Maybe this was because of time limits, or maybe the framing of tasks in the school context directed the attention to get the tasks done, more than to in-depth reading. In the cases I studied, the learning objectives were directed towards subject knowledge in Social Studies, rather than to developing the students' specialized literacy for reading digital graphs. As pointed out in my introduction, the curriculum encourages a focus on literacy integrated in other learning outcomes. Amid everyday classroom demands this double focus seems hard to maintain. This points to a need for special attention towards literacy even in secondary schooling, including basic skills in using visual, numeric, and verbal resources as well as digital media (Norwegian Directorate for Education and Training, 2013).

My discussion of best, weaker, and failing practices should not be taken as authoritative universals; each literacy event must be understood in context. In a different situation the objectives of reading or the data visualizations read may justify a more critical, or even subversive, literacy practice. Some of the experiences from my classroom observations may still be transferable, such as the time it takes to get acquainted with the graph and the digital options it affords; the need to question underlying data; and the challenge of contextualizing what is being displayed. Considering the increased use of data visualization in society, the curriculum's ambitions to teach students search strategies, in combination with the ability to evaluate the objectives and relevance of one's sources, seems vital to lifelong learning.

Acknowledgements

This research has been funded by Research Council of Norway through the INDVIL project. A special thanks to my colleague, Professor Anne Løvland, for taking part in developing the research design and assisting in collecting the data.

References

Allen, W. L. (2018). Visual brokerage: Communicating data and research through visualisation. *Public Understanding of Science*. Advance online publication. https://doi.org/10.1177/0963662518756853

Barton, D. (2007). *Literacy: An introduction to the ecology of written language.* Malden: Blackwell Publishing.

Bezemer, J., & Kress, G. (2016). *Multimodality, learning and communication: A social semiotic Frame*. London & New York: Routledge.

Cairo, A. (2016). *The truthful art: Data, charts, and maps for communication*. Berkeley, CA: New Riders.

Chevalier, F., Henry Riche, N., Alper, B., Plaisant, C., Boy, J., & Elmqvist, N. (2018). Observations and reflections on visualization literacy in elementary school. *IEEE Computer Graphics and Applications, 38*(3), 21-29. http://doi.org/10.1109/MCG.2018.032421650

Hasan, R. (1996). Literacy, everyday talk and society. In: R. Hasan & G. Williams (Eds.), *Literacy in society*. (pp. 377-424). London: Longman.

Gapminder (2018). Downloaded April 26, 2018 from https://www.gapminder.org/tools/#_data_/_lastModified:1524739828963&lastModified:1524739828963;&chart-type=bubbles

Google Public Data (2018). *Unemployment rate in Europe* [Data Set]. Downloaded April 26, 2018 from https://www.google.com/publicdata/explore?ds=z807pt6rd5uqa6_

Kress, G. (2003). *Literacy in the new media age*. London & New York: Routledge.

Kress, G., & van Leeuwen, T. (2006). *Reading images: The grammar of visual design* (2nd ed.). London: Routledge.

Norwegian Directorate for Education and Training (2013). Social Studies Subject Curriculum (SAF1-03 / Samfunnsfag). *The Knowledge Promotion Reform*. Revised 2013. Retrieved April 23, 2018 from https://www.udir.no/kl06/SAF1-03

NDLA (2018). Gapminder. Retrieved August 1, 2018 from https://ndla.no/nb/node/7660?fag=36

Prince, R., & Archer, A. (2014). Exploring academic voice in multimodal quantitative texts. *Literacy and Numeracy Studies*, 22(1), pp. 39-57. https://doi.org/10.5130/lns.v22i1.4178

Roe, A. (2008). *Lesedidaktikk—etter den første leseopplæringen. [Reading didactics—After the initial teaching of reading]*. Oslo: Universitetsforlaget.

van Leeuwen, T. (2005). *Introducing social semiotics*. London & New York: Routledge

About the author

Elise Seip Tønnessen is Professor in the Department of Nordic and Media Studies, University of Agder, Norway. Her research interests range from literacy across modes and media, to reception studies in educational as well as leisure time settings. She has published on literacy, children's media culture, and children's literature.

13. Data visualization literacy: A feminist starting point

Catherine D'Ignazio and Rahul Bhargava

Abstract
We assert that visual-numeric literacy, indeed all data literacy, must take as its starting point that the human relations and impacts currently produced and reproduced through data are unequal. Likewise, white men remain overrepresented in data-related fields, even as other STEM (Science, Technology, Engineeering and Medicine) fields have managed to narrow their gender gap. To address these inequalities, we introduce teaching methods that are grounded in feminist theory, process, and design. Through three case studies, we examine what feminism may have to offer visualization literacy, with the goals of cultivating self-efficacy for women and underrepresented groups to work with data, and creating learning spaces where, as Philip et al. (2016) state, 'groups influence, resist, and transform everyday and formal processes of power that impact their lives'.

Keywords: Data literacy; Feminism; Community; Inequality; The arts

Introduction

There is a growing body of literature arguing that working with data is a key modern skill (Letouzé et al., 2015; Wolff et al., 2016). And yet, while highly valued as a precursor to evidence-driven insight, data are expensive—to collect, maintain, and mobilize. Corporations, governments, and elite universities are the primary institutions which have the resources to undertake this work. Within those institutions, white men remain overrepresented in data-related fields such as computer science, engineering, and artificial intelligence, even as other STEM fields like biology have managed to narrow their gender gap (Corbett & Hill, 2015; Neuhauser, 2015; West, Whittaker,

Engebretsen, M. and H. Kennedy (eds.), *Data Visualization in Society.*
Taylor & Francis Group, 2020
DOI 10.5117/9789463722902_CH13

& Crawford, 2019). This has resulted in a growing literature around bias in data collection (Angwin et al., 2016), algorithmic decision-making (O'Neil, 2016), and machine learning training sets (Buolamwini & Gebru, 2018). Acknowledging these basic inequalities in the ecosystem—that data and skills to work with them are in the possession of groups that are already privileged in society—lays the groundwork for how educators can start to discuss data literacy more broadly.

We assert that visual-numeric literacy, indeed all data literacy, must take as its starting point that the human relations and impacts currently produced and reproduced through data are unequal. Thus, educators are faced with a choice. They may either 'integrate the younger generation into the logic of the present system' or teach learners how to 'participate in the transformation of their world' (Freire, 1968, p. 16) through data-driven inquiry. The vast majority of data science programmes, trainings, and tools choose the former. This choice may not be nefarious or intentional, but rather because alternatives may not be readily apparent. This chapter explores an emancipatory approach to data visualization literacy based in feminist scholarship and pedagogy in an attempt to chart an alternate course.

A body of work that owes its emergence to the women's suffrage movements of the nineteenth century, feminist theory encompasses a range of ideas about how identity is constructed, how power is assigned, and how knowledge is generated, as well as how a range of intersectional forces such as gender, race, class, and ability combine to influence the experience of being in the world. It is important to note that while feminist scholarship uses gender as a starting point for considerations of social inequality, a feminist approach is not only about cis and trans women, nor only about gender. Deployed as a tool for critical inquiry, feminist thinking seeks to situate knowledge in specific human bodies and to 'unmask universalism' (Davis, 2008)—to show how things that appear to be neutral or objective are in fact biased towards the bodies that hold power—typically male, white, abled, heterosexual, and well-educated. For example, the quintessential feminist critique of data visualization is Donna Haraway's characterization of it as 'the gaze from nowhere' (1989, p. 581). Because the view is not situated in a body or a perspective, it has the aura of neutrality. But, of course, the view from nowhere is always the view from somewhere—more often than not it is the view from a dominant location of power over people whose views and knowledge are suppressed and silenced (Collins, 2009, p. 251; Eubanks, 2018; Noble, 2018; Walter & Andersen, 2013).

For this chapter, we draw specifically on prior work (D'Ignazio & Klein, 2016) that connects feminist theory to the design of data visualizations. Our goal is to demonstrate the relevance of feminist concerns with gender, social difference, and power in relationship to the teaching and learning of data visualization. Data visualization is sometimes taught with the idea that data are neutral and objective; visualizations are methods for depicting those data; and the right method of depiction can be found by understanding the basics of human visual perception and cognition—which are sometimes imagined to be universally the same across contexts, culture, and history (Kennedy et al., 2016a). Instead, we wish to craft an alternate starting point that acknowledges the social and political context in which data are collected and communicated, cultivate self-efficacy in women, people of colour, and other minoritized groups to work with data and visualization, and focus learners' attention on what happens in the world as a result of an act of data communication.

The editors of this volume ask, '*What does literacy mean when it comes to data visualization, and how can visual-numeric literacy be enhanced?*' (this volume). We assert that because visualizations are outputs of a process, visual-numeric literacy is part of a larger process of data literacy, which itself draws on other approaches such as statistical literacy, numeracy, and critical information literacy. In earlier work, we proposed that data literacy '*includes the ability to read, work with, analyze and argue with data as part of a larger inquiry process*' (D'Ignazio & Bhargava, 2016, p. 1).

While this definition makes it sound like it is an individual ability, integrating feminist thinking opens up questions as to the nature of literacy itself. Is data literacy a set of autonomous skills acquired by an individual? Or, following bell hooks's notion of an 'open learning community' (1994) and proponents of new literacy studies (Street, 1994), is data literacy a set of social practices, learned and practised in and through a social context such as an organization or community? Or, following feminist computer scientist Lynette Kvasny (2006), is teaching about data a site of ideological transmission, a place where, if we are not careful, we risk reinscribing structural oppression? From our teaching practices we would answer these questions 'yes' and 'yes' and 'yes'.

From this complicating ground, then, data literacy cannot begin with technical skills like making and interpreting bar charts and network diagrams. It necessitates a starting point grounded in higher-order critical thinking and making skills that connect data back to the social and political reality from which they were produced.

But what might this look like in practice? In the remainder of this chapter, we outline three short cases of data visualization learning that come from our practice as educators and then analyse them in relation to the design principles laid out by one of us, Catherine, and Lauren Klein (2016), in our paper 'Feminist Data Visualization'. In that paper, we created six preliminary principles of feminist data visualization, drawing from work in feminist science and technology studies, feminist human-computer interaction, feminist digital humanities, and critical cartography & GIS. The principles are: (1) Rethink binaries, (2) Embrace pluralism, (3) Examine power and aspire to empowerment, (4) Consider context, (5) Legitimize affect and embodiment, and (6) Make labour visible. Due to space considerations as well as the exploratory nature of this work, we focus on analysing (2), (4), and (5) as they relate most directly to data visualization literacy.

Three cases of data visualization learning

The Groundwork Somerville data mural

Our first case study focuses on the process of working with a community group—Groundwork Somerville (GW)—and local youth to design and paint a data-driven story as a community mural. This example of a 'data mural' documents one approach to an action-oriented, community-situated activity that builds various data literacies.

GW focuses on empowering participants to improve environmental, economic, and social well-being, specifically through nature-focused programmes. One of their main programmes involves youth to create, plant, and maintain gardens as well as sell the produce that results. Immigrants and low-income families are the main beneficiaries. Additionally, many of the vegetables planted are chosen to reflect the immigrant makeup of the community. GW was interested in working with us in order to reinforce their goals of youth development, to beautify an urban garden, and to tell a story about their impact.

The collaboration followed a process which moved from identifying data, finding a story, collaboratively designing a visual to tell that story, painting the mural, and finally hosting an unveiling event (see Bhargava et al., 2016 for details). GW shared qualitative and quantitative data with our team, and we worked together to narrow in on data to include in a multi-page handout for the youth. In terms of demographics, there were six young women of colour, seven young men of colour, two young white women and two young white men on the GW team. With the data in hand, we hosted a brainstorming session with youth to analyse these data and generate a story and visuals to tell it. Based on these handouts, the participants identified

Figure 13.1. Some of the sketches students created. Photos by Emily Bhargava. Printed with permission.

a number of facts and quotes in order to tell the GW story. Inspired by some data visualizations we shared with them, the youth then sketched visuals for telling the story. This invitation to sketch concretized the data and helped them bridge into a narrative structure more readily. The youth were responsible for the visual language of the narrative, which was a key pedagogical goal of ours.

The resulting narrative arc told a story about the GW 'winning formula' and how it was benefiting the community through 'together livin' better'. The visual designs were stitched together by Emily Bhargava into a consistent mural design. Painted on the large metal fence behind one of the converted lots, roughly 80 feet long and 10 feet tall, it showcases the GW impact story at the site of one of the reclaimed urban farms; literally telling a story about the space, in the space itself.

At the unveiling, viewers and participants alike commented on the impacts. One attendee said, 'What strikes me is that you've managed to tell a story with an equation and very simple images'. Others commented on the visual encodings and symbolic language—'The bike sticks in my mind'. Validating our goal of increasing data literacy with the youth participants, one commented that 'I learned that by pictures you can also send out a message', and another said, 'I learned how to take data and make a story'.

This example highlights that people who don't 'speak' data or self-identify as 'geeks' or 'techies' can be effectively involved in data analysis and storytelling by focusing on an arts-based, socially-oriented invitation. Our goals centred on building the confidence of youth to engage with data and enhancing the built environment in an impactful way. The choice of a mural as the medium leveraged the long history of public murals being used to comment on and change the public discourse about a topic. The

Figure 13.2. The data mural. Photo by Rahul Bhargava. Printed with permission.

participatory, youth-driven process offers an example of how to engage people in a collaborative meaning-making process to amplify their understanding of how data can communicate. In this case, data analysis and visualization are methods for connecting more deeply to the community, not modelled as end points in themselves.

'Asking questions' with WTFcsv

For the second case of data visualization learning, we introduce an activity called 'Asking questions' from the DataBasic.io suite of tools and activities which we built. DataBasic.io consists of simple, web-based tools for beginners that introduce concepts of working with data ranging from quantitative text analysis to network analysis. For the purposes of this case, we focus on the tool WTFcsv and its accompanying learning activity 'Asking questions'.

WTFcsv helps learners analyse a comma-separated-values (CSV) file to look for potential data-driven stories to tell. The software analyses each column from a spreadsheet file uploaded by the learner and returns a data visualization that summarizes the patterns in each column (Figure 13.3).

Newcomers often approach data thinking of it as consisting only of numbers. Two of the primary learning goals for WTFcsv are that 1) learners understand that data have many types, including numbers, categories, text, and dates, and 2) that different types of visualizations are appropriate for summarizing different types of data. For example, temporal data are shown as a line-chart histogram on a time-series axis. Numeric data are shown as a histogram, with buckets created linearly. Text data are shown as a column chart if there are only a few types (categorical data), or a word cloud if there are many entries (i.e. open text).

'Asking questions' is the learning activity that accompanies the WTFcsv tool, based on Tactical Technology Collective's notion of 'asking your data some questions' (Tactical Tech, 2014). While newcomers to spreadsheet analysis often attribute some wizardry to the data analysis process, this activity tries to introduce them to a simple, inquiry-based process for getting acquainted with a dataset.

Figure 13.3. The WTFcsv results screen. Printed with permission.

Learners break into small groups, choose one of the sample datasets, examine WTFcsv's summary visualizations, and brainstorm questions that they want to ask the data.[1] Facilitators encourage learners to use the visualizations to generate many types of questions, including context questions ('What's the source of these data? Why did they collect it? Who uses it?'), ethical questions ('Is it OK to publish people's full names? How did sex end up as a binary variable?'), quality questions ('Are these data complete? How were they acquired?'), data formatting questions ('What does the "Parch" column mean?'), as well as data analysis questions ('Did women survive at a higher rate than men?'). After 10 minutes of brainstorming questions and sources for connected data, learners share their most interesting question back to the group. The debrief conversation focuses on how rich questions often involve multiple data sources, the variety of questions that can come from one dataset, and the critical process of recognizing any bias in identified questions.

We have run more than 30 workshops with WTFcsv. In an evaluation of the WTFcsv activity, learners responded well to these choices. One participant commented that the activity 'helps you from the beginning to understand the possibilities of your spreadsheet'. The fact that they framed 'possibilities' as plural is meaningful, in the sense that it is important for newcomers to understand the role of *exploratory data visualization*—the way in which visual aggregation can serve to provoke important questions and next steps towards the formulation of a knowledge claim. Another commented that the tool was 'different because usually there are just text and numbers, not lots of images and graphs and the ability to look at them all right away'. This fulfils our primary goal that learners understand that data visualization can play an important role in the exploration and meaning-making process.

'ConvinceMe' with the Data Culture Project

The third case study is an activity we developed to enable people to practice the skill of making arguments with data to convince people to take action, called 'ConvinceMe'. Many definitions of data literacy focus on the ability to read and write with data; fewer include the idea of arguing with data as a core skill. We believe that this is critical for putting data into action in the real world. Without this, data end up divorced from the fact that

1 Current options for English-language speakers include Titanic Passengers, UFO Sightings, and Dogs of NYC. When viewed in other languages, the tool offers different culturally and geographically appropriate examples.

they often represent real people, and are used to make decisions about them. This activity is the final one in our Data Culture Project (see http://dataculturproject.org), a lightweight self-service curriculum available for free to any organization that wishes to work on building an inclusive data culture.

The activity itself is simple to run, and relies on the creativity of the participants. It begins with a room of at least 10 people. Facilitators share a printed data visualization about some topic. We use one about water conservation, which argues that choosing to consume beef has an extraordinarily high 'water cost'. After talking through the graphic, we ask the group to identify 3-5 key stakeholder groups that can influence that topic. For example, in the case of water conservation, stakeholders might be farmers, policymakers, a shopping family member, or a restaurant owner. Volunteers are solicited to role-play those stakeholders, and invited to stand in the front of the group. The rest of the group is asked to make a short data-driven argument to specific stakeholders, asking them to change some behaviour. For example, this could consist of telling a 'shopping parent' stakeholder about the high water cost of meat, and invite them to try a vegetarian diet by joining the 'meatless Monday' movement.

The primary learning goal for ConvinceMe is for learners to practice making data communication decisions in a situated environment, with a specific audience in mind. For newcomers, data often appear neutral and abstract, but to practice making arguments with data you have to re-concretize them. The role-playing stakeholders embody those being asked to take action—they physically step forward if they are convinced by the arguments and step backwards if they are not. Another learning goal for ConvinceMe is to situate data visualizations and data-driven arguments as tools for advocacy and social transformation. The goal for the activity is not to create a picture, but rather to move a particular stakeholder towards a desired action. The act of inviting participants to make a persuasive argument with data breaks down the narrative of data as neutral.

A group of 25 non-profit organizations participated in the first cohort of the Data Culture Project in Fall 2017 and ran the ConvinceMe activity, with positive feedback. One noted that their arguments 'used a lot of shame and guilt', leading them to reconsider how they frame their calls to action. A mid-sized non-profit valued most the 'importance of identifying stakeholders and trying to understand their interests'. The act of practising the arguments with real people impacted their approaches to picking stories to tell, and how to tell them. Another group began rethinking their whole approach to data storytelling, and decided to do this activity 'before anyone even collects any data', so they could 'think up front about who should be convinced about what'.

Learning and feminist data visualization

While mainstream data visualization teaching often starts by instruction in chart types, or learning a software tool, or learning about human perception, we argue that we need to begin with a wider lens before zooming into technical specifics. In this section, we will consider the three cases of data visualization learning described above in relationship to three design principles of feminist data visualization outlined by D'Ignazio and Klein (2016)—'Consider context', 'Legitimize affect & embodiment', and 'Embrace pluralism'—in order to explore their implications for a feminist starting point for visual numeric literacy.

Consider context

One of the central tenets of feminist epistemology is that knowledge is 'situated' (Haraway, 1989, p. 581). What this means is that context matters—What are the social, cultural, historical, and material conditions in which knowledge is produced? What are the identities of the humans making the knowledge? A feminist perspective advocates for connecting datasets and data visualizations back to their context, to better understand their limitations and ethical obligations, and, ultimately, the ways in which power and privilege may obscure truths.

Situating datasets and data visualizations for learners is a particular challenge, since the conventions of both spreadsheets and precise graphics make them appear objective (Kennedy et al., 2016a), particularly for non-technical newcomers. The 'Asking questions' activity illustrates one way to situate data and visualizations. Instead of trying to 'find stories', the position of asking questions helps model for learners a process of inquiry and exploration where meaning is not something to be 'found' hidden in the dataset, but rather produced through an iterative process of investigation that involves many bits of information that are not included in the dataset itself. And encouraging many types of questions, including questions about trust in sources, missing data, and data formatting, helps learners start to connect the data back to the institutional and historical context where they were collected, emphasizing that those things also matter deeply to any meaning that comes from patterns observed in the data. The fact that learners use visualizations to ask those questions matters as well—this demonstrates the value of visualization not as the definitive, objective word on a subject, but rather as a helpful exploratory step in a process of building meaning.

ConvinceMe functions in a similar way to draw connections back to situated human experience, and grounds data visualization in acts of communication between different types of stakeholders. While data visualization research within technically oriented disciplines often focuses on time to task metrics, such as how well an individual can decode the meaning of a particular chart, there is very little research on how data visualizations or data-driven arguments help move groups from different positions and/or different cultures to action (or not). If you know that your data visualization needs to move farmers to use less water, then you will make different decisions about what data to highlight and what format to use than if you need to convince parents. ConvinceMe creates a lightweight, bounded playground in which people can begin to understand visualization as acts of situated communication.

Legitimize affect & embodiment

This principle of feminist data visualization derives from the argument by feminist thinkers that experiences that derive from sensation and emotion have been systematically devalued over quantitative methods of knowing. Patricia Hill Collins notes in her articulation of Black feminist epistemology that in an ideal knowledge situation 'neither emotion nor ethics is subordinated to reason' (2009, p. 266). There has been work on the rhetorical function of data visualization in narratives (Hullman & Diakopoulos, 2011) and persuasion (Pandey et al., 2014). But the role of emotion in data visualization has been understudied, with the exception of work by Kennedy, Hill, Aiello, and Allen (2016) and several chapters in this book (Gray, this volume; Hill, this volume; Simpson, this volume). In contrast to learning about data which involves abstraction and distance from the subject matter (or from the subjects themselves), how might we acknowledge embodied and affective experiences in the data visualization learning process? How might an intimate, emotional connection to data be considered an asset to the analysis, visualization, and learning process?

The GW data mural legitimates affect and embodied experience in several ways. First, the data that the youth analysed and used to tell a story are data about themselves and their organization. Data visualization techniques are often discussed as though the subject matter of the data is interchangeable and neutral (Kennedy, Hill, Allen & Kirk 2016). Many teaching examples use so-called 'classic' datasets like *mtcars* (Kosara, 2018), but, as D'Ignazio has stated in prior work, 'Cars. Who cares?' (2017, p. 8), people will learn better and more deeply from data that they have an experiential understanding of and an emotional connection to (Kennedy, Hill, Aiello & Allen, 2016). The GW project began with data that was about something intimate and emotionally connected to the youth and their community. Likewise, the project ended with a data visualization that was emotionally connected to them—the input and the output were deeply situated.

Both the GW data mural and the ConvinceMe activity also use embodied, arts-based ways of knowing for learner engagement and think beyond the screen in terms of data visualization. In the case of GW, the youth not only came up with the iconography but also painted it onto a giant mural in the garden. The act of assembling the data visualization was itself an embodied, social act, undertaken in community. And ConvinceMe uses performance and narrative to construct a humorous social situation where peers have to use a data graphic to convince each other to shift their behaviour. Data visualization is important to each of these cases, but a 2D screen-based graphic is not modelled as the end point. Significantly, ConvinceMe seeks to value personal testimony in addition to data-driven graphics. Indeed, the graphic is the jumping off point, but needs the embodied personal testimony (the speaker) and custom-tailoring to an audience (the speaking situation) in order to move them to action. Legitimizing affect and embodiment may mean seeking the appropriate form for the appropriate audience, as well as modelling in learning activities how visualizations fit in to an embedded advocacy or community-building process.

Embrace pluralism

The design principle 'Embrace pluralism' comes from feminist scholarship's long history of challenging claims of objectivity, neutrality, and universalism, emphasizing instead how knowledge is always constructed within the context of a specific subject position as well as within a community of knowers (Harding, 1991). Black feminist scholars like Patricia Hill Collins have demonstrated how discourses of objectivity systematically exclude the voices of women and people of colour, among others, with the burden of oppression most heavily borne by those who sit at the intersections of the 'matrix of domination' such as Black women (Collins, 2008). A key contribution of this line of feminist thinking has been to recognize how a multiplicity of voices, rather than one single loud, magical, or technical voice, often results in a more complete picture of the issue at hand.

Embracing pluralism in data visualization learning can counter the media narrative that sometimes constructs data scientists as 'unicorns' or 'wizards'—solo technical geniuses who can use advanced analytical or artistic techniques to tame large datasets into insights. Learning activities for newcomers often have to counteract these preconceptions by modelling an alternative meaning-making process that is social and dialogue-based rather than individual and technical. While skill development is necessary for data visualization, we would argue that starting with those things reinforces the naive notion that data are about solo technical mastery. We choose to model a process that communicates that answers are best found in dialogue with a community of knowers, who approach a topic area with many perspectives.

It is significant that all three of the cases discussed above model a process of valuing different voices and producing knowledge through dialogue in group. Rather than students being positioned as individual learners in front of computer screens, the learning situation is social. In the case of the data mural, youth worked in small groups to contextualize the GW numbers and in one large group to paint the mural. In the case of the WTFcsv activity, learners work in groups of three to brainstorm questions—people learn from their peers' questions. And ConvinceMe is an activity conducted in a larger group which intentionally centres the idea of multiple stakeholder voices as well as formulating appeals to those particular standpoints. You 'win' by making a data-driven appeal to one of those stakeholders, convincing them to move towards the speaker. While the first two activities model a social, pluralistic process for deriving meaning from data and their visualization, this last activity embraces pluralism on the reception side of data communication, helping learners understand that different audiences may be moved by different narrative and visual arguments.

Conclusion

As stated at the beginning, one of the reasons that a feminist approach to data is useful and necessary is because of the power differentials and inequalities in the data ecosystem. Resources to collect, store, and analyse data are not distributed equally, nor are the technical skills to work with data. The cases we have discussed, along with the feminist design principles that guide them, represent a starting point for visual-numeric literacy. This is an area for further research and evaluation: does a feminist-informed learning programme lead to increased self-efficacy around data and its visualization for more women, people of colour, and other minoritized groups?

Relatedly, while many of our learning activities *model a feminist process*, they do not explicitly tackle issues of power, structural inequality, and bias in the *content of what is taught*. What might learning activities for the same audience (adult, non-technical newcomers) look like that specifically address concerns of gender and racial bias, the political economy of data, and so on? In order to integrate these conversations into the learning situation, Philip, Olivares-Pasillas, and Rocha (2016, p. 365) argue that we need to consider cultivating racial literacy and gender literacy side-by-side with data literacy. They write, 'Spaces must be facilitated for students to engage with the structural and ideological contexts of data visualizations if these tools are to authentically engage them in democratic deliberations—spaces where they grapple with how groups influence, resist, and transform everyday and formal processes of power that impact their lives'.

References

Angwin, J., Larson, J., Mattu, S., Kirchner, L. (2016, May 23). Machine bias: There's software used across the country to predict future criminals. And it's biased against blacks. *ProPublica*. Retrieved from https://www.propublica.org/article/machine-bias-risk-assessments-in-criminal-sentencing

Bhargava, R., Kadouaki, R., Bhargava, E., Castro, G., & D'Ignazio, C. (2016). Data murals: Using the arts to build data literacy. *Journal of Community Informatics, 12*(3). Retrieved from http://ci-journal.org/index.php/ciej/article/view/1276

Buolamwini, J., & Gebru, T. (2018). Gender shades: Intersectional accuracy disparities in commercial gender classification. *Proceedings of Machine Learning Research: Proceedings of the Ist Conference on Fairness, Accountability, and Transparency, 81*, 77-91.

Collins, P. H. (2009). *Black feminist thought: Knowledge, consciousness, and the politics of empowerment* (2nd ed.). New York: Routledge.

Corbett, C., & Hill, C. (2015). *Solving the equation: The variables for women's success in engineering and computing.* American Association of University Women. Retrieved September 18, 2019 from https://www.aauw.org/research/solving-the-equation/

D'Ignazio, C. (2017). Creative data literacy: Bridging the gap between the data-haves and data-have nots. *Information Design Journal*, 23(1), 6-18. https://doi.org/10.1075/idj.23.1.03dig

D'Ignazio, C., & Bhargava, R. (2016). DataBasic: Design principles, tools and activities for data literacy learners. *The Journal of Community Informatics*, 12(3), 83-107. Retrieved from http://ci-journal.org/index.php/ciej/article/view/1294

D'Ignazio, C., & Klein, L. (2016, October). *Feminist Data Visualization*. Paper presented at and published in the workshop proceedings from the Workshop on Visualization for the Digital Humanities at IEEE VIS Conference, Baltimore, Maryland.

Davis, K. (2008). Intersectionality as buzzword: A sociology of science perspective on what makes a feminist theory successful. *Feminist Theory, 9*(1), 67-85. https://doi.org/10.1177/1464700108086364

Eubanks, V. (2018). *Automating inequality: How high-tech tools profile, police, and punish the poor.* New York: St. Martin's Press.

Freire, P. (2015). *Pedagogy of the oppressed* (50th Anniv.). New York: Bloomsbury Academic.

Haraway, D. J. (1988). Situated knowledges: The science question in feminism and the privilege of partial perspective. *Feminist Studies, 14*(3), 575-599.

Harding, S. G. (1991). *Whose science? Whose knowledge? Thinking from women's lives.* Ithaca: Cornell University Press.

Hooks, B. (1994). *Teaching to transgress: Education as the practice of freedom.* New York & London: Routledge.

Hullman, J., & Diakopoulos, N. (2011). Visualization rhetoric: Framing effects in narrative visualization. *IEEE Transactions on Visualization and Computer Graphics, 17*(12), 2231-2240.

Kennedy, H., Hill, R. L., Aiello, G., & Allen, W. (2016). The work that visualisation conventions do. *Information, Communication and Society, 19*(6), 715-735. https://doi.org/10.1080/1369118X.2016.1153126

Kennedy, H., Hill, R. L., Allen, W., & Kirk, A. (2016). Engaging with (big) data visualizations: Factors that affect engagement and resulting new definitions of effectiveness. *First Monday, 21*(11). https://doi.org/10.5210/fm.v21i11.6389

Kvasny, L. (2006). Social reproduction and its applicability for community informatics. *The Journal of Community Informatics, 2*(2). Retrieved from http://ci-journal.org/index.php/ciej/article/view/342

Kosara, R. (2018, March 21). How to get excited about standard datasets [Blog post]. Retrieved from *eagereyes.org*.

Letouzé, E., Bhargava, R., Deahl, E., Noonan, A., Sangokoya, D., & Shoup, N. (2015). *Beyond data literacy: Reinventing community engagement and empowerment in the age of data.* Retrieved from http://datapopalliance.org/item/beyond-data-literacy-reinventing-community-engagement-and-empowerment-in-the-age-of-data/

Neuhauser, A. (2015, June 29). 2015 STEM Index Shows Gender, Racial Gaps Widen. *US News & World Report.* Retrieved from https://www.usnews.com/news/stem-index/articles/2015/06/29/gender-racial-gaps-widen-in-stem-fields

Noble, S. U. (2018). *Algorithms of oppression: How search engines reinforce racism.* New York: New York University Press.

O'Neil, C. (2016). *Weapons of math destruction: How big data increases inequality and threatens democracy.* New York: Crown.

Pandey, A. V., Manivannan, A., Nov, O., Satterthwaite, M., Bertini, E. (2014). The persuasive power of data visualization. *IEEE Transactions on Visualization and Computer Graphics, 20*(12), pp. 2211-2220.

Philip, T. M., Olivares-Pasillas, M. C., & Rocha, J. (2016). Becoming racially literate about data and data-literate about race: Data visualizations in the classroom as a site of racial-ideological micro-contestations. *Cognition and Instruction, 34*(4), 361-388. https://doi.org/10.1080/07370008.2016.1210418

Street, B. (2003). What's 'new' in New Literacy Studies?: Critical approaches to literacy in theory and practice. *Current Issues in Comparative Education, 5*(2), 77-91.

Tygel, A. F., & Kirsch, R. (2016). Contributions of Paulo Freire for a critical data literacy: A popular education approach. *The Journal of Community Informatics, 12*(3), 108-121.

Walter, M., Andersen, C. (2013) *Indigenous statistics: A quantitative research methodology.* Walnut Creek, CA: Left Coast Press.

West, S. M., Whittaker, M., & Crawford, K. (2019). *Discriminating systems: Gender, race and power in AI.* AI Now Institute. Retrieved from https://ainowinstitute.org/discriminatingsystems.html.

Wolff, A., Gooch, D., Cavero Montaner, J., Rashid, U., & Kortuem, G. (2016). Creating an understanding of data literacy for a data-driven society. *The Journal of Community Informatics, 12*(3), 9-26. Retrieved from http://ci-journal.net/index.php/ciej/article/view/1286

About the authors

Catherine D'Ignazio is Assistant Professor of Civic Media and Data Visualization at Emerson College, a senior fellow at the Emerson Engagement Lab and a research affiliate at the MIT Media Lab. catherine_dignazio@emerson.edu

Rahul Bhargava is a Research Scientist at the MIT Center for Civic Media and MIT Media Lab. rahulb@media.mit.edu

14. Is literacy what we need in an unequal data society?

Lulu Pinney

Abstract

Having the skills and awareness to make sense of data visualizations has become a contributing factor in determining who gets to participate in our data-driven society. Initiatives that seek to enable people to make sense of some aspect of our digital, datafied worlds are often described in terms of literacy. However, taking a closer look at different usages of literacy across academia, policy, and practice reveals different notions of power embedded in different populations' implicit understanding of the term. Situated in the emerging field of critical data studies, the field that is concerned with understanding data's role in reproducing and creating social inequalities, this is a conceptual chapter that asks how useful literacy is in this context.

Keywords: Know-how; Expertise; Everyday; Data justice; Datafication; Participation

Introduction

In this digital age, information and data are presented to us more and more often, on a range of subject matter, from many sources, across a variety of different channels, in different formats, relating to most aspects of our lives. This presents us with many things of which we need to make sense. Correspondingly, you do not have to look far to find a project or initiative offering to teach us how to make sense of some aspect of our digital, datafied worlds. Often these projects' descriptions include the term literacy, and examples can be found across academia, practice, and policy. One definition of literacy, from the United Nations Educational, Scientific and Cultural Organization (UNESCO), is as follows:

Engebretsen, M. and H. Kennedy (eds.), *Data Visualization in Society.*
Taylor & Francis Group, 2020
DOI 10.5117/9789463722902_CH14

> Literacy is defined as the ability to identify, understand, interpret, create, communicate and compute using printed and written materials associated with diverse contexts. Literacy involves a continuum of learning in enabling individuals to achieve their goals, develop their knowledge and potential and participate fully in community and society. (UNESCO, 2005)

However, as a concept with a long history, literacy has come to mean different things to different people in different contexts.

The emerging academic field of critical data studies is concerned with social inequalities that are created and reproduced as a consequence of the widespread production, circulation, and uses of data. With data increasingly being 'mobilized graphically' (Gitelman & Jackson, 2013, p. 12), the relationship between power and data visualizations in society is also of critical interest, including the inequality that results from not being able to make sense of data visualizations.

This chapter is a conceptual one. It explores different notions of power embedded in implicit understandings of terminology used in projects that seek to enable people to make sense of the data society they live in, in the context of the inequalities that result from that same data society. In doing this it has been helpful to distinguish between literacy as a concept and literacy as a term. A dictionary definition (OED Online, 2018) of both words is provided here:

> Concept: a general idea or notion, a universal; a mental representation of the essential or typical properties of something, considered without regard to the peculiar properties of any specific instance or example.
> Term: A word or phrase used in a precise sense in a particular subject or field, or by a particular group of people; a technical expression; a piece of jargon.

I argue that literacy is useful as a concept because it enables those affected by inequality to ask critical questions. However, as a term, I find its use for engaging populations is limited.

The concept of literacy is useful for researching an unequal data society

Referring to the world we live in as a 'data society' is to acknowledge not only the ubiquitous presence of data in society but also that these data have an impact on our worlds and our experiences of living in them. The widespread use of digital technology means we are creating data much of the time, with examples including what we discuss on social media and with whom, how many kilometres we run and where, or where we use our credit cards and what we buy. The impacts of data like these on our lives depend on the assumptions, biases, methods, and motivations of the organizations and individuals who collect and use our data (boyd & Crawford, 2012). Whoever collects the data and decides how they are going to be used is in a position of power, whether or not they realize it, relative to the people whose data are collected. This has led scholars to ask questions about the relationship of digital data to issues such as surveillance, privacy, exploitation, discrimination, and exclusion that can result from such a data society (e.g. boyd & Crawford, 2012; Eubanks, 2018; Noble, 2018). These issues are the focus of the field of critical data studies, which interrogates data's contribution to social inequality either through reproducing existing inequalities or creating new ones (Kennedy, 2018).

Data visualizations also contribute to inequality in our data society. Boyd and Crawford (2012) argue that, alongside having access to data, having the skills to work with data is also a factor in determining the societal divides that emerge. Gitelman and Jackson (2013) observe that 'data are mobilized graphically' (p. 12), that is to say that to be useful to us, data are usually represented visually. Therefore, whether or not individuals and institutions have the skills to work with data *visualizations* also impacts who gets to participate in a data society (Kennedy & Hill, 2017).

What is known about skills for working with data visualizations is that they are diverse and, in the context of social inequality, must include critical awareness as well as practical dimensions. Though there is no definitive list of skills for working with data visualizations, doing so involves the following steps, each one contributing to the end-to-end production of any data visualization: data creation, processing, and distribution; visual representation and design of data; and finally, data visualization distribution and then viewing. In addition to the practical skills needed to perform each of these steps, the importance of critical skills is highlighted by Gray et al. (2016), who illustrate the social and cultural factors that lead to mediation in every step in the production of a data visualization. These factors include the people, institutions, infrastructure, tools, methods, usage, aesthetics, and contexts involved, all of which are shaped by human decision-making, bias, and assumptions. A further factor to consider is that this mediation is obscured by both the seemingly simple outward appearance of data visualizations and the popular belief that data visualizations, like the data on which they are based, are objective (Kennedy & Hill, 2016). It can be seen that on top of diverse practical skills, the social and cultural provenance of data visualizations demands a critical awareness of data visualization as a practice (Gray et al., 2016; Kennedy & Hill, 2016) without which Kennedy and Hill (2016) argue, in summary, data visualizations will continue to 'privilege certain viewpoints, perpetuate existing power relations and create new ones' (p. 5). It is also for this reason that researchers interested in the role of data visualizations in society often talk about the skills and awareness to *make sense of* data visualizations, inviting consideration of both the practical and critical dimensions, rather than *working with* data visualizations, a notion which is more commonly associated with practical, operational, or technical skills.

There is a significant body of work around how to make sense of data visualizations cognitively and perceptually (Card, Mackinlay, & Shneiderman, 1999; Ware, 2012) and this knowledge underpins much popular, practical guidance for working with data visualizations (Cairo, 2013; Few, 2013; Ware, 2012). However, this work often treats data visualizations as isolated texts that are independent of their provenance, the person who is looking at them, and the context in which this happens. Overlooking these sociocultural factors limits the possibility of finding out how the skills and awareness needed for making sense of data visualizations impact power relations and participation in a data society. This is where the concept of literacy can provide a useful framework, and there are two key features that make it so, which I discuss next: firstly, literacy as a social practice; and, secondly, literacy as an enabler of social change.

Thinking of literacy as a social practice is to understand that literacy is relevant to everyone, practised in different aspects of our everyday lives, and dependent on both context and individual (Barton & Hamilton, 1998; Street, 1984). This concept, developed by New Literacy Studies scholars over the last 40 years, has displaced the 'autonomous model' (Street, 1984), a traditional approach to literacy that Street criticized for characterizing literacy as a set of neutral, technical, decontextualized skills that, if an individual was in possession of them—or literate—could be deployed regardless of time, place, or purpose. The usefulness of thinking of literacy as a social practice for research into the skills and awareness needed for making sense of some aspect of modern society against a backdrop of inequality can be illustrated from literatures on visual literacy, media literacy, information literacy, data literacy, and digital literacy, five literacy fields that relate to data visualization literacy. The common goal of literacy initiatives across these fields is enabling people to be both active users and producers of visuals, media, information, data, or digital media respectively. However, they all also emphasize the importance of doing these things critically, by taking into account the broader contexts in which users and producers are operating. This results in a recognition of the need for a wide range of multidimensional skills, critical as well as practical. The extent of any individual's performance of such skills can be modelled on a continuum and will fluctuate depending on the context in which the skills are being drawn (Avgerinou & Pettersson, 2011; Bassett, Fotopoulou, & Howland, 2013; Letouzé et al., 2015; Potter, 2005; SCONUL, 2011). Understanding literacy as a social practice in this way allows researchers to account for the influence of sociocultural factors on the skills and awareness needed for making sense of data visualizations in an unequal data society, that is to say to include the critical dimension necessary for raising questions around power in the context of inequality. In this way the concept of literacy is useful to researchers.

The second useful feature of the concept of literacy is as an enabler of social change. This emancipatory dimension of the concept of literacy is exemplified by the work of Paolo Freire, who understood literacy as the ability to make sense of the world in which we live (Freire, 1996). He believed and practised the idea that it is only by enabling people to identify the power structures regulating their lives that they can challenge them. This is an approach that values and builds on the knowledge and experience of those adversely affected by inequality. This understanding of literacy builds on the first by also raising questions around power, but then goes further by also understanding it as enabling those affected by power imbalances to ask critical questions. In this way the concept of literacy has the potential to also be useful to those who experience inequality in a data society.

Many scholars share this emancipatory understanding of the concept of literacy because it is useful for researching how sense is made of some aspect of society in the context of social inequality. Examples relevant to the field of critical data studies can be illustrated through the work of several authors of chapters in this book: D'Ignazio and Bhargava do ongoing *data literacy* work with communities, including their introductory web tool DataBasic (D'Ignazio & Bhargava, 2016) and the Data Culture Project (Bhargava, 2018); Tønnessen is researching *visual-numeric literacy* in secondary schools (this volume) through the Innovative Data Visualization and Visual-Numeric Literacy (INDVIL) project of which she and this book are a part; Gray et al. (2018) recently called for *data infrastructure literacy* as 'the ability to account for, intervene around and participate in the wider socio-technical infrastructures through which data are created, stored and analysed' (p. 1); Archer and Noakes are researching data visualization's role in *academic literacies* in higher education (this volume); Kennedy, Hill, and colleagues' Seeing Data project (Kennedy et al., n.d.) was designed around developing the general public's *visualization literacy*; and Feigenbaum and colleagues' (2016) Datalabs project sought to 'establish a sustainable training model for *data literacy*, data-driven research and data storytelling' in journalism education (p. 62). All of these scholars, to a greater or lesser degree, have understood the concept of literacy as both a social practice and an enabler of social change. That is to say, this is the understanding they implicitly associate with literacy, and is why literacy is useful in their work. What none of them do, however, is consider that their implicit understanding of literacy is not necessarily the same as everyone else's.

Literacy does not mean the same thing to everyone

Literacy, both as a term and as a concept, is widely used beyond the examples just given, in academia as well as in practice, policy, and everyday life. This includes usage as a term in its own right, *literacy*, as well as part of compound terms, for example *digital literacy* or *visualization literacy*. Some scholars from the academic disciplines of information, computer, and cognitive science research *visualization literacy* (Boy, Rensink, Bertini, & Fekete, 2015; Lee et al., 2016). In practice, *data literacy* initiatives are emerging all the time, with online examples including datatothepeople.org and dataliteracy.fit. There are policies for developing *media, information,* and *digital literacies* at UK national and European scales (Department for Digital, Culture, Media, & Sport, 2017; Vuorikari, Punie, Carretero, & Van Den Brande, 2016), with *data literacy* beginning to be talked about in the context of the UK government's own use of data (Duhaney, 2018; Knight, 2018). In the news, Kate Winslet is warning of the 'shame' of *illiteracy*' for young women who cannot read or write (Coughlan, 2018). In my inbox, a former client recently asked if I can recommend any *data literate* graphic designers.

While the term literacy is widely used, the implicit understandings that different people associate with it vary. To consider its different meanings it is helpful to refer back to the UNESCO definition quoted at the start of this chapter. All of the elements in this definition align with the emancipatory understanding of literacy as a concept, as already presented. However, as a concept with a long history and a term with wide usage, its meanings when used in the other examples given are narrower than the UNESCO definition. In everyday usage, literacy is often taken to mean simply the ability to read and write. For many people it is also associated with school. In its usage as a compound term, literacy often acts as a metaphor for technical or operational skills, with the widespread emergence of 'digital literacy' policy initiatives as an example (Bassett et al., 2013; Knobel & Lankshear, 2007). While these usages reflect some elements of the UNESCO definition, none of them encompass an understanding of literacy's potential to enable social change and, in this way, they indicate a traditional implicit understanding of literacy, one that derives from the 'autonomous model' (Street, 1984). This sits in direct opposition to the emancipatory understanding of literacy, a contradiction that has fuelled much academic debate (Cook Gumperz, 2006; Gee, 2015). Where the application of the emancipatory understanding of literacy has the potential to empower those affected by social inequality by positioning them and their knowledge at the centre of a process of learning and change, embedded in the traditional understanding of literacy is the notion that power lies, and remains, with those who already have it. This is a consequence of literacy's primary usage in the context of schooling, where what is taught is defined, tested, and standardized by those in positions of power (Cook Gumperz, 2006; Gee, 2015). As such, applications of this understanding of literacy are not concerned with addressing inequality in society in the emancipatory sense, even though acquiring literacy, understood in a traditional way, can indeed be empowering. Lastly, there can also be negative connotations implicit in the term literacy in its everyday usage. These have their origin in the historical reification of literacy which equated it with the well-being of society. This led to the popular belief that 'literate people are [...] more intelligent, more modern, more moral' (Gee, 2015, p. 67). The continued currency of this belief today is evident in the pejorative inference that any use of the term illiterate carries with it. Thus, having considered a range of instances where people use literacy as a term or as a concept, it can be seen that there are different notions of power embedded in different population's implicit understandings of it.

When terms do not mean the same thing to everyone, there is an impact in everyday life. This is something Bassett et al. (2013) researched empirically in the context of computer use in community organizations. The researchers were interested in, amongst other things, what uses of digital technology, and expectations of use, result from the two terms literacy and expertise. They did this through interviewing and observing both professional and new users of digital technologies within community projects that either focused on enabling marginalized groups to access digital technologies or used digital technologies to undertake cultural activities. The researchers also reviewed literacy discourses in policy documents. They found that the widely used term 'digital literacy' was not ambitious enough for under-served populations: the term was understood reductively as a set of actions undertaken to avoid risk and harm; it encouraged passive, not active, participation; and it did not foster any ambition in the creative use of digital technologies, focusing just on access instead. Conversely, they found that thinking with the term expertise meant that participants expected more in terms of their own digital media skills. This shows that the implicit understandings of terminology have an impact in everyday life, in this instance influencing the kinds of use, users, skills, and expectations that result.

Literacy can be understood by different populations in multiple and divergent ways, with different notions of power embedded within different understandings, and this has implications when working in the context of inequality. When associated with an emancipatory understanding, literacy is a useful concept for framing initiatives that seek to address inequality in marginalized communities. However, while widespread in certain academic fields, this understanding is not popularly shared. Instead, with a variety of other understandings of the term more common in everyday usage, those same communities might be confused, insulted, or just think that a literacy initiative is not relevant to them. At worst, literacy is a term that carries implications of the social domination that emancipatory literacy initiatives seek to counter. This is why it is important to consider alternative terms.

How useful are the alternatives to literacy?

Other academic concepts that are used for thinking about the skills and awareness needed to make sense of aspects of society include competence, skill, know-how, and expertise, so it is these that I have considered as alternatives to literacy. Like literacy, as well as being concepts, these are all also terms that are used in the everyday. Having explored why the emancipatory understanding of literacy is useful for researching social inequality, as well as the reasons that its multiple and divergent implicit meanings are problematic, two criteria emerge for assessing alternatives. Firstly, the emancipatory understanding of literacy is useful because, as a concept, it enables both researchers and populations affected by inequality to ask critical questions around power. This provides one criterion that any alternative to literacy, as a concept, should also meet. However, one of the key features of emancipatory literacy research is that it is informed by those who might be affected by the issue being researched, that is to say the research is situated in their everyday lives. This is where the term literacy, with different notions of power embedded within different understandings of it, has the potential to cause problems when used in public initiatives that seek to address social inequality. Therefore, the second criterion is that, as a term in everyday usage, any alternative to literacy should not cause problems as a result of differential understandings of the relationship between the term, its meanings, and the power relationships in which it is embedded and which its use seeks to address.

Competence is a concept researched primarily in educational psychology and management studies. While there is no simple definition, it can be usefully thought of as the knowledge, skills, and attitudes—or cognitive competences, functional competences, and behavioural competencies (intentional change of spelling in that last instance) respectively—that are learned at work, post-education, to meet the demands of a particular occupation (Le Deist & Winterton, 2005). Le Deist and Winterton (2005) make the case for developing a common typology of competence across vocational education and work-based learning, as well as across occupations and locations, ultimately to promote greater transparency and mobility. However, they also note that interrogation from a sociocultural perspective of existing efforts to standardize competence, for example with certificates or assessments, or of the influence of context and culture on understandings of competence, has been neglected. From this point of view the concept of competence does not provide a useful lens for thinking about power. There is no evidence to report on the everyday usage of the term, although its antonym, incompetent, is also popularly used. It is not hard to imagine that, like the term illiterate, the inference of deficiency associated with such a term would not be welcomed by anyone at whom it was directed.

Skill is a word that, in its everyday usage, can be found describing factors that contribute to all of the other terms being considered here, and vice versa. However, it is also a concept in its own right. It has no simple definition, but it is understood as an ability, with both mental and physical dimensions, that can be applied at different levels ranging from competent at one extreme to expert at the other (Attewell, 1990). Academic interest in skill derives from thinking about where skills are situated and how they are described, acquired, transferred, and measured in relation to their value in the labour market, particularly since the advent of technology. The concept of skill is therefore used in a range of fields including economics, psychology, and sociology (Attewell, 1990; Green, 2011). It is the latter that is of interest here, as sociological research has investigated the notion of skill as a social construction highlighting gender and class inequalities in particular (Green, 2011). To my knowledge, skill has not been researched as a term in everyday usage. However, it is noticeable that literature discussing skill-related issues, in any field, uses the term low-skilled rather than unskilled. Like illiterate and incompetent, unskilled is a term in everyday usage that has the potential to infer a deficiency.

Know-how is a further alternative to literacy, a concept often used in management and organizational studies concerning our knowledge about using technology. One definition is 'our ability to perform skills without being able to articulate how we do them' (Collins, 1992, p. 56). It is also known as tacit knowledge and is based on a set of social skills, sitting in contrast to knowledge that can be modelled (Collins, 1992). However, while it is helpful to consider that there are different types of knowledge that go into making sense of situations, and there is acknowledgement that tacit knowledge is dependent on social context, only considering one type of knowledge will not provide insight into the full range of skills and awareness needed for making sense of data visualizations. Know-how is a concept that Pols (2014) has used to privilege patients' knowledge in the field of medical research, where traditionally it is lay people who support, rather than inform, medical knowledge. In her case study of people with severe lung disease, Pols developed the concept of 'know-Now' (2014, p. 88)—a variation on 'know-How' specifically for interpreting new situations—to explain how patients articulate the knowledge that they develop and use in their daily lives and make it transferable and useful to others. This adaptation of the concept of know-how does enable voices to be heard that usually are not. As a term, to the best of my knowledge, know-how does not have troubling notions of power embedded within it.

Expertise is the last alternative to literacy being considered here, a concept that is discussed in science and technology studies (STS) literature, in the context of public understanding of science, where the value of lay versus expert knowledge is debated. The difficulty of trying to define expertise relates to identifying and describing where the boundary lies between expert and lay knowledge (Collins & Evans, 2002). In his case study about the interactions between scientific experts and the sheep farmers whose livelihoods were negatively impacted by the radioactive fall-out from the Chernobyl nuclear accident, Wynne (1996) found that the perspective of the scientists 'was just as socially grounded, conditional and value-laden' (p. 38) as that of the farmers. It is through the recognition that expertise is socially situated that the concept invites questions to be raised around power. Elsewhere, feminist STS scholars in particular have challenged male dominance over what counts as technical knowledge and expertise (Ford & Wajcman, 2017). Research has highlighted opposing effects of implicit understandings of the term expertise when used in community projects. Bassett et al. (2013) found that the term expertise, in contrast to literacy, meant 'to raise the bar' (p. 28) in relation to people's expectations about their own computer skills, and in what they can produce. However, Rey-Mazón et al. (2018), reporting on three distinct community projects that had all used open-source technologies for collecting data, found that when certain people or institutions are recognized to have expertise, the value of other people's contributions to collective inquiry and knowledge is diminished, and as such the term expert was seen to 'bolster imbalances in power' (n.p.).

In summary, four concepts—competence, skill, know-how, and expertise—are alternatives to literacy which may provide a lens to think about how inequality intersects with people's abilities for making sense of their worlds. Measured against the first criterion that, as a concept, any alternative should enable those affected by inequality to ask critical questions around power, with the exception of competence, all concepts enable this. Like literacy, skill, know-how, and expertise are all acknowledged to be socially and culturally situated, a perspective which invites critical questions. However, all four alternatives can be found in most accounts of literacy. While the alternatives all relate to one or more aspects of literacy, individually they are each smaller in scope than literacy. As such, these alternative concepts may not be as useful as literacy for researching the skills and awareness needed for making sense of data visualizations in an unequal data society. Against the second criterion that, as a term, any alternative to literacy should not cause problems as a result of differential understandings of the relationship between the term, its meanings, and the power relationships in which it is embedded and which its use seeks to address, know-how seems to be the least problematic term. It does not have multiple meanings, nor does it have an antonym that infers deficiency.

Conclusion

In asking whether literacy is what we need in an unequal data society, is has been useful to consider it both as a concept and as a term, as well as thinking about four alternatives. I conclude that, *as a concept*, literacy is the most useful. Not only does it enable both researchers and populations affected by inequality to ask critical questions around power, it also offers the broadest scope for researching the skills and awareness needed for making sense of data visualizations in an unequal data society. However, *as a term* for engaging with populations, know-how provides the best alternative to literacy, not having notions of power embedded in any implicit understandings associated with it.

References

Attewell, P. (1990). What is skill? *Work and Occupations, 17*(4), 422-448. https://doi.org/10.1177/0730888490017004003

Avgerinou, M. D., & Pettersson, R. (2011). Toward a cohesive theory of visual literacy. *Journal of Visual Literacy, 30*(2), 1-19. https://doi.org/10.1080/23796529.2011.11674687

Barton, D., & Hamilton, M. (1998). *Local literacies: reading and writing in one community.* London: Routledge.

Bassett, C., Fotopoulou, A., & Howland, K. (2013). Expertise: A scoping study. *Working Papers of the Communities & Culture Network+.*

Bhargava, R. (2018). Launching the Data Culture Project. Retrieved September 24, 2018 from https://medium.com/mit-media-lab/launching-the-data-culture-project-58fc2dbb17ed

Boy, J., Rensink, R. A., Bertini, E., & Fekete, J. (2015). A principled way of assessing visualization literacy. *IEEE Transactions on Visualization and Computer Graphics, 20*(12).

boyd, d., & Crawford, K. (2012). Critical questions for big data. *Information, Communication and Society, 15*(5), 662-679. https://doi.org/10.1080/1369118X.2012.678878

Cairo, A. (2013). *The functional art: An introduction to information graphics and visualization.* Berkeley, CA: New Riders.

Card, S. K., Mackinlay, J. D., & Shneiderman, B. (1999). *Readings in information visualization: Using vision to think.* San Francisco, CA: Morgan Kaufmann Publishers.

Collins, H. M. (1992). Replicating the TEA-Laser: Maintaining scientific knowledge. In: H.M. Collins (Ed.), *Changing order: Replication and induction in scientific practice* (2nd ed.). (pp. 51-78). Chicago and London: The University of Chicago Press.

Collins, H. M., & Evans, R. (2002). The third wave of science studies: Studies of expertise and experience. *Social Studies of Science, 32*(2), 235-296. https://doi.org/10.1177/0306312702032002003

Cook Gumperz, J. (Ed.). (2006). *The social construction of literacy* (2nd ed.). Cambridge: Cambridge University Press.

Coughlan, S. (2018, September 7). Kate Winslet warns of 'shame' of illiteracy. *BBC News.* Retrieved from https://www.bbc.co.uk/news/education-45435973

D'Ignazio, C., & Bhargava, R. (2016). DataBasic: Design principles, tools and activities for data literacy learners. *The Journal of Community Informatics, 12*(3), 83-107.

Department for Digital, Culture, Media & Sport. (2017). 2. Digital skills and inclusion—giving everyone access to the digital skills they need. Retrieved August 16, 2018 from https://www.gov.uk/government/publications/uk-digital-strategy/2-digital-skills-and-inclusion-giving-everyone-access-to-the-digital-skills-they-need

Department for Education. (2014). National curriculum in England: Framework for key stages 1 to 4. Retrieved September 8, 2018 from https://www.gov.uk/government/publications/national-curriculum-in-england-framework-for-key-stages-1-to-4/the-national-curriculum-in-england-framework-for-key-stages-1-to-4

Duhaney, D. (2018). Data literacy—improving conversations about data. Retrieved August 16, 2018 from https://dataingovernment.blog.gov.uk/2018/02/21/data-literacy-improving-conversations-about-data/

Eubanks, V. (2018). *Automating inequality: How high-tech tools profile, police, and punish the poor.* New York: St. Martin's Press.

Feigenbaum, A., Thorsen, E., Weissmann, D., & Demirkol, O. (2016). Visualising data stories together: Reflections on data journalism education from the Bournemouth University Datalabs Project. *Journalism Education, 5*(2), 59-74.

Few, S. (2013). Data visualization for human perception. In: Interaction Design Foundation (Ed.), *The encyclopedia of human-computer interaction* (2nd ed.). Retrieved July 13, 2017 from https://www.interaction-design.org/literature/book/the-encyclopedia-of-human-computer-interaction-2nd-ed/data-visualization-for-human-perception

Ford, H., & Wajcman, J. (2017). 'Anyone can edit', not everyone does: Wikipedia's infrastructure and the gender gap. *Social Studies of Science, 47*(4), 511-527. https://doi.org/10.1177/0306312717692172

Freire, P. (1996). *Pedagogy of the oppressed.* London: Penguin Group.

Gee, J. P. (2015). *Social linguistics and literacies: Ideology in discourses* (5th ed.). New York: Routledge.

Gitelman, L., & Jackson, V. (2013). Introduction. In: L. Gitelman (Ed.), *Raw data is an Oxymoron.* (pp. 1-14). Cambridge, MA: MIT Press.

Gray, J., Bounegru, L., Milan, S., & Ciuccarelli, P. (2016). Ways of seeing data: Toward a critical literacy for data visualizations as research objects and research devices. In: S. Kubitschko & A. Kaun (Eds.), *Innovative methods in media and communication research.* (pp. 227-251). Cham: Palgrave Macmillan.

Gray, J., Gerlitz, C., & Bounegru, L. (2018). Data infrastructure literacy. *Big Data & Society, 5*(2), 1-13. https://doi.org/10.1177/2053951718786316

Green, F. (2011). *What is skill? An inter-disciplinary synthesis.* Centre for Learning and Life Chances in Knowledge Economies and Societies. Retrieved September 10, 2018 from https://www.llakes.ac.uk/sites/default/files/Green%20-%20What%20is%20Skill%20-%20final.pdf

Kennedy, H. (2018). Living with data: Aligning data studies and data activism through a focus on everyday experiences of datafication. *Krisis, 1*, 18-30.

Kennedy, H., & Hill, R. L. (2016). The pleasure and pain of visualizing data in times of data power. *Television & New Media, 18*(8), 769-782. https://doi.org/10.1177/1527476416667823

Kennedy, H., & Hill, R. L. (2017). The feeling of numbers: Emotions in everyday engagements with data and their visualisation. *Sociology, 52*(4), 830-848. https://doi.org/10.1177/0038038516674675

Kennedy, H., Hill, R. L., Allen, W., & Kirk, A. (n.d.). Original seeing data research—Seeing data. Retrieved September 24, 2018 from http://seeingdata.org/original-seeing-data-research/

Knight, M. (2018). A data literacy scale? Retrieved August 16, 2018 from https://digitalblog.ons.gov.uk/2018/04/05/a-data-literacy-scale/

Knobel, M., & Lankshear, C. (Eds.). (2007). *A new literacies sampler.* New York: Peter Lang.

Le Deist, F. D., & Winterton, J. (2005). What is competence? *Human Resource Development International, 8*(1), 27-46. https://doi.org/10.1080/1367886042000338227

Lee, S., Kim, S. H., Hung, Y. H., Lam, H., Kang, Y. A., & Yi, J. S. (2016). How do people make sense of unfamiliar visualizations?: A grounded model of novice's information visualization sensemaking. *IEEE Transactions on Visualization and Computer Graphics, 22*(1), 499-508.

Letouzé, E., Bhargava, R., Deahl, E., Noonan, A., Sangokoya, D., & Shoup, N. (2015). Beyond data literacy: Reinventing community engagement and empowerment in the age of data. *Data-Pop Alliance White Paper Series*. Data-Pop Alliance (Harvard Humanitarian Initiative, MIT Media Lab, & Overseas Development Institute) & Internews.

Noble, S. U. (2018). *Algorithms of oppression: How search engines reinforce racism.* New York: New York University Press.

Concept [Def. 2]. (2018). In: *OED Online.* Oxford University Press. Retrieved October 10, 2018 from http://www.oed.com/view/Entry/38130?rskey=t1pjzZ&result=1

Term [Def. 11a]. (2018). In: *OED Online.* Oxford University Press. Retrieved October 10, 2018 from http://www.oed.com/view/Entry/199409?rskey=eGVZob&result=1

Pols, J. (2014). Knowing patients: Turning patient knowledge into science. *Science, Technology, & Human Values, 39*(1), 73-97. https://doi.org/10.1177/0162243913504306

Potter, W. J. (2005). *Media literacy* (3rd ed.). Thousand Oaks, CA & London: Sage.

Rey-Mazón, P., Keysar, H., Dosemagen, S., D'Ignazio, C., & Blair, D. (2018). Public lab: Community-based approaches to urban and environmental health and justice. *Science and Engineering Ethics, 24*(3), 971-997.

SCONUL. (2011). *The SCONUL seven pillars of information literacy.* SCONUL Working Group on Information Literacy. Retrieved April 15, 2017 from https://www.sconul.ac.uk/sites/default/files/documents/coremodel.pdf

Street, B.V. (1984). *Literacy in theory and practice.* Cambridge: Cambridge University Press.

UNESCO. (2005). *Aspects of literacy assessment: topics and issues from the UNESCO Expert Meeting.* Retrieved from http://unesdoc.unesco.org/images/0014/001401/140125eo.pdf

Vuorikari, R., Punie, Y., Carretero, S., & Van Den Brande, L. (2016). DigComp 2.0: The digital competence framework for citizens. Update Phase 1: The Conceptual Reference Model. *JRC Science for Policy Report*. Luxembourg Publication Office of the European Union.

Ware, C. (2012). *Information visualization: Perception for design* (3rd ed.). Waltham, MA: Morgan Kaufmann.

Wynne, B. E. (1996). Misunderstood misunderstandings: Social identities and public uptake of science. In: A. Irwin & B. E. Wynne (Eds.), *Misunderstanding science?: The public reconstruction of science and technology*. (pp. 19-46). Cambridge: Cambridge University Press.

About the author

Lulu Pinney is a PhD student in the Department of Sociological Studies at the University of Sheffield. She is researching non-experts' experiences of working with data and its visualization in community organizations. She comes to this research from professional practice where she designed data visualizations as well as running introductory data visualization training.

15. Multimodal academic argument in data visualization

Arlene Archer and Travis Noakes

Abstract

This chapter investigates the semiotic and rhetorical strategies for realizing argument in data visualizations produced by second-year journalism students. The semiotic strategies include use of colour, typography, graphics, and the rhetorical strategies include establishing credibility and the use of citation. The effect of the underlying basis for comparison of data on the argument is examined, as are the selection and processing of data. The chapter investigates the semiotic encoding of ideational material and the ways relationships are established within the discourse communities constructed in the data visualizations. This way of looking at academic argument has important implications for teaching these text-types in higher education in order to produce critical citizens; both in terms of production and critical analysis.

Keywords: Academic argument; Social semiotics; Data visualization; Higher education

Introduction

In an age in which more and more data are produced and circulated visually and digital environments make the production of data visualizations increasingly accessible, it is important to develop critical tools for people to engage with these kinds of texts. Data can be represented through a range of modes (such as writing, visuals, and numbers) and different data visualizations (from tables to graphs). There are design choices to be made in the production of these texts in terms of size, shape, colour, and composition in order to represent a particular argument to a particular

Engebretsen, M. and H. Kennedy (eds.), *Data Visualization in Society.*
Taylor & Francis Group, 2020
DOI 10.5117/9789463722902_CH15

audience in the most apt way. It is useful to analyse how this repertoire of semiotic resources works together, especially in terms of fulfilling specific functions in academic argument. These functions range from establishing logical relationships to constructing hypotheses. There is a need to develop a pedagogy that takes into account the functions of academic argument, as academic discursive conventions can serve as 'gatekeepers' in terms of student access (Prince & Archer, 2008; Lea & Street, 1998). This chapter explores a framework for looking at argument in data visualizations, and applies this to students' texts in a second-year journalism course. Student texts can highlight the constructed nature of academic argument through inconsistencies and disjunctures, thus exposing the normative practices of the discipline. Our analysis aims to identify semiotic signifiers of argument within data visualizations, in order to assist both in the production and critique of these kinds of texts.

A social semiotic approach to argument in data visualizations

Our approach to exploring argument is multimodal social semiotics, where meaning-making is seen as a social practice (Martinec & van Leeuwen, 2008; van Leeuwen & Jewitt, 2001). This approach has been shown to be productive in analysing a range of professional and pedagogical texts, such as technical drawings (Simpson, 2016), infographics (Prince & Archer, 2014; Bateman, Wildfeuer, & Hiippala, 2017), and PowerPoint slides (Zhao, Djonov, & van Leeuwen, 2014). The assumption underpinning this approach is that meaning-making is informed by context. Also, meaning potentials are understood to be constructed through the selection and configuration of semiotic modes through the interests of the producer of the texts (Jewitt, 2009, p. 15). 'Mode' refers to socially and culturally shaped resources for making meaning, such as written language, spoken language, and visual representation (Kress, 2010). According to Halliday (1978), every sign performs three kinds of functions. The ideational function represents the world, concepts, and processes. The interpersonal function indexes the stance that the meaning-maker is taking towards audiences and the represented content. The textual function refers to the ways in which complexes of signs form coherent texts. In this metafunctional view, data visualization texts can represent a state of affairs, a relationship between abstract 'participants', and also indicate a particular relationship with an audience. These ideational and interpersonal aspects are realized through the textual organization or composition of the texts.

A social semiotic approach can also help understand how writing and visuals work together in academic argument. A data visualization could, for instance, be used as evidence, it could be a part of an argument (the proposition, for instance), or it could constitute 'restatement' in a different mode. Sometimes, the communicative work of different modes in a text can complement each other, but they can also contradict each other. Where different modes realize different aspects in a complementary way, textual coherence may be achieved. However, where there is a disjuncture between the messages of the modes, coherence could be compromised. Recognizing the communicative work that modes are performing in a text is important in order for students to both critique and produce argument.

Academic argument is a semiotic practice which engages with ideas in the world and with the existing positions and conventions of a discipline (Coffin, 2009, p. 513). Broadly speaking, argument is a logical set of ideas that is supported by evidence. Evidence can be the existing accepted material that an 'arguer' agrees with, or resists, but nonetheless draws on to establish a position. Previously, one of us identified some underlying ways of organizing knowledge in multimodal argument, including narrative, induction, contrast, and comparison (Archer, 2016). Firstly, narrative structures can be used to represent sequencing in time, but also change from one state to another (Kress & van Leeuwen, 2006). Secondly, induction or theorizing the relation of the particular to the general is another important structure. This can be both descriptive (backward-looking) and predictive (forward-looking). In other words, one can generalize from the specific instance, as well as make predictions about specific cases based on the general. Thirdly, academic argument can be realized through comparisons (both similarities and differences) and through contrast (where a juxtaposition sets up a tension between components of the text). Both comparisons and contrasts are based on underlying classifications according to specific categories. Clearly, this list is not exhaustive, but is useful in that it identifies structures of argument that work across modes. In general, we see argument as a textual form that produces 'difference' rather than closure (Kress, 1989). In foregrounding difference, argument can open the space for reconsideration, for a shift in values and attitudes, and for an extension of thought, producing 'new cultural values and knowledge' (Kress, 1989, p. 12). This notion of 'difference' offers insight into how tension is established in argument, particularly those based on comparison and contrast.

The chapter now moves on to propose a way of looking at argument in academic data visualization texts. It first investigates the semiotic encoding of ideational material in argument. Here the focus falls on the basis for comparison; the underlying discourses that are drawn on, and the ways these are realized using the resources of composition, colour, font, size, and shape. Secondly, the chapter explores the ways relationships are established within the discourse communities constructed in data visualizations, focusing on how credibility is established, and the use of citation.

Semiotic encoding of ideational material in data visualizations

The first ideational aspect to look at is the basis for comparison in an argument or the underlying classification for comparison. Selection and classification are always ideological. What is selected as well as the chosen basis for comparison is often as important as what is omitted. A second ideational aspect to explore is the normative discourses and practices that shape data visualizations. The analytical focus should thus fall on the analysis of semiotic resources (such as composition, shape, and colour) as located within the discourses, practices, and technologies that regulate the use of these resources. In recent social theory, discourse is understood to refer to the ways social institutions define and regulate the practices within those institutions. Discourses are 'socially constructed knowledges of (some aspect of) reality which give expression to the meaning and values of an institution or social grouping' (Kress & van Leeuwen, 2001, p. 4). Discourses regulate the practices within those institutions through the use of language or other semiotic modes. Ledin and Machin (2016), for example, draw attention to the way 'strategic' diagrams recontextualize agents, processes, and causalities, and are often embedded in neoliberal discourses. It is also worthy of note that in some instances, normative discourses are built into the latest visual communication technologies, such as Microsoft PowerPoint or Excel (Zhao et al., 2014).

Data visualizations can be conceptualized in terms of van Leeuwen's (2008) 'new writing' as they encapsulate some of the principles of contemporary integrated design enabled by digital technology. This form of writing is 'at once more visual than the old "page" media, and less pictorial than the old "screen" media' (van Leeuwen, 2008, p. 132). Ideational content is encoded through the resources of composition, colour, font, size, and shape, as well as the choice of data visualization. For instance, bar charts compare quantities, pie charts show proportions of the whole, and line graphs show quantities over time.

In addition to looking at choices of semiotic resources, it is also important to look at the relations between image and writing. Many multimodality theorists have attempted to systematically describe visual/verbal relations (cf. Martinec & Salway, 2005; Unsworth, 2006). In general, the relations between data visualization and writing can be thought of in terms of similarity, complementary, and opposition relations. In similarity relations, one could look at how one mode exemplifies the other. In complementary relationships, what is represented graphically and in writing may be different, but complementary. In opposition relations the content of the written text contrasts with that of the data visualization. Ideational and, at times, interpersonal content, are encoded through the relations between visual and written aspects, as well as through the semiotic choices of font, colour, size, shape, and compositional choices (of positionality and directionality).

Establishing social relations in data visualizations

The type of graphic chosen can establish credibility in argument and lend more authority to the numbers. Trimbur and Press (2015) argue that truth values ascribed to various modes are shaped by a struggle for rhetorical authority within the means of representation. Data visualizations tend to be assigned credibility as the assumptions underlying the numbers are generally hidden and numerical representations are often regarded as more factual and objective than other kinds of evidence (Porter, 1995; Zhou & Hall, 2018). Whilst conveying results from questionnaires through statistical means enables participant anonymity, the existence of the participants then becomes masked by a number. The selection of data behind data visualization texts is of course subjective and every number can and should be interrogated. For this reason, it is important to cultivate a critical awareness (Kennedy et al., 2016) amongst both educators and students.

In academic writing, credibility in argument is often established through tentative assertions which are realized through discourse markers such as 'hedging' and 'emphatics' (Hyland, 1999, p. 104). Hedges indicate the writer's decision to withhold complete commitment, whereas emphatics construct certainty. What is of interest for data visualizations is the ways that credibility is established across the written and visual modes, and whether there is any 'hedging' or tentativeness in the representation of the data and the argument.

In Prince and Archer (2014) we argued that uncertainty about a 'point statistic' is often provided through the use of a 'confidence interval'. A confidence interval is a resource for representing a certain parameter and a range of possible values. This enables the presentation of findings in a

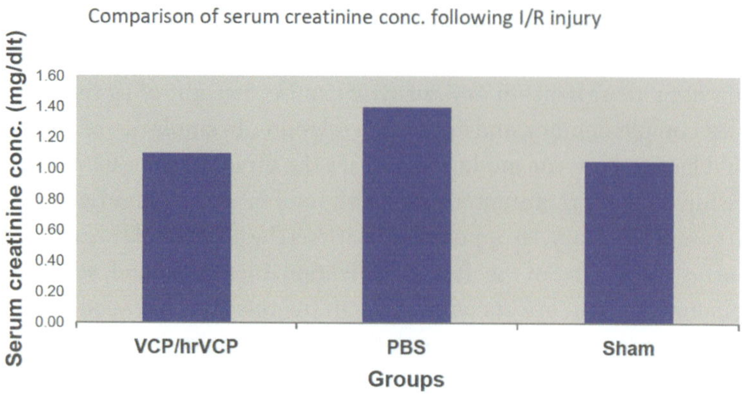

Figure 15.1a. Visual data hedging through the use of a confidence interval. Illustration by A. Archer & T. Noakes.

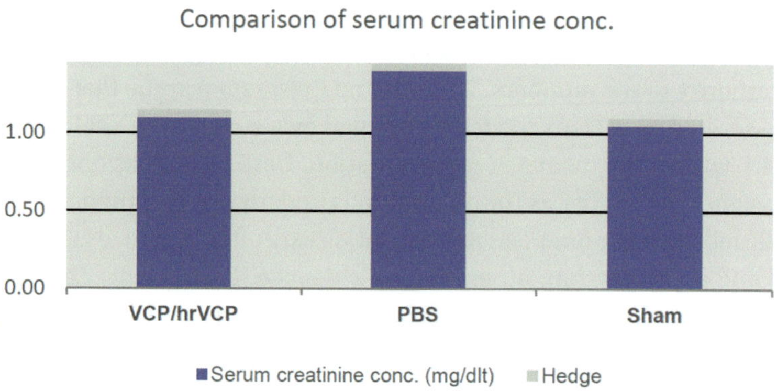

Figure 15.1b. An alternate visual form of hedging with maximum y-axis. Illustration by A. Archer & T. Noakes.

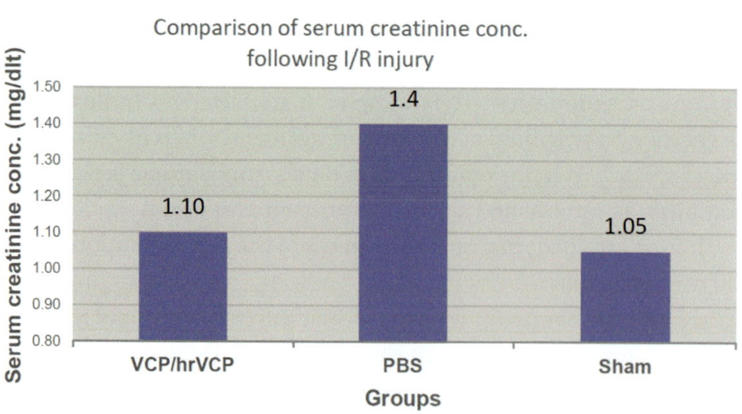

Figure 15.1c. Another visual form of hedging with the maximum and minimum values labelled. Illustration by A. Archer & T. Noakes.

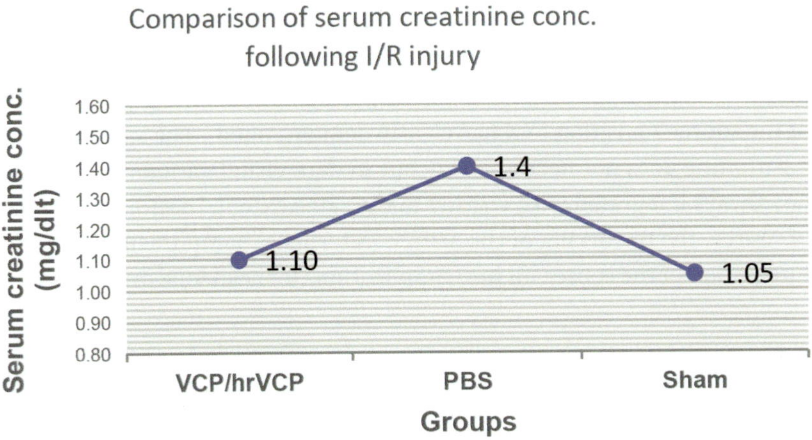

Figure 15.2. Maximum and minimum values indicated using separate line graphs. Illustration by A. Archer & T. Noakes.

more tentative way, which can be likened to 'hedging' in academic discourse (see Figure 15.1a). Here the resources for indicating uncertainty include a line that spans the given point (both above and below it) to indicate the possible range of values. Other visual alternatives for indicating a confidence interval could include adding a second box chart (see Figure 15.1b), entering the range with data labels (see Figure 15.1c), or using separate line graphs to suggest the maximum and minimum ranges (see Figure 15.2).

Another way of establishing relations is the use of citation for appropriating a source into an argument and using the arguments of others to negotiate a position in a particular discourse community. All texts are always positioned in relation to a network of other texts. Choices about the integration of sources include the selection of material from the source, the form of the citation, and some kind of framing. There are a number of options for citation in data visualization. The data could be generated empirically by the researcher and thus no citation of external sources is necessary. A second possibility is the integration of a researcher's own data with cited data. Thirdly, data can be compiled from multiple sources within one information graphic. For example, data sources can be cited in a list at the bottom of a data visualization.

The placing of the in-text references in data visualizations is of importance. The source could be more foregrounded if placed in the label rather than the caption, for instance. In data visualizations, the words 'taken from…' could indicate the graphic is a reproduction or a 'quote' from the original source with all the deferment of authority that this entails. If the source is introduced as 'adapted from…' it indicates some kind of paraphrasing or reworking. The way in which the source is introduced is thus of importance. In writing, the 'reporting' verbs used to introduce a citation can be neutral (as in XX states), or sceptical (XX would have us believe), or strongly supportive of the source's position (XX has clearly demonstrated). The same kind of positioning does not necessarily occur in data visualizations as they are not always integrated with writing to the same extent. In sum, the choice of source, the reworking of the source in terms of paraphrasing, and the integration of the source all have implications for academic argument.

Data visualization in a second-year journalism course

We will now employ the framework outlined above to look at the semiotic and rhetorical strategies for realizing argument in data visualizations produced by second-year journalism students at a university in South Africa. The students were required to design a poster using data visualizations that focused on educational inequalities in two geographical areas in Cape Town (Noakes, 2017). Each student was taught to contrast up to three aspects of inequality within a poster design and to export the resulting text for blog publication. For the purposes of this chapter, all projects were reviewed as a convenience sample (Ferber, 1977) that would allow us to learn from the struggles of inexperienced students experimenting with data visualization design. Our research has provided input towards improving this pathfinder curriculum by incorporating ideas of multimodal argument.

As the task required, all the arguments in the student produced texts are based on comparisons, showing the differences in levels of education obtained in different geographical areas. Some students chose to focus on social issues (such as pregnancy, poverty, single- or no-parent households) and others on access issues (internet, home language, unemployment, income). Here we look at two of the posters produced. The students gave permission for their work to be discussed and they understand that we are drawing attention to both the successful and unsuccessful aspects of argument through visualization. The first poster highlights the difficulty of producing argument rather than description, and the second one showcases the struggle between causation and correlation in these types of texts.

Description versus Argument

The data visualization poster in Figure 15.3 compares two areas in Cape Town that feature vastly different living circumstances, namely Nyanga

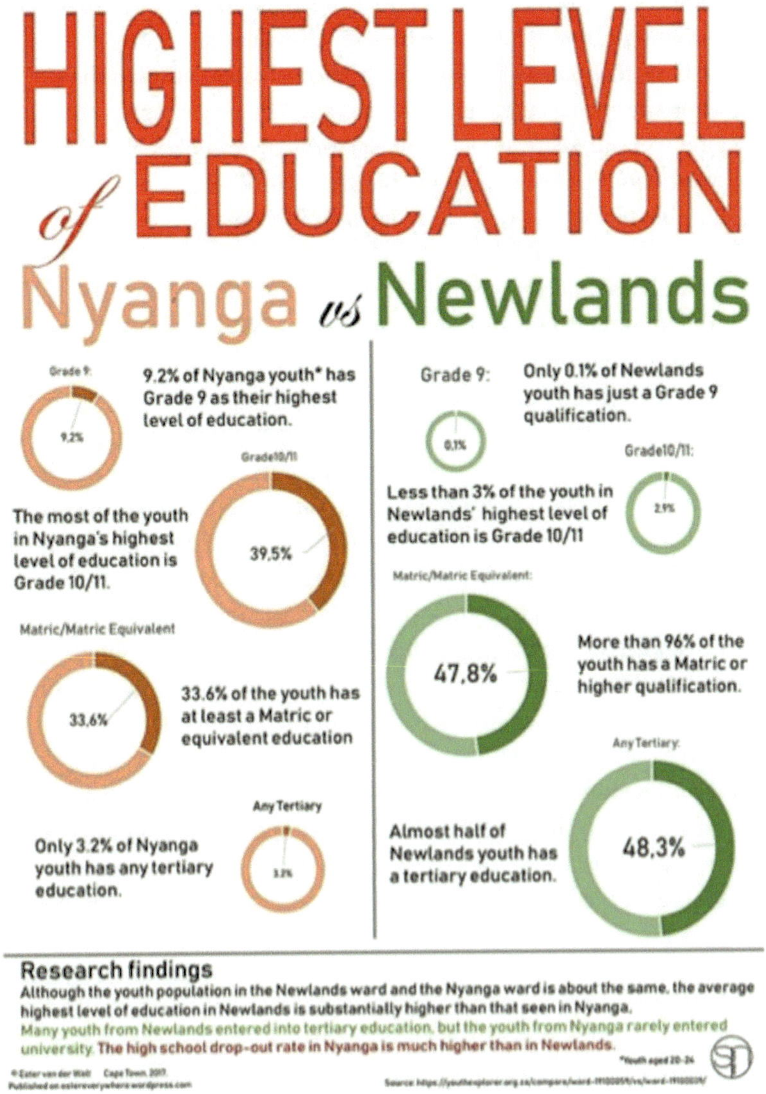

Figure 15.3. Nyanga versus Newlands. Poster by E. van der Walt, 2017. Reprinted with permission.

and Newlands. Nyanga is a predominantly 'black' township situated about 26 km from the city centre. It is one of the poorest areas in Cape Town and has a high unemployment rate. Newlands, on the other hand, is an upmarket suburb located at the foot of Table Mountain. It has many good schools and sports and recreation facilities. The poster compares the highest level of education achieved by the youth in each area (Grade 12 refers to the final year of schooling). The poster tends towards description rather than argument, as it does not identify possible contributing factors for the difference in educational performance in Nyanga and Newlands.

The basis for contrast is stated as: 'the youth population in the Newlands ward and the Nyanga ward is about the same'. In fact, according to the youth explorer website, there are 4,648 youth in Nyanga and 3,765 in Newlands, so the word 'about' is rather a larger qualifier. This discrepancy aside, making the number of youths the underlying basis for contrast could serve to efface and neutralize the vast differences between the areas. What is not stated, for example, is that Nyanga is 1.2 km squared whereas Newlands is 44.1 km squared, making the youth per square km in Nyanga 4,018.3 as compared to 85.4 youth per square kilometre in Newlands (youthexplorer.org.za).

The underlying structure of the argument is a binary where two aspects are juxtaposed. The contrast is set up visually through two main resources, namely layout and colour. In terms of layout, the poster is divided by a vertical line into two sections, Nyanga on the left and Newlands on the right. In terms of colour, van Leeuwen (2008) points out that colour can be used both for its connotative potential and to signify textual cohesion. The poster employs colour to signify particular features of the two areas as well as to establish the contrast. The title 'Nyanga' and the data related to Nyanga are depicted in a 'rusty red' or orange, emphasizing the dryness (dust), less development, and poor infrastructure of the area. This is opposed to the green of Newlands which points to the notion of the 'leafy suburb', as well as natural beauty (the forest and nature of this high rainfall area). Other design choices in the poster include one simple graph type throughout, a 'donut' chart. The poster uses size as a semiotic resource in argument: sizing the graphs in accordance with their percentage values, and the font sizes get bigger for larger percentages.

A citation is placed at the bottom right corner of the poster in the form of a URL, the 'youth explorer' website (https://youthexplorer.org.za). Such a citation may lack credibility in not actually citing the originating source of the data (the Western Cape Education Department). A similar issue occurs where students have attributed image sources in their presentations to 'Google', the search engine. A critical reader would expect to be able to use the link to directly access the attributed images, much as website database references should refer directly to their sources, not to aggregators or search engines that are intermediaries.

The poster establishes credibility by employing the academic discourse conventions of hedging ('about') and qualified emphatics ('substantially'). Credibility is also established by presenting the 'research findings' dispassionately as facts:

> Although the youth population in the Newlands ward and the Nyanga ward is *about* the same, the average level of education in Newlands is *substantially* higher than that seen in Nyanga. Many youth from Newlands entered into tertiary education, but the youth from Nyanga rarely entered university. The high school drop-out rate in Nyanga is much higher than in Newlands.

However, it would appear that information is presented here, rather than argued, as the underlying causes for this vast disparity are not identified. This objective presentation of 'facts' could serve to erase the people and the hardships of the area (from overcrowding to crime).

Correlation versus causation

In Figure 15.4, the data visualization text attempts to make an argument. It hypothesizes the underlying causes for the differentials in education levels achieved in the two chosen areas that roughly encompass Camps Bay (Ward 54) and Hout Bay (Ward 74). The poster attempts to look at factors for academic exclusion. It provides an explanation for the potential exclusionary role of language, and a link is made between internet access and academic throughput.

The student claims to have chosen these two areas as they are 'neighbouring' wards in close proximity to each other. Whereas Camps Bay is a more affluent area along the Atlantic seaboard, Hout Bay is a somewhat demographically mixed area. Once a more homogenously upmarket area, Hout Bay now includes a large informal settlement, Imizamo Yethu, which was established about 25 years ago. The area houses approximately 33,600 people in high density living. However, the complexity of the demographics and history of Hout Bay cannot be reflected here in a simplistic contrast with Camps Bay, as was the task brief.

The poster is divided into three sections, using a band of colour in the middle to separate the sections. Colour is employed predominantly as an organizational feature, rather than for its connotational affordances, and shades of purple and orange dominate. Blue is used for Ward 74 and orange for Ward 54 in the charts. While the student has chosen these colours to tie in with the pallet of her painter logo, the orange, blue, and white colours used may upset viewers who could perceive this as a reference to the colour of the Apartheid-era South African flag. However, as a non-South African student, it is unlikely that she was aware of the

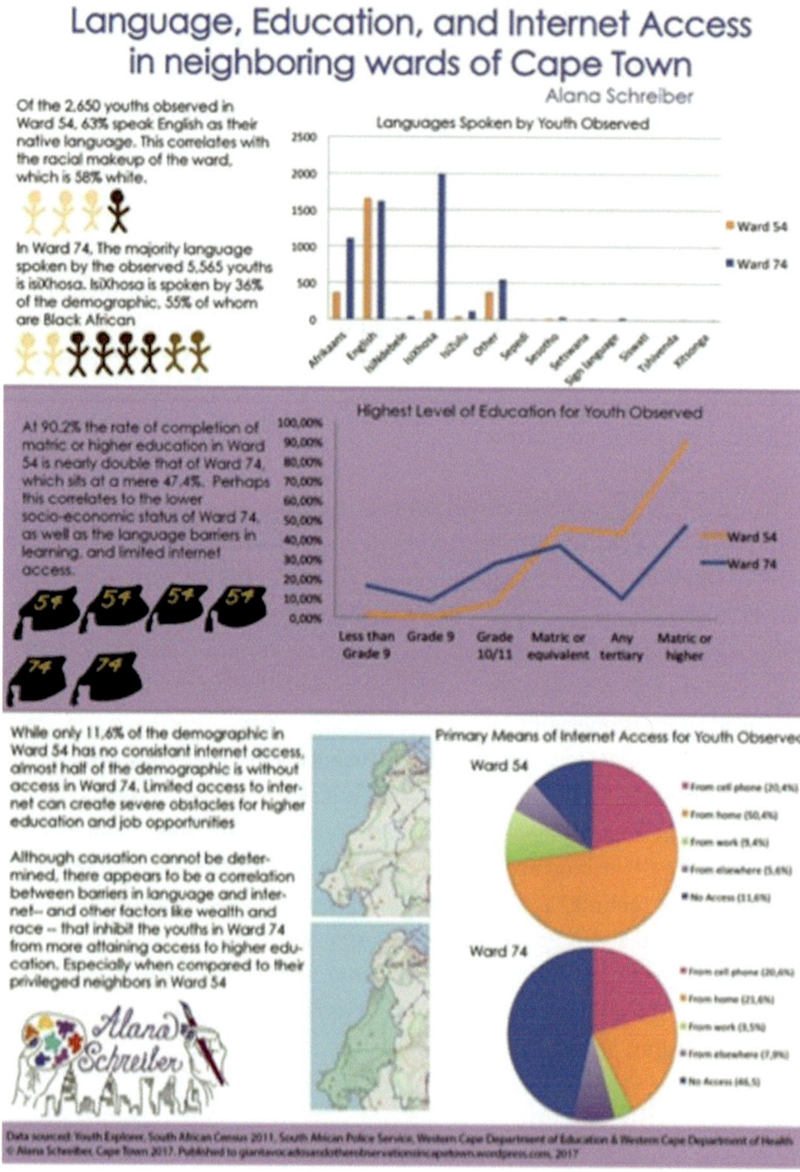

Figure 15.4. Language, Education and Internet Access in neighbouring wards of Cape Town: Camps Bay versus Hout Bay. Poster by Alana Schreiber, 2017. Reprinted with permission.

negative connotations of this choice. The stark contrast in achievement of higher education (90.2 percent versus 47.4 percent) is indicated visually by the blue and orange lines. This is echoed in the writing on the left side, which states:

> At 90.2% the rate of completion of matric or higher education in Ward 54 is nearly double that of Ward 74, which sits at a mere 47.4%. Perhaps this correlates to the lower socio-economic status of Ward 74, as well as the language barriers in learning, and limited internet access.

The writing here establishes credibility through the use of hedges ('nearly') and tentative statements ('perhaps') which is in accordance with conventions of academic discourse. However, it is not a dispassionate representation of 'facts' as indicated by a 'mere 47.4%' which indicates surprise or some outrage at the low figures.

A simplified visual is used to restate the written argument (namely, four 'ward 54' graduation hats placed above two 'ward 74' graduation hats). These graduation hats indicate the notions of half and double without being statistically accurate. The use of the graduation hat graphics alongside the line chart in the poster could be seen to be the equivalent of first and third person in data visualization texts. While third person writing often characterizes research that uses both qualitative and quantitative methods, Zhou and Hall suggest incorporating more of the first person, as they claim this 'adds to the subjective experience as part of the evidence for the author's claims and makes the author's perspective and constructive role in creating meaning in a study more visible' (2018, p. 2). As with the graduation hats, the rows of figures at the top left of the poster show the demographics of each area, but not in any statistically accurate way. These kinds of simplified representations may be a way of reintroducing narrative and experience into data visualization.

The bottom band of the poster compares the internet access of the two areas. While the student accedes that internet access is not causal in educational achievement, she suggests that there is a correlation.

> Although causation cannot be determined, there appears to be a correlation between barriers in language and internet—and other factors like wealth and race—that inhibit the youths in Ward 74 from more attaining access to higher education. Especially when compared to their privileged neighbours in Ward 54.

The explanation for the internet's role assumes that the internet is used during school assignments. However, this is seldom the case, especially at government schools that first-language isiXhosa speakers would typically attend. The academic argument here is mostly made through written text, and not through data visualization. The poster samples quite disparate datasets ('level of education', 'language', and 'internet access'), which require written explanation. The writing thus serves to make links between the graphs, and then to summarize the argument. The argument is realized effectively, however, through the juxtaposition of pie charts, creating 'difference', and through the writing, which communicates a sense of outrage by using words like 'inhibit' and 'privileged'. Here, Kress's (1989) notion that argument foregrounds difference is pertinent, as this juxtaposition opens the space for reconsideration, for a shift in values and attitudes, even if correlation and causation become slightly blurred in the representation.

Discussion and implications

The past three years in South African higher education have seen a growing movement known as '#feesmustfall', which was unprecedented in its scale and violence (Jansen, 2017) in calling for free, decolonized education (Ndlovu, 2017). Disparities in access to higher education as residual effects of the apartheid system and a slow and disproportionate throughput of students are part of the reasons for the #feesmustfall movement and the call to decolonize higher education. This forms part of the circulating discourses and contexts in which the students produced these data visualization texts. In fact, at the time the students were to present their posters, lectures were cancelled due to protest action, and students were given the option to submit a PowerPoint presentation together with an audio file. This context highlights the need for students to become critical users, assessors, and producers of scientifically grounded data visualizations, with an understanding of the surrounding discourses.

Despite access to higher education being of topical importance, the type of data provided via youthexplorer.org.za does not currently support students to design posters that might contextualize these issues. For example, the site does not provide data on academic access in local universities, nor drop-out rates in degrees. Youthexplorer.org.za's data come from a particular governmental discourse with particular values that organizes and selects 'important' data on youths made available to site users. It is important to understand what a discourse focused on the state's role for improving the wellbeing of young people neglects, such as promoting equitable access to tertiary education.

On a more specific level, the students tended to confuse correlation and causation, such as, assuming that internet access supports a high pass rate. Rather, in the South African context, internet access is a marker of privilege that is often linked to households that can afford better schooling. In addition, the data visualizations did not really allow for blurred categories as each data point needs to be assigned to a separate category. Data visualizations tend to simplify qualitative complexity into a number. For example, studying or a 'gap year' or any other reason for not working often fell under the category of 'unemployment' in the data visualizations. Data visualization is thus often a simplification of complexity. To enable students to read and produce argument, we need to develop an awareness of processes of simplification in order to inculcate a critical perspective on meaning-making.

This chapter has highlighted what Kennedy and Hill (2017) call the 'complex entanglement' of aspects of data visualization: knowing how to physically create these texts; the underlying discourses and ideological work of data visualizations; and the pleasure and aesthetics of data visualization. We have presented an analysis of two data visualization texts in an attempt to explore how academic argument is constructed through the interplay between multiple semiotic resources. Written language, visual representation, and the representational choices we make (like using a bar rather than a point on a graph) all contribute to academic argument. We have shown how, in composing data visualizations, students encode ideational material and establish relationships within the discourse community, both through citation and establishing credibility. A way of looking at academic argument such as the one explored here could be useful to facilitate awareness and analysis of data visualization texts for students in order to enable access to their invisible norms and conventions. The discourses that shape data visualizations are expressed through choice, such as the type of representation and the composition of the representation. Producing these data visualizations can facilitate recognition of the social provenance of texts, namely that data visualizations are often seen as objective and neutral, rather than ideological. We have fed these insights back into this particular journalism course. However, this chapter has argued that this has important implications for the way we teach these text types in higher education in general in order to produce critical citizens, both in terms of production and critical analysis.

Acknowledgements

The research is based upon work supported by the British Academy Newton Advanced Fellowship scheme.

References

Archer, A. (2016). Multimodal academic argument: Ways of organizing knowledge across writing and image. In: A. Archer & E. Breuer (Eds.), *Multimodality in Higher Education.* (pp. 93-113). Leiden: Brill.

Bateman, J., Wildfeuer, J., & Hiippala, T. (2017). *Multimodality: Foundations, research and analysis—A problem-oriented introduction.* Berlin: De Gruyter Mouton.

Coffin, C. (2009). Contemporary educational argumentation: A multimodal perspective. *Argumentation, 23*(4), 513-530. https://doi.org/10.1007/s10503-009-9161-z

Ferber, R. (1977). Research by convenience. *Journal of Consumer Research, 4*(1), 57-58. http://dx.doi.org/10.1086/208679

Halliday, M. A. K. (1978). *Language as social semiotic: The social interpretation of language and meaning.* London: Arnold.

Hyland, K. (1999). Disciplinary discourses: Writer stance in research articles. In: C. N. Candlin & K. Hyland (Eds.), *Writing: Texts, processes and practices.* (pp. 99-121). London & New York: Longman.

Jansen, J. (2017). *As by fire: The end of the South African university.* Cape Town: Tafelberg.

Jewitt, C. (Ed.). (2009). *The Routledge handbook of multimodal analysis.* Abingdon & New York: Routledge.

Kennedy, H., Hill, R. L., Allen, W., & Kirk, A. (2016). Engaging with (big) data visualizations: Factors that affect engagement and resulting new definitions of effectiveness. *First Monday, 21*(11). https://doi.org/10.5210/fm.v21i11.6389

Kress, G. (2010). *Multimodality: A social semiotic approach to contemporary communication.* London & New York: Routledge.

Kress, G. (1989). Texture and meaning. In: R. Andrews (Ed.), *Narrative and argument.* (pp. 9-21). Milton Keynes: Open University Press.

Kress, G., & van Leeuwen, T. (2006). *Reading images: The grammar of visual design* (2nd ed.). London: Routledge.

Kress, G., and van Leeuwen, T. (2001). *Multimodal discourse: The modes and media of contemporary communication.* London: Arnold.

Lea, M. R., & Street, B. (1998). Student writing in Higher Education: An academic literacies approach. *Studies in Higher Education, 23*(2), 157-172. http://doi.org/10.1080/03075079812331380364

Ledin, P., & Machin, D. (2016). Strategic diagrams and the technologization of culture. *Journal of Language and Politics, 15*(3), 322-336. https://doi.org/10.1075/jlp.15.3.06led

Martinec, R., & Salway, A. (2005). A system for image-text relations in new (and old) media. *Visual Communication 4*(3), 337-371. https://doi.org/10.1177/1470357205055928

Martinec, R., & van Leeuwen, T. (2008). *The language of new media design: Theory and practice.* London & New York: Routledge.

McCandless, D. (2009). *Information is beautiful.* London: Collins.

Ndlovu, M. (2017). *# FeesMustFall and youth mobilisation in South Africa: Reform or revolution?* Abingdon: Routledge.

Noakes, T. (2017). Designing infographics on educational inequalities in Cape Town's wards—a new #UCT Media Studies project. Retrieved August 31, 2018 from http://www.travisnoakes.co.za/2017/11/designing-infographics-on-educational.html

Porter, T. M. (1995). *Trust in numbers: The pursuit of objectivity in science and public life.* Princeton, NJ: Princeton University Press.

Prince, R., & Archer, A. (2008). A New Literacies approach to academic numeracy practices in Higher Education in South Africa. *Literacy and Numeracy Studies, 16*(1), 63-75.

Prince, R., & Archer, A. (2014). Exploring academic voice in multimodal quantitative texts. *Literacy and Numeracy studies, 22*(1), 39-57.

Simpson, Z. (2016). 'Drawn Writing': The role of written text in civil engineering drawing. In: A. Archer & E. O. Breuer (Eds.), *Multimodality in Higher Education.* (pp. 241-255). Leiden: Brill.

Trimbur, J., & Press, K. (2015). When was multimodality? Modality and the rhetoric of transparency. In: A. Archer & E. Breuer (Eds.), *Multimodality in writing: The state of the art in theory, methodology and pedagogy.* (pp. 17-42). Leiden: Brill.

Unsworth, L. (2006). Towards a metalanguage for multiliteracies education: Describing the meaning-making resources of language-image interaction. *English Teaching: Practice and Critique, 5*(1), 55-76.

van Leeuwen, T. (2008). New forms of writing, new visual competencies. *Visual Studies, 23*(2), 130-135. https://doi.org/10.1080/14725860802276263

van Leeuwen, T., & Jewitt, C. (Eds.). (2001) *Handbook of visual analysis.* London: Sage Publications; New Delhi: Thousand Oaks.

Youth Explorer. (2011). About Youth Explorer. Retrieved August 31, 2018 from https://youthexplorer.org.za/about

Zhao, S., Djonov, E., & van Leeuwen, T. (2014). Semiotic technology and practice: A multimodal social semiotic approach to PowerPoint. *Text & Talk, 34*(3), 349-375.

Zhou, X., & Hall, J. (2018). Mixed methods papers in first-person and third-person: Writing voices in dialogue. *Journal of Mixed Methods Research, 12*(3), 344-357. https://doi.org/10.1177/1558689816652755

About the author

Arlene Archer, PhD, is an Associate Professor in Applied Linguistics. She is the director of the Writing Centre at the University of Cape Town and leads the South African Multimodality in Education group.

Travis Noakes is an Honorary Research Associate at the Centre for Film and Media, University of Cape Town. His research interest is describing young adults' diverse strategies and productions as online content creators.

Section IV

Data visualization semiotics and aesthetics

16. What we talk about when we talk about beautiful data visualizations

Sara Brinch

Abstract

'Beautiful' is an adjective often used in descriptions of well-designed data visualizations. How the concept is used, however, reveals that it is applied to characterize a variety of qualities. Going beyond mere descriptions, the use of the concept also lays bare a certain ambivalence among scholars and practitioners towards *how* beauty matters, and which means it serves in data visualization. Interrogating 'beautiful' as a characterizing word, combined with a study of cases of 'best practice' used as examples of beautiful visualizations in various discourses, this chapter presents an analysis of what is regarded as beautiful within the field of data visualization design. This, in turn, can inform the understanding of what beauty means in visualizing data, in the purpose of facilitating the viewer's comprehension and engagement.

Keywords: Beautiful data; Aesthetics of data visualization; Anti-sublime; Data visualization and design

Introduction

> Beautiful /ˈbjuːtɪfʊl,ˈbjuːtɪf(ə)l/ Pleases the senses and mind aesthetically.

Beauty is to be found everywhere, including the field of data visualization. Stunning in the various ways designers use colours, forms, and lines to turn data into imagery and graphical information, data visualizations can attract attention to their aesthetically pleasing expressions. But is it this kind of aesthetic endeavour that is appreciated or discussed in books, articles, and talks which address beautiful data visualizations? When authors like Julie Steele and Noah Lillinsky, or Edward Tufte, choose titles like *Beautiful*

Engebretsen, M. and H. Kennedy (eds.), *Data Visualization in Society.*
Taylor & Francis Group, 2020
DOI 10.5117/9789463722902_CH16

Visualizations and *Beautiful Evidence* for their books, what, exactly, does that imply? And, when David McCandless initiates the *Information is Beautiful* award along with Kantar, what exactly is being awarded?

Visualizations of big data have brought along a growing consciousness about their potentials for forming knowledge (e.g. Gray, Bounegru, Milan, & Ciuccarelli, 2016; McCosker & Wilken, 2014). In the field of digital culture and new media studies, data visualization as visual communication and cultural expression has become an object of study (e.g. Gray et al., 2016; Manovich, 2014). In addition, beyond academia, attention directed towards the artistic and aesthetic aspects of visualization is noticeable through a variety of museum exhibitions, art projects, contests, and awards. The beauty of an original and well-designed data visualization seems to matter—but how, and what is being appreciated when a visualization is said to be beautiful? This study addresses the rather general question of how beauty matters in the field of data visualization, making an argument for a shared understanding of what the purpose of an aesthetically pleasing visualization is. Based on a selection of textbooks, publications, and contests, this chapter focuses on how the concept of beauty and the beautiful are being used in various discourses related to data visualization: in instructional texts, in critical or evaluating notes, as well as in the community of practitioners discussing 'best practice'. Based on these texts I present a selection of five types of beautiful data visualizations that can be derived from these same discourses.

The ancient Greeks pointed out the challenge of discussing the nature of beauty: 'while we know with relative ease what a beautiful horse or a beautiful man or possibly even a beautiful pot is', the author of *Greater Hippias* writes, 'it is much more difficult to say what "Beauty" unattached to any objects is' (cited in Scarry, 2011, p. 9). One can even claim, like Elaine Scarry does, relying on Immanuel Kant, that there is no sense in trying to explain beauty without pointing to something particular. I therefore elaborate on the types of beautiful data visualizations with examples of best practice: award-winning data visualizations and visualizations often referred to in studies, textbooks, and discussions among practitioners.

The aesthetics of data visualization

Data visualizations represent matters of the world through graphic interpretations of the data generated from investigating or researching these matters. The data visualization designer performs his or her work based on skills in how to relate semiotic properties of data visualizations, like types of representations (different graphs, models, imagery), and aesthetic qualities (the use of modalities and combinations of them, suitable for the purpose), to the means the data visualization aims to serve and the context in which it is to be included. Aesthetics is to be understood as a way of thinking not only about the design and appearance of the visualization, but also about how the viewer or reader perceives the data visualization, based on the sensory experience that springs from it. The viewer's perception has both a cognitive and an emotional side, both of which are activated when encountering the data visualization and the context in which it is found (newspaper article, textbook, etc.). This resonates with Jay Lemke's thoughts that feelings and meanings are part of the same material processes (2015, p. 589). What's more, the form and expression of a data visualization will be of importance in the way the viewer or reader perceives and comprehend the matter being communicated. A data visualization that generates particular interest and engagement at first glance could be a visualization that exemplifies a double meaning of 'aesthetics'—both sensory experience *and* beauty.

Beauty can be a philosophical concept, and an aesthetic category belonging to the field of art criticism or other professional disciplines such as design, in which it forms an ideology governing the aesthetic principles being applied. However, beauty can also be thought of as something with a more formative aspect: it can evoke the feeling of good mood or enjoyment (Eco, 2010, p. 10), to contemplate it, or to replicate it (Scarry, 2011, p. 3). Ideas about beauty are historical constructs, changing over time, as well as over fields of interest. When addressing beauty, and the beautiful within the field of data visualization, one has to address it in relation to the dominant understanding of what defines data visualization, and what its main purpose or function is. When Andy Kirk defines data visualization as '[t]he representation and presentation of data to facilitate understanding' (2016, p. 19), this also has implications for how to think about the matter of beauty in data visualizations, and what will be regarded as beautiful visualizations.

The matter of beauty and the beautiful

Making use of 'beautiful' when describing or characterizing data visualizations, the adjective is always used to underline that the particular visualization differs from others, in a positive way. When included in textbooks or articles, however, 'beautiful' refers to a variety of qualities, depending on the visualization in question. Furthermore, attempts to articulate an understanding of the value of 'beauty' in data visualization is never oriented towards beauty as such, but ends up as arguments about why it is secondary to other aspects of the design, or why it cannot be aligned with other virtues. For example, in *Data Visualisation: A Handbook for Data Driven Design*, Kirk starts out by explaining how the book's argument is based on several dichotomic distinctions, one of them being 'Useful vs. Beautiful': 'this book is not intended to be seen as a beauty pageant', he explains (2016, pp. 6-7). The works discussed should be regarded as useful examples, from which we can learn something, not only pleasant expressions to be looked at. He talks rather about *elegant* design. In his handbook, elegance is one of three principles of good design (the other two being 'trustworthy' and 'accessible'), used as pointers for understanding, for an aspiring designer.

Elegance is explained as a quality that will attract the viewers' attention and make an impression on them, but as with beauty never something that should be thought of or planned for in itself: 'When working on a problem, I never think about beauty. I think only how to solve the problem. But when I am finished, if the solution is not beautiful, I know it is wrong' (Richard Buckminster Fuller, cited in Kirk, 2016, p. 43). This quote is used by Kirk to suggest a shared ambivalence towards beautiful design among practitioners: on the one hand it should not be sought deliberately nor be the prevailing principle in the design process. On the other, it is seen as a result of a successful design process, as the sum of essential or key elements related to visualizations, like Steele and Iliinsky argue (2010), as well as a way of getting the viewers' or readers' attention, which can be turned into an engagement with the topics or issues being visualized.

Then again, there are designers and design companies which embrace beauty, making it one of their leading principles for how to engage and motivate the viewer. One such design company is Accurat, co-funded by designer Giorgia Lupi, stating the following on their website:

> We pursue beauty:
> Beauty is not a frill. We know how to engage and motivate people to dig deeper and take time to explore the intricacies of a visual data analysis. We deploy our rigorous methods to achieve the ideal balance between familiar visual motifs and unexpected aesthetics, a powerful combination that leverages studies on perception to trigger curiosity and interest, and creates indelible images in the minds of users. (Accurat, 2018)

To Lupi, beautiful design is a trigger to get people curious to explore the contents the visuals convey. 'I like the idea of making people say "Oh that's beautiful! I want to know what this is about!"' (Lupi, cited in Kirk, 2016, p. 46).

In *The Truthful Art* (2016), Cairo lists 'beautiful' as a quality associated with great visualizations, along with 'truthful', 'functional', 'insightful', and 'enlightening' (2016, p. 45). Together, they constitute the framework upon which the book is built, Cairo writes, but they are not flawless: every one of them is 'dangerously polysemic' (p. 45). To Cairo, beauty always consists of a balanced mix of sensual and intellectual pleasures, and I believe he, like Giorgia Lupi, sees the formative power of a beautiful expression when he paraphrases one of Donald A. Norman's reflections from *Emotional Design* (2003): 'beauty matters because attractive and pleasing things work better. They put us in a good mood, and so they invite us to invest some effort in understanding how to operate them' (2016, p. 56). 'Beauty is, thus,' Cairo writes 'not a thing, or a property of objects, but a measure of the emotional experience of awe, wonder, pleasure, or mere surprise that those objects may unleash' (p. 45). It is an emotional experience that *can* be turned into an intellectually-oriented interest in looking more closely into things.

So elegance can be regarded as an expression of beauty, but so can efficiency and simplicity (Cairo, 2016). This is a kind of beauty that Edward R. Tufte discusses. In *The Visual Display of Quantitative Information* (2001) he praises the ways the best designs draw the viewer into what he calls the 'wonder of the data', either by narrative power, immense detail, or by elegant presentations. Neither of these imply decorations or ornamentations of any kind (Tufte, 2001, p. 121). To him there is a danger of turning visualization into 'chart junk' or 'redundant data-ink' through decoration of any kind—be it by ways of graphically highlighting the information (explained as 'Unintentionally Optical Art' and 'The Grid'), or by letting an idea for appearance govern the design process, resulting in 'Self-Promoting Graphic' (p. 116).

This way of thinking about beauty as something that is found in clean and simplifying design can also be found in an early theoretical discussion of data visualization aesthetics, Lev Manovich's essay 'Data visualisation as new abstraction and anti-sublime' from 2002. Drawing on Immanuel Kant's notion of the sublime, Manovich characterizes much data visualization art as 'anti-sublime'. The reason for this, he argued, was the inherent purpose of visualization practice: to make 'phenomena that are beyond the scale of human senses into manageable visual objects' (2002). Manovich's neologism has later been expanded upon to the field of data visualization as such (not art exclusively) in which anti-sublime can be understood as 'that which can be easily understood' or as 'user friendly' (Sack, 2007). More recently, Anthony McCosker and Rowan Wilken note the limitations of Manovich's notion, and instead introduce the potential for the *diagrammatic capacities* of visualizations of big data, through which the complexity of big datasets can be communicated (2014, pp. 159-163). However, as Manovich's later works visualizing cultural data exemplify, his initial reflections on the anti-sublime have to be seen in relation to their historical context, and in regard to the technical limitations governing digital data visualizations at the turn of the millennium.

Beauty of various types, by various means

Despite their variations, these reflections share a view of what the purpose of beauty is: to get attention or evoke a sentiment in order to make the viewer or reader interested in investing time and effort in understanding the information the data visualization communicates. But *how* this attention or sentiment can be produced varies. The variations, however, can be sorted into types. The ones I take into consideration are the designer, the visualizations themselves, the viewer or reader, and the social reception of the visualization. The examples are all award-winning or highly acclaimed data visualizations.

Beautiful by expressing fine craftsmanship

In the first paragraph of his book *Beautiful Evidence*, Edward Tufte gives an example of what he regards as beautiful evidence, by quoting Federico Ceci, a fellow of Galileo Galilei. When commenting upon Galileo's 38 hand-drawn images of sunspots, Ceci found these drawings a 'delight both by the wonder of the spectacle and the accuracy of expression' (Tufte, 2006, p. 9). Here, accuracy of expression reflects both the images as detailed representations of sunspots, as well as the work the scientist/artist has performed. In addition, the visualization also made a natural phenomenon apparent for the human eye to perceive.

A data visualization can be valued as beautiful by the way it represents the work of a highly skilled designer who develops expertise in the means and techniques of turning data into visual imagery in extraordinary ways. Charles Joseph Minard was such a designer, with his 'Figurative Map' of the successive losses of the French army in the Russian Campaign 1812-1813 (Figure 16.1).

Hailed by Tufte in several of his books (including a chapter-long analysis of Minard's analytical design in *Beautiful Evidence* (2006)), and often mentioned by others (Cook, 2013; Friendly, 2002; Tableau, 2018) for the way the map

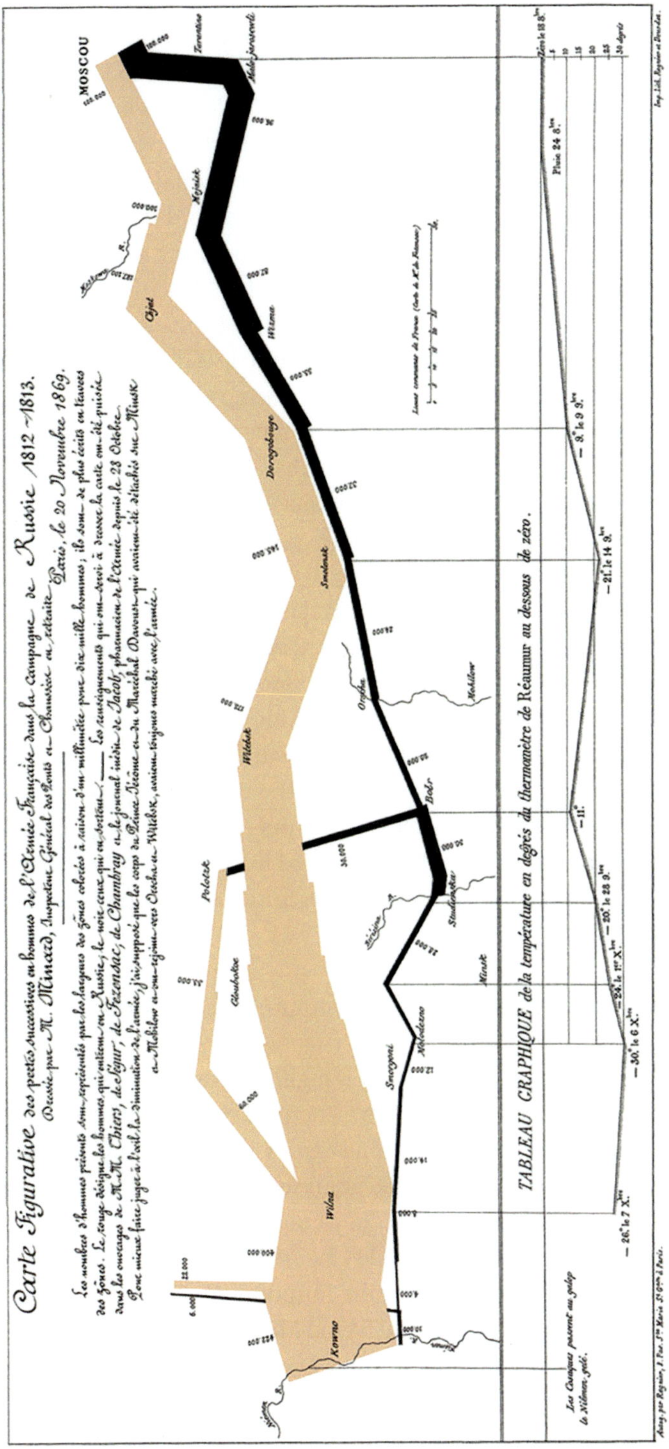

Figure 16.1. Charles Joseph Minard's map of Napoleon's Russian Campaign 1812-1813. C. J. Minard (1869). (https://en.wikipedia.org/wiki/Charles_Joseph_Minard#/media/File:Minard.png). Public domain.

communicates the tragedy of Napoleon's attempt at invading Russia so clearly, Minard's work belongs to a canon of data visualization and infographics works. The dominant data are the numbers of soldiers, visualized as a band following the path eastward from Paris towards Moscow, with the number of soldiers indicated by its width. As an ideational representation of the historical facts, the band becomes narrower and narrower in its direction east, before turning even narrower when pointing west, following the army's withdrawal out of Russia, after its defeat in Moscow. It is precise, and in some ways quite simple, but also a visually stunning presentation of how an army of over 420,000 men ended up with less than 10,000 during the campaign. A narrative is formed by the narrowing band, contextualized with the geographical representations of a map, as well as information about temperature, dates, and the distances the soldiers walked. Minard's precision in his analytical design and condensed compositional meaning is the reason for the map's status as a highly valued example of statistical graphics.

Being beautiful by expressing fine craftsperson-ship can be seen as a way of giving status to someone highly skilled within the praxis of visualizing data graphically. This can, in turn, be expressed as a distinction of a particular designer or studio: expecting something out of the ordinary from designers and studios that has already been awarded and hailed for their previous work. By this, the personal touch or the creative sign of a studio will influence as well as affirm what is to be understood as the politics of beauty within the field of data visualizations and infographics. The category 'Studio of the Year' in the Kantar Information is Beautiful Awards contest, is an example of this.

Beautiful by presenting complexity in a comprehensible form

Accuracy of expression is something that is highly cherished when complex matters or calculations are visualized in a comprehensible representation, or when huge datasets are given a form and expression communicating with clarity and accessibility. With digitally analysed datasets, data visualizations are described as beautiful when they manage to express complex or massive amounts of data into very detailed, but still elegant and clean visual expressions. One example of this can be found in Fernanda Viégas and Martin Wattenberg's highly acclaimed 'Wind Map' from 2012 (Figure 16.2).

The visualization is discussed in Cairo's *The Truthful Art* in relation to creativity and innovation in data visualization and infographics (2016, pp. 351-352), and in Kirk's *Data Visualisation* as an example of elegance and

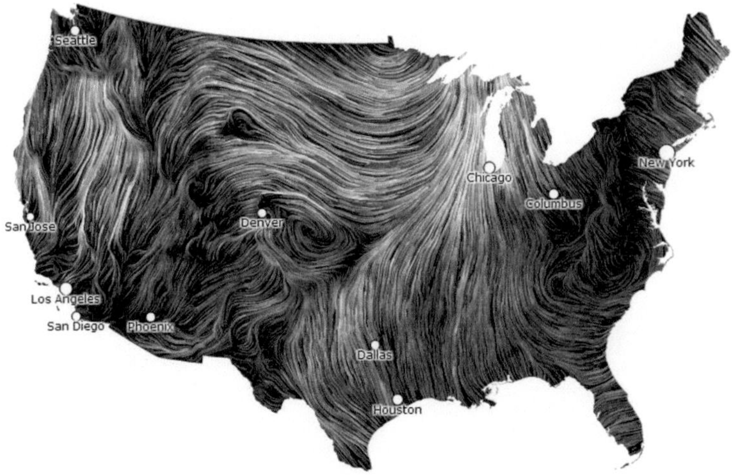

Figure 16.2. From 'Wind Map' by F. Viégas and M. Wattenberg (2012). (http://hint.fm/wind). Copyright 2012 by F. Viégas and M. Wattenberg. Reprinted with permission.

beauty when it comes to the designers' choice of concentrated form and colour (2016, p. 289). Where Minard's expertise could be seen in the way he managed to present the multivariate analysis as a tragic narrative of the French invasion unfolding over time and in space in one single, multimodal expression, Viégas and Wattenberg use a single framing (the outline of USA's geographical border) as a means of presenting a real-time visualization of an ever-changing phenomenon: wind. Wind streams of course do not stop at the border, but the way the borderline effectively isolates the geographical area coloured in a monochromatic palette of steel grey, the phenomenon (animated in real time based on data derived from a digital forecast database) is visualized with great clarity—and beauty.

Other examples of turning complexity into a single expression which makes us aware of aspects that we cannot see without the aid of the visualization include: Jaz Parkinson's 'Color Signatures' (2013, see http://jazparkinson.tumblr.com/), presenting quantitative analysis of colours found in famous novels in poster-like images in which the colours are distributed as lines of various width, and Charlie Clark's *The Colors of Motion* (2014, see https://thecolorsofmotion.com/), visualizing movies as a multi-lined, interactive image in which each frame of a film is displayed as a line in the frame's average colour. Each line of the image is clickable, revealing the frame the colour is calculated from (Cook, 2015, pp. 140-141).

Figure 16.3. 'Blade Runner' from the project 'The Color of Motion' by C. Clark (2014). (https://thecolorsofmotion.com/). Copyright 2014 by C. Clark. Reprinted with permission.

While these are visualizing single works, another of Viégas and Wattenberg's visualizations, Flickr Flow, presents a single phenomenon in its variations over time at one single place: Boston Common Park and its variations in colours over the year (see http://hint.fm/projects/flickr/). All these cases exemplify a visual method of capturing phenomena and making them perceptible in a unified visual expression, accessible and understandable with only a minimum of verbal explanation or support.

As Kennedy, Hill, Aiello, and Allen argue, one highly operative convention within the field of data visualization is the use of a clean layout, which can rhetorically obscure the complexity of the data and the matters being represented (2016, p. 729). The cases presented here exemplify that the same convention is applied when linearly distributed information is being turned into data that is visualized in simultaneous expressions.

Beautiful by letting complexity keep its complex character

Data visualizations can also be cherished for presenting a complex phenomenon or matter in all its complexity. *On Broadway,* a large-scale data compilation and visualization project designed by Daniel Goddemeyer, Moritz Stefaner, Dominikus Baur, and Lev Manovich, exemplifies this (see http://on-broadway.nyc). *On Broadway* is both a portrayal of the street in New York, based on demographic info (household income), transport data (taxi drop-offs and pick-ups), and various smartphone-based activities visitors and inhabitants perform along the 13.5-mile long street, such as snapshots on Instagram and Twitter messages. These data have been analysed and turned into a complex composite representing the area, with which the viewer can interact. This visualization project, awarded Silver in the category Kantar Most Beautiful Award in 2015, makes the complexity of the various data compiled in the visualization comprehensible, by applying the geographical location along the street as the main compositional principle. This gives the viewer the chance to orient him- or herself along and between horizontal lines of data gathered from Manhattan, with the street of Broadway giving what Andy Kirk calls a 'continuous narrative' (2016, p. 299). *On Broadway* is also an example of how to design comprehensible *yet* sublime data visualization, a visualization that initially seems to contradict Manovich's own argument about data visualization as anti-sublime (2002). At first encounter, the visualization seems far too detailed for any viewer's perception or cognition to comprehend, expressing instead a myriad of cultural artefacts, human activities generating digital information, combined with statistic variables. However, by using street geography as an organizing principle for combining various strata of data, as well as the possibility of interacting with the installation (such as zooming in on the information), the visualization invites the viewer to engage with the data and the strata. Through the means of an interactive digital medium, it turns something of sublime character into a comprehensible representation of vibrant city life.

Another example of presenting complexity in its complex character is Stefanie Posavec's projects *Writing Without Words*, which visualizes works of literature as 'literary organism', a method of displaying both the author's style of writing as well as the structure of the book down to its very sentences (see http://stefanieposavec.com/writing-without-words/). Her visualization of Walter Benjamin's essay 'Art in the age of mechanical reproduction' has caught the attention of professional data visualization designers (Kirk, 2016, p. 279), as has her work both on Jack Kerouac's *On the Road* (Cairo, 2013, pp. xvi, 243-250; Popova, 2009) and on Charles Darwin's six editions of *On the Origin of Species*, together with Greg McInery under the title '(En)tangled Word Bank' (Cairo, 2013, p. 348). They share a similar tree structure with branching lines representing chapters, ending up in leaf-shaped fades of sentences, colour coded according to various styles of writing. When explaining the beauty of Posavec's work, Cairo identifies it as a combination of the appearance, the typeface, and palette of colours (so does Kirk, 2016, p. 278), as well as in the way the visualization presents the viewer the opportunity of extracting multiple different readings (Cairo, 2016, pp. xvi-xvii)—in other words, its complexity.

Beautiful by activating

'When the eye sees something beautiful, the hand wants to draw it', Ludwig Wittgenstein once stated (Scarry, 2011, p. 4). A dominant way of understanding how beauty matters within the field of data visualization is—as discussed above—as a means to help people engage emotionally or cognitively with the visualization, study it in more detail, or even take action from it. 'At its best, [a data visualization] plants the seed for a moral inclination to do something to nudge [the] world a little bit closer to how it should work', as cultural critic Maria Popova puts it. She continues: 'Therein lies the quality that sets the great and the mediocre apart: information design that merely informs or simply delights fails to move this moral dial' (cited in Cook, 2015, p. x).

In the award-winning interactive visualization project 'Poppy Field' (Figure 16.4), designed by Valentina D'Efilippo, a clean layout is combined with emotionally involving visual metaphors and the possibility for the reader/viewer to interact with the data, displaying numbers of casualties of wars worldwide, in the century spanning from 1914 to 2014. A short digital

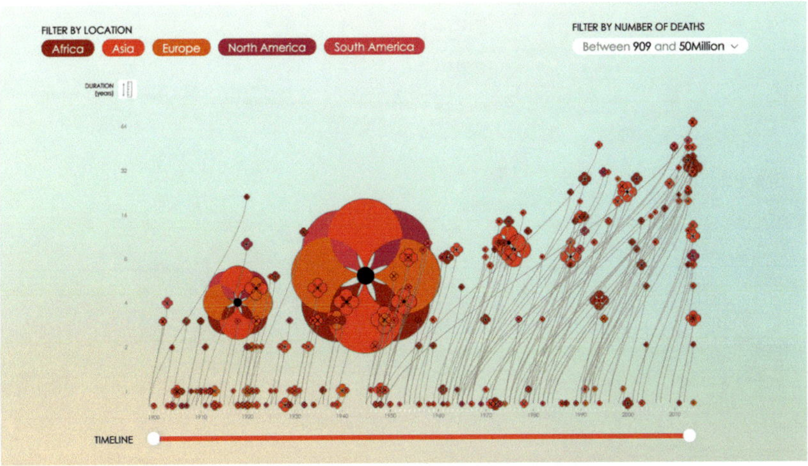

Figure 16.4. 'Poppy Field' by V. D'Efilippo (2014): (http://www.poppyfield.org/). Copyright 2014 by V. D'Efilippo. Reprinted with permission.

narrative introduces the topic for the visualization: The First World War, initially called the Great War, with more than 10 million casualties, was thought of as the war to end all wars. A hundred years, and numerous wars later, people are still dying in wars. This is visualized by a field of poppies, each representing a war and expressing the number of casualties in its size.

The poppy, being a symbol of commemorating military deaths in wars, is a cultural convention that may be unfamiliar to some, and therefore also provocative. The reader can explore the field, zooming in on each flower and extracting information. The interactive aspects of the visualizations invite the reader/viewer to get involved in the digital story, experiencing for themselves the scale and devastating amounts of war casualties.

And finally: Beautiful by being a work of art

Some data visualizations are as much an artistic project as an explanatory one. The artworks *Herald/Harbinger* (Ben Rubin & Jer Thorp, 2018, see https://jerthorp.com/herald-harbinger), and *Bloom* (Ken Goldberg et al., 2013, see http://hint.fm/projects/bloom/), building on natural and seismic data, respectively, are just two examples of this. Even though being a work of art does not make the data visualization beautiful as such, a project combining original ways of thinking about topics relevant for analysis and visualization, innovative ways of applying design principles, and a fresh take on the various possibilities of materializing the visualization is more likely to be recognized or interpreted as a work of art than others.

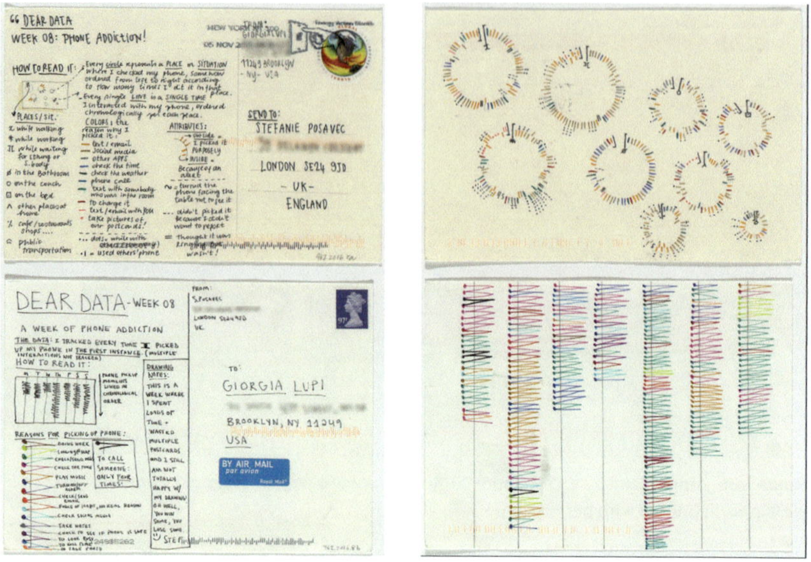

Figure 16.5a and b. Front and backside of Week 8 (Phone Addiction / A Week of Phone Addiction). From *Dear Data by Giorgia Lupi and Stefanie Posavec,* 2014 (http://www.dear-data.com/theproject). Copyright 2014 by G. Lupi and S. Posavec. Printed with permission.

When asked if design was 'an expression of art', designer Charles Eames replied: 'I would rather say that it is an expression of purpose. It may (if it is good enough) later be judged as art' (cited in Cairo, 2016, p. 59). These words seems to apply to the visualization project 'Dear Data' (see Figure 16.5), initiated and executed by Giorgia Lupi and Stefanie Posavec as a way of getting to know each other, almost like pen pals. But only almost: the project was governed by rules making it a systematic investigation as well as a creative endeavour: during 52 weeks starting in September 2014, the two turned everyday life phenomena into something that should be observed, mapped, analysed, and visualized—ending with 104 postcards sent and received, two for each topic. Initiating as an analogue, private, one-to-one hand-drawn visualization experiment questioning how we can learn more about ourselves and other people through the methods of collecting and mapping personal data with pen and paper, it became a project that grew in reputation. Through dissemination in various media (including the project's own website), a book version of the project was released in September 2016 and two months later it was included in MoMA's collection. The project has travelled along various tracks: from private postcard correspondence, net-based dissemination, printed mass media, and then exhibition objects in an art institution's collection.

There are several aspects that make the data visualizations presented in *Dear Data* art, the most important being how the project invites us to reflect upon what data visualization and information design are, and even prior to that: what data are, how data can be a personal matter, and how the practice of data visualization is governed by a set of conventions. Not through questions directed towards us, but by the choices the two designers made: what to observe and collect data about, by which purpose, how to express it, and finally how to communicate it. *Dear Data* became an instant classic, winning the Kantar Information is Beautiful Award special prize 'The Most Beautiful Project' in 2015, being included as an example in Cairo's *The Truthful Art* (2016), Kirk's *Data Visualization* (2016) and Cook's *The Best American Infographics* (2016) among others. However, within the context of this article, the project's most beautiful aspect is its potential for making scholars, critics, and the data visualization community reflect upon the basics of data visualization. Explained in social semiotic terms, the project exemplifies the art of visualizing data: its ideational meanings (what to represent), its interpersonal meaning (who is addressing whom through the visualization's representation), and its compositional meaning (how is it designed and given an expression).

What *do* we talk about when we talk about beautiful data visualizations? A conclusion

When seen in relation to the field of data visualization, Oxford Dictionary's explanation of the term 'beautiful' as something that 'pleases the senses and mind aesthetically' also explains how a well-designed visualization can work on us. We contemplate in expressions or objects we find pleasing, either by their appearances or by displaying something that is already of interest to us. Information and knowledge are beautiful, David McCandless (2012) emphasizes, and the human is a curious creature, seeking information by nature. However, living under conditions overloaded with information of every kind, we need guidance in orienting our attention and focus. Beautiful design can do that. However, even though many of the examples of data visualizations included here are widely acclaimed and discussed because of the ways they present information, and by that constituting something close to an (Anglo-American) canon of beautiful data visualizations, each example can be regarded differently by an individual reader.

We can conclude that a shared opinion within the field of data visualization is that beauty serves a distinct purpose in making us engage in finding out what the visualization communicates. However, if we find a data visualization beautiful, and acclaim it for its aesthetic qualities, it is at the same time germane to encounter the visualization and its context with a critical eye and mind. As Helen Kennedy et al. argue, power is always at work in data visualizations, even though they are designed following conventions that make them seem neutral (2016, p. 716). The same rings true for beautiful visualizations. In that regard, one of the greatest values of projects such as *Dear Data* is the way they make us critically reflect on what data are, and what the meaning of data visualizations can be.

References

Accurat. (2018). *About*. Retrieved September 9, 2018 from https://www.accurat.it/about/

Cairo, A. (2013). *The functional art: An introduction to information graphics and visualization*. Berkeley, CA: New Riders

Cairo, A. (2016). *The truthful art: Data, charts and maps for communication*. Berkeley, CA: New Riders.

Clark, C. (2014). *The color of motion*. Retrieved September 9, 2018 from https://thecolorsofmotion.com/detail/blade-runner-1982

Cook, G. (Ed.) (2016). *The best American infographics 2015*. Boston & New York: Mariner Books.

D'Efilippo, V. (2014). Poppy Field. Retrieved September 9, 2018 from http://www.poppyfield.org/

Eco, U. (2010). *On beauty: A history of a Western idea* (2nd ed.). London: MacLehose Press.

Friendly, M. (2002). Visions and re-visions of Charles Joseph Minard. *Journal of Educational and Behavioral Statistics, 27*(1), 31-51.

Gray, J., Bounegru, L., Milan, S., & Ciuccarelli, P. (2016). Ways of seeing data: Toward a critical literacy for data visualizations as research objects and research devices. In: S. Kubitchsko & A. Kaun (Eds.), *Innovative methods in media and communication research*. (pp. 227-251). Cham: Palgrave Macmillan.

Kennedy, H., Hill, R. L., Aiello, G., & Allen, W. (2016). The work that visualisation conventions do. *Information, Communication and Society, 19*(6), 715-735. https://doi.org/10.1080/1369118X.2016.1153126

Kirk, A. (2016). *Data visualisation: A handbook for data driven design*. London: Sage.

Lempe, J. (2015). Feeling and meaning: A unitary bio-semiotic account. *International handbook of semiotics*. (pp. 589-616). New York & London: Springer.

Lupi, G., & Posavec, S. (2014). *Dear Data*. Retrieved September 9, 2018 from https://www.moma.org/collection/works/216733?artist_id=67122&locale=en&page=1&sov_referrer=artist

Lupi, G., & Posavec, S. (2016). *Dear data: A friendship in 52 weeks of postcards.* New York: Princeton Architectural Press; London: Particular Books.

Manovich, L. (2002). *Data Visualisation as New Abstraction and Anti-Sublime.* Retrieved September 9, 2018 from http://manovich.net/index.php/projects/data-visualisation-as-new-abstraction-and-anti-sublime

Manovich, L. (2014). Visualization methods for media studies. In: C. Vernallis, A. Herzog, & J. Richardson (Eds.), *Oxford handbook of sound and image in digital media.* (pp. 253-278). Oxford: Oxford University Press.

McCandless, D. (2012). *Information is beautiful* (new ed.). London: William Collins.

McCosker, A., & Wilken, R. (2014). Rethinking 'big data' as visual knowledge: the sublime and the diagrammatic in data visualisation. *Visual Studies, 29*(2), 155-164. http://doi.org/10.1080/1472586X.2014.887268

Minard, C. J. (1869). *Napoleon's Russian Campaign 1812-1813.* Retrieved September 9, 2018 from https://en.wikipedia.org/wiki/Charles_Joseph_Minard#/media/File:Minard.png

Popova, M. (2009). *Writing without words: Visualizing Jack Kerouac's On the Road.* Retrieved September 9, 2018 from https://www.brainpickings.org/2009/05/07/writing-without-words/

Sack, W. (2007). *Aesthetics of information visualization.* Retrieved September 9, 2018 from http://virus.meetopia.net/pdf-ps_db/WSack_esthetics-of-informations.pdf

Scarry, E. (2011). *On beauty and being Just.* London: Duckworth.

Steele, J., & Iliinsky, N. P. N. (2010). *Beautiful visualization.* Sebastopol: O'Reilly.

Tableau. (2018). *Data is beautiful: 10 of the best data visualization examples from history to today.* Retrieved September 9, 2018 from https://www.tableau.com/learn/articles/best-beautiful-data-visualization-examples

Tufte, E. R. (2001). *The visual display of quantitative information* (2nd ed.). Cheshire, CT: Graphics Press.

Tufte, E. R. (2006). *Beautiful evidence.* Cheshire, CT: Graphics Press

Viégas, F., & Wattenberg, M. (2012). *Wind map.* Retrieved September 9, 2018 from https://www.moma.org/collection/works/163892

About the author

Sara Brinch is an Associate Professor in visual communication at the Norwegian University of Science and Technology (NTNU). She has been teaching students within the field of film, media studies, and visual culture for the past 20 years, and received her PhD in Media Studies in 2004. Her main area of research is knowledge production and audiovisual media.

17. A multimodal perspective on data visualization

Tuomo Hiippala

Abstract

This chapter discusses the multimodality of data visualizations, that is, how they combine multiple modes of expression, such as written language, photographs, diagrammatic elements, and illustrations, in various printed and digital media. Because the medium in which a data visualization is presented determines the modes of expression available, the chapter shows how different media can be pulled apart for multimodal analysis. The proposed approach is illustrated by analysing static information graphics, non-interactive, and interactive dynamic data visualizations.

Keywords: Multimodality; Media; Data visualization; Information graphics; Interactive media

Introduction

The current interest in multimodality, or how multiple modes of communication cooperate and interact, has opened up new opportunities for theoretical reflection and practical application within several fields. Thus we find in linguistics increasingly widespread statements that language should be seen as just one form of communication among many other, equally important expressive resources; visual communication begins to consider aspects of language; art history takes in the moving image; human-computer interaction design is extended to include tactile and gestural communication rather than just language, and so on. Research on data visualization, however, has not yet fully benefitted from the interdisciplinary perspective that defines most of the current research on multimodality. Previous research has established principles for visualizing information

Engebretsen, M. and H. Kennedy (eds.), *Data Visualization in Society*.
Taylor & Francis Group, 2020
DOI 10.5117/9789463722902_CH17

(Tufte, 1983, 1997), mapped the repertoires of visual expression (Bertin, 1981, 1983, 2001; Engelhardt, 2002) and explored the perception and reception of data visualizations (Holsanova, Holmberg, & Holmqvist, 2009; Ware, 2012; de Haan et al., 2018).

Another stream of research has recently called for attention to how data visualizations may privilege certain perspectives or appear objective (Dick, 2015; Kennedy, Hill, Aiello, & Allen, 2016) and identified inequalities in access to the kinds of literacies needed for making sense of data visualizations (D'Ignazio, 2017). These contributions have provided a much-needed critical perspective to complement the design- and reception-oriented approaches introduced above. At the same time, however, attempts to describe the multimodality of data visualizations have been relatively few (exceptions include Engebretsen & Weber, 2017; Bateman, Wildfeuer, & Hiippala, 2017), although Ledin and Machin (2018) have argued that any form of critical inquiry that targets contemporary forms of communication, such as data visualizations, must now be supported by a robust theory of multimodality. Ledin and Machin's (2018) call for increased support from theory resonates with the oft-cited quote from Halliday, who observed that:

> A discourse analysis that is not based on grammar is not an analysis at all, but simply a running commentary on a text. (1994, p. xvi)

Although the notion of 'grammar' has been suggested as problematic for multimodal analysis, because it relies on strong assumptions about *form*, a property that visual modes of expression do not necessarily respect (Bateman, 2014a, 168), the need for a theory of multimodality that can reveal structural regularities and explain the choices made within specific modes of expression remains crucial for making systematic observations about multimodal discourse. Multimodal analyses are not only highly valuable on their own right for increasing our knowledge about multimodality as a phenomenon, but can also support critical perspectives on data visualizations by placing these analyses on a robust, multimodally-informed foundation.

A multimodal approach focuses on how meaningful *combinations* of written language, illustrations, photographs, diagrams, maps, layout, and other modes of expression emerge in data visualizations. How such combinations are supported multimodally and become interpretable across a wide range of different media remains a wide and open research question. For data visualization, answering these questions requires a theory capable of engaging with all forms of media in which data visualizations appear.

In this chapter, I aim to show how the framework we proposed in Bateman et al. (2017) can be used as a foundation for analysing the multimodality of data visualizations. The proposed approach starts with a detailed account of the media in which data visualizations are presented, the modes of expression provided by the medium, and what kinds of engagement the combination of media and modes demands from those interacting with them. By doing so, I attempt to show how a comprehensive theory of multimodality can be used to identify the detail that is needed in critical inquiry (cf. Ledin & Machin, 2018).

Media and their canvases

Data visualizations are used in different kinds of communicative situations across a wide range of media. They are presented on websites, printed on newspapers, shared on social media feeds, and projected on public displays, to name just a few examples (Lima, de Castro Andrade, Monat, & Spinillo, 2014; Bounegru, Venturini, Gray, & Jacomy, 2017; Amit-Danhi & Shifman, 2018). For this reason, identifying the medium in which the data visualization is presented is a natural first step for their analysis, which has far-reaching consequences for a description of their multimodality. However, if media are characterized purely on the basis of their physical or technological characteristics, for instance, by setting up dichotomies such as 'print' or 'digital', we risk oversimplifying the medium in which data visualizations appear (see e.g. Elleström, 2010; Bateman, 2017).

In order to break down the abstract concept of 'media' (or medium) and prepare it for multimodal analysis, Bateman et al. (2017, pp. 86-87) adopt the notion of a *canvas* to describe any potential carrier of semiotic modes that may be taken up for interpretation. The notion of a canvas places very few demands on the underlying materiality—almost anything capable of carrying intentionally-produced signs will do. Thus a note scribbled on a napkin is just as interpretable as a daily menu written on a chalkboard, because the presence of semiotic modes signals that the canvas in question is offered up for interpretation. Multimodality research conceptualizes semiotic modes as socially-shaped resources for making and exchanging meanings, and just like semiotic modes, the canvases provided by a medium come to be by virtue of being embedded within a community of users (Bateman, 2011; Kress, 2014).

Bateman et al. (2017) propose that physical or technical media may be characterized as recognizable 'bundles' of canvases defined by patterns of production and consumption. To exemplify, the medium of news broadcast often allocates parts of the screen to news tickers, stock and weather information, and other overlays, in addition to the audiovisual broadcast (Tan, 2011). These parts of the medium differ from each other in terms of their characteristics: the notion of a canvas allows picking them out for description. From the consumer's perspective, identifying these canvases not only generates expectations about what kinds of communicative situations may take place on them, but also anticipates the semiotic modes most likely to be encountered in a particular communicative situation. In order to characterize the properties of a canvas, Bateman et al. (2017, p. 104) propose accounting for several material properties: space (2D or 3D), temporality (static or dynamic), transience (permanent or fleeting), and how the user is positioned with respect to the canvas (distanced observer or immersed participant).

Because some of these affordances are inherited from the materiality of the medium, this is also where differences begin to emerge between canvases. Nevertheless, all canvases that carry data visualizations must have an inherent *spatial* (2D) extent, which provides access to expressive resources provided by layout (Waller, 2012). First differences emerge within the *temporal* extent: spatial canvases without temporal extent are considered *static*, whereas their counterparts with a temporal extent can be characterized as *dynamic*. Dynamic canvases may be either *immutable* or *mutable*, which also determines their degree of interactivity (Weber, 2017, pp. 246-247). In most cases, these canvases are also *designed*, which in this context implies that the content (or underlying data) cannot be altered by the user.

For the multimodal analyst, being able to pick out canvases and their properties for closer analysis is crucial for making sense of how the underlying medium is used to support a data visualization. This is necessary for establishing differences between data visualizations presented on their own dedicated websites (see e.g. Zambrano & Engelhardt, 2008; Bounegru et al., 2017) and those embedded in social media feeds, or for capturing the differences between information graphics in printed newspapers and their counterparts in digital media (see e.g. Lima et al., 2014). In both cases, the theoretical apparatus must be capable of taking on the complexity of the communicative situation in which a data visualization is mobilized, as opposed to hiding it from view. This is why the following section introduces an additional perspective, which attempts to capture the kinds of interaction demanded by canvases, shifting the attention from the production of canvases to their consumption.

Exploration and composition

Dynamic data visualizations and static information graphics have been suggested to demand different kinds of engagement from their viewers (Lima et al., 2014; Weber, 2017). Bateman et al. (2017, p. 105) characterize this engagement in terms of *ergodic work*, redefining the concept originally developed by Aarseth (1997) for multimodality research. The concept of ergodic work seeks to characterize a communicative situation in terms of the effort required from those participating in the situation. More specifically, the concept emphasizes how participants co-construct the communicative situation they are interacting with/in and to what extent the participants may manipulate the situation (cf. Bucher & Niemann, 2012). Because communicative situations can take place on canvases embedded within one another, different canvases may demand different forms of ergodic work.

As a form of ergodic work, engaging with interactive data visualizations may be broadly characterized as *exploration* (Bateman et al., 2017, p. 108). Exploration involves substantial ergodic work on behalf of the viewer, in the form of interacting with the visualization, for instance, by choosing which parts of the underlying data are rendered by manipulating the data visualization via an interface. The extent to which the visualization may be manipulated is naturally determined by its degree of interactivity (Weber, 2017, pp. 246-247). What remains beyond the user's reach, however, is the underlying data. In other words, the presentation of the data may be altered, but not the data themselves. For this reason, the communicative situation of engaging with an interactive data visualization may be characterized as *ergodic* yet *immutable* (Bateman et al., 2017, p. 108).

Another form of ergodic work required for interpreting data visualizations is that of *composition*, which requires the viewer to determine how the information presented on a 2D canvas is to be put together. The ergodic work of composition involves selective visual perception and interpretation, which may be revealed using methods such as eye-tracking, as Holsanova et al. (2009) and de Haan et al. (2018) have shown for data visualizations in printed and digital newspapers (Bateman et al., 2017, pp. 107-108). It should be noted, however, that as forms of ergodic work, composition and exploration are not mutually exclusive. In fact, exploring a data visualization must necessarily involve ergodic work in the form of composition, as interpreting an interactive data visualization involves making sense of information rendered on the screen at a given point in time. These embedded forms of ergodic work emerge naturally from canvases embedded within each other (p. 109).

To summarize, the concept of ergodic work draws attention to the different forms of engagement demanded by data visualizations. As the following analyses will show, differences in ergodic work may be traced back to the properties of the physical/technical medium in which the data visualization is realized.

Three example analyses

In this section, I demonstrate how the procedure set out in Bateman et al. (2017, p. 228) can be used to identify canvases in three different data visualizations, in order to lay a foundation for their multimodal analysis. All three examples discussed below are on the topic of sustainability, such as biological conservation, global warming, and marine pollution. The examples feature contributions from various semiotic modes in the form of written language, photography, diagrams, and graphic elements. For current purposes, I do not seek to pursue a detailed analysis of their structure and functions, but characterize them rather broadly. The same applies to any discourse relations that hold between them. In contrast, by focusing on the canvases I seek to provide the means for increased analytical control, laying a foundation for more detailed analyses.

Static information graphics

Figure 17.1 shows an information graphic produced by Graphic News, a London-based agency that produces news graphics for media outlets around the world. Whilst not a data visualization, this infographic contains visualized data alongside other elements, and as such the framework under discussion applies here. The information graphic combines several modes of expression—written language, photography, maps, two-dimensional illustrations, and diagrammatic elements—which are organized on several overlapping canvases. The wealth of semiotic modes present exemplifies why information graphics may be conceptualized as a *composite semiotic mode*, which provides the 'glue' necessary for joining together contributions from individual semiotic modes (Bateman et al., 2017, p. 289). This 'glue' may be traced back to a specific form of *discourse semantics* that supports the interpretation of such composite units, which uses the layout space to set up potential relations between elements that make up the composite unit (p. 264).

For this reason, interpreting information graphics requires ergodic work in the form of composition. The viewer must identify the semiotic modes,

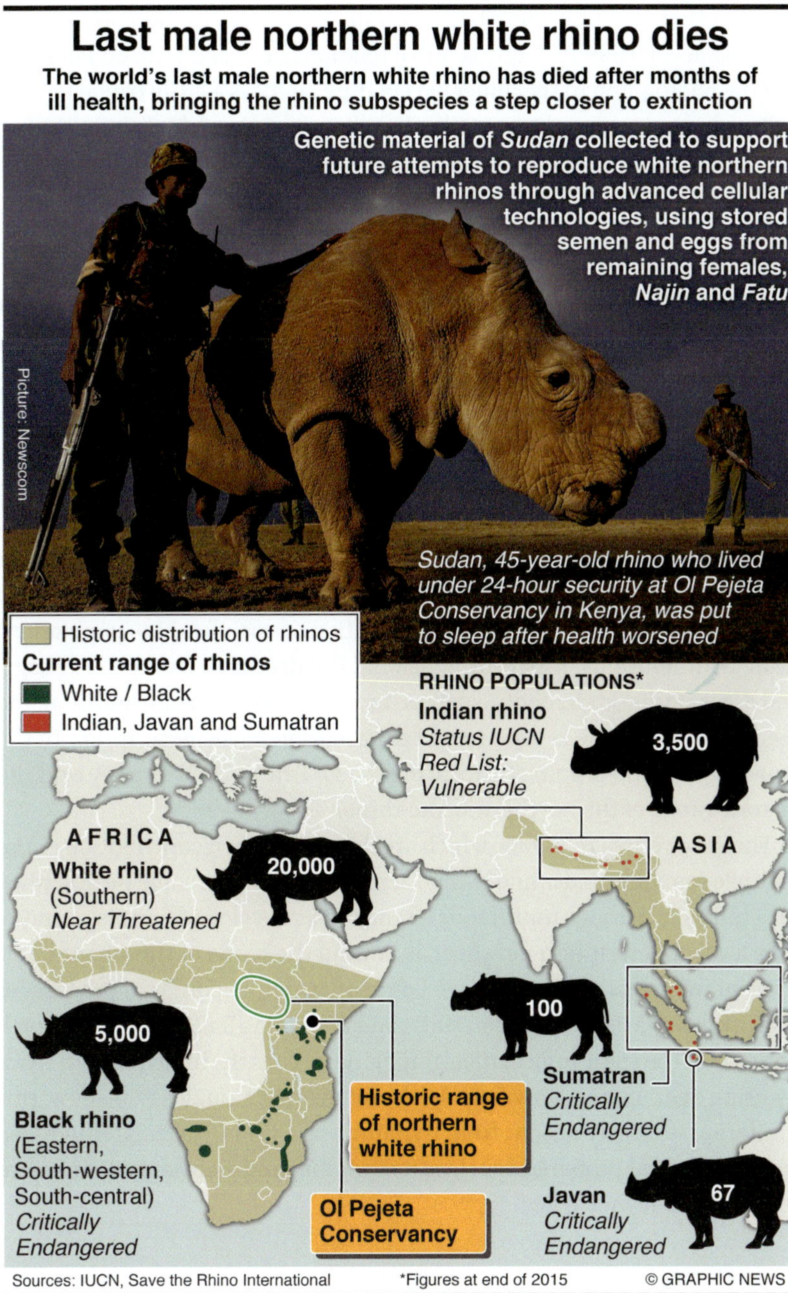

Figure 17.1. A static information graphic reporting on the death of the last male northern white rhino. Produced by *Graphic News*. Copyright 2018 by Graphic News. Printed with permission.

consider their specific contributions, and relate them to each other in the layout space. Resolving the discourse relations between written language and the photograph in the upper part of the graphic is a fairly trivial task, as written language is used to provide a headline, background information, and to identify the rhino in question. Making sense of the lower part, by contrast, may prove more challenging due to discourse relations that hold between multiple semiotic modes presented on several overlapping canvases, which are a common feature of information graphics (Bateman et al., 2017, p. 291).

The lower part of the visualization features a map that shows the current and historic geographical distributions of rhino populations. MacEachren (2004, p. 317) notes that maps use overlays to present complex phenomena in space and time, but required processes of attribution—assigning meaning to the overlays—are often dependent on other modes of expression. This process of attribution is exemplified in Figure 17.1 by the accompanying legend, which uses coloured graphic elements and written language to group together different species of rhino and establishes their current and historic spatial distributions. Laid out on top of the map is another canvas, which provides additional information on specific rhino populations using combinations of two-dimensional illustrations, written language, and diagrammatic elements, such as lines and containers. In addition, the diagrammatic mode is used to add information to the description in the upper part of the graphic by locating the Ol Pejeta Conservancy and the historic range of the northern white rhino.

These discourse relations, which are drawn between contributions from multiple semiotic modes and extend across the canvases, could be described in detail using various multimodal frameworks developed for this purpose. This level of description, however, is beyond the scope of this chapter (for a recent overview of this area, see Bateman, 2014b). In order to prepare for drawing comparisons between static and dynamic 2D canvases, it is worth noting how the static information graphic negotiates the limited layout space by using overlapping canvases. As the following examples will show, this limitation is largely absent from dynamic data visualizations, which can exploit material properties such as temporality and transience to increase the available layout space.

Non-interactive dynamic data visualizations

Figure 17.2 features four screenshots captured from a dynamic data visualization, which illustrates temperature anomalies by country between 1900 and 2016. Created by Antti Lipponen, a researcher at the Finnish Meteorological

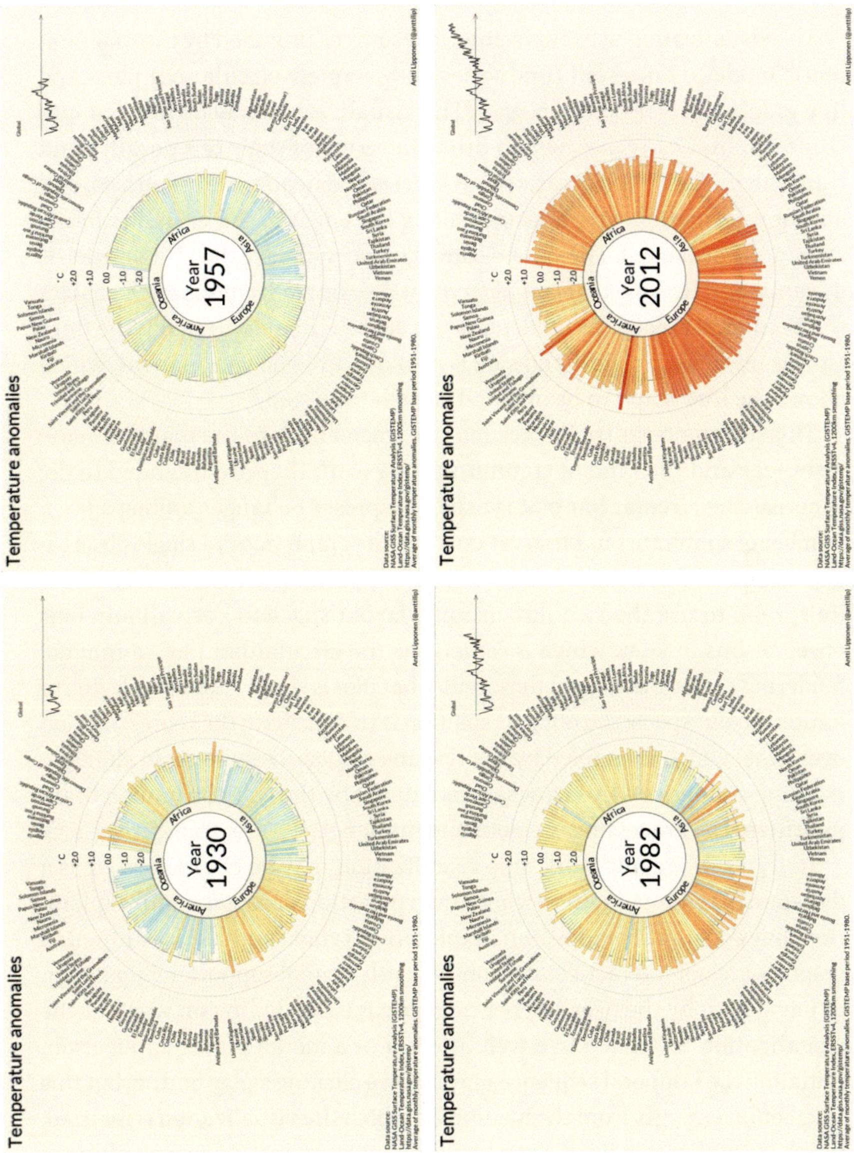

Figure 17.2. Four screenshots from the non-interactive dynamic data visualization 'Temperature anomalies arranged by country 1900–2016' showing temperature anomalies arranged by country 1900–2016. By Antti Lipponen (CC BY 2.0).

Institute, this 2D visualization is centred around a circular structure with the data—represented by coloured bars—and their respective labels laid out on concentric circles. The labels (countries) and bars (observations) are organized along the concentric circles according to geographical location. In addition, the top right-hand side of the visualization features a line graph which shows the global average temperature for each year, summarizing the individual observations presented using the circular bar plot.

The visualization uses two semiotic resources provided by the diagrammatic mode to represent time series data, namely circular bar plots and line graphs (Tversky, 2017, p. 350). The visualization may be divided into three distinct canvases, which differ in terms of their temporality and transience. The first canvas, which carries the header, data sources, and author information, all positioned along the edges of the visualization, is static and permanent. The second canvas in the middle of the visualization is dynamic and fleeting in terms of *temporality* and *transience*, which enables the circular bar plot to be rendered again at each time step. Finally, the third canvas on the top right-hand side is also dynamic but permanent, which allows the line graph to be updated at each time step.

This difference in transience may be traced back to the diagrammatic resources and the kind of communicative work they are intended to do. Whereas the circular bar plot is used to represent changes among a large number of simultaneous observations, the line graph tracks a single observation over time to summarize the trend. The permanent canvas allows the line graph to use the two-dimensional layout space to keep all previous observations in view, which is something the circular bar plot cannot do: rendering each time step on the circular bar plot is simply not feasible due to limited layout space; the obvious solution is to distribute the representation over time, which is enabled by the fleeting canvas. Despite rapid changes, tracking changes on this canvas is facilitated by the way the human brain prioritizes the processing of colour and line length (Ware, 2012, pp. 154-155).

In terms of ergodic work, this visualization may be characterized as a dynamic data visualization, whose interpretation does not entail exploration, but constant composition. Exploration is not required, as animated graphics are not interactive, and consequently cannot support manipulation or navigation by the user (Weber, 2017, p. 247). Depending on whether the visualization is opened in a web browser or a media player application, initiating the temporal sequence may involve clicking a play button, but this interaction emerges from the medium in which the visualization is realized, not the visualization itself. Such low-level interactions are commonly used for imposing control over embedded dynamic canvases in digital media, and should not be confused with interactivity inherent to the data visualization (cf. Hiippala, 2017, pp. 424-425), which are taken up for discussion below.

Interactive dynamic data visualizations

The final example in Figure 17.3, *The Seas of Plastic*, is an interactive dynamic data visualization created by Dumpark, a design agency based in Wellington,

New Zealand, which visualizes how plastic pollution accumulates into large circulating gyres in oceans. To do so, the visualization provides two distinct views—designated as map and source views, respectively—which are both presented on their own canvases. These canvases may be rendered for viewing via the interface in the top-right corner of the embedding canvas, which remains constantly visible to the user.

In addition to the interface for exploring the visualization, additional levels of interactivity are introduced on the two canvases. The map view, shown in the upper part of Figure 17.3, features a 2D representation of a globe that may be freely rotated by clicking and dragging. A legend, positioned in the lower left-hand corner, is used to attribute meaning to the overlays rendered on the globe, which bears close resemblance to the discourse relations in the information graphic in Figure 17.3. The user may also select a specific gyre on the right-hand side interface, which rotates the globe into a position that shows the selected gyre. Selecting a source or a gyre in the source view highlights coloured bands that show the source or destination of plastic pollution. Multiple sources or gyres may also be selected simultaneously for drawing comparisons between them.

Together, multiple user interfaces on several canvases lend this data visualization a high degree of interactivity. According to Weber (2017, p. 247), this entails that the users are allowed to explore the visualization, interact with the data, and influence its representation, which corresponds closely with what Bateman et al. (2017, p. 108) characterize as ergodic work in the form of *exploration*. At the same time, the contributions from various semiotic modes and the discourse relations that hold between them closely resemble those found in static information graphics and non-interactive dynamic data visualizations: written language provides additional information on graphics, legends accompany cartographic representations, colour creates distinctions, etc., which must be decomposed and put back together for interpretation. In other words, making sense of the interactive data visualization requires ergodic work both in the form of exploration and composition, a feature which separates it from static information graphics and non-dynamic data visualizations.

The need for exhaustive analyses

Engebretsen and Weber (2017, p. 289) have recently argued that a multimodal account of data visualization must move beyond identifying which semiotic modes are used and for what purpose, and move towards a closer analysis of

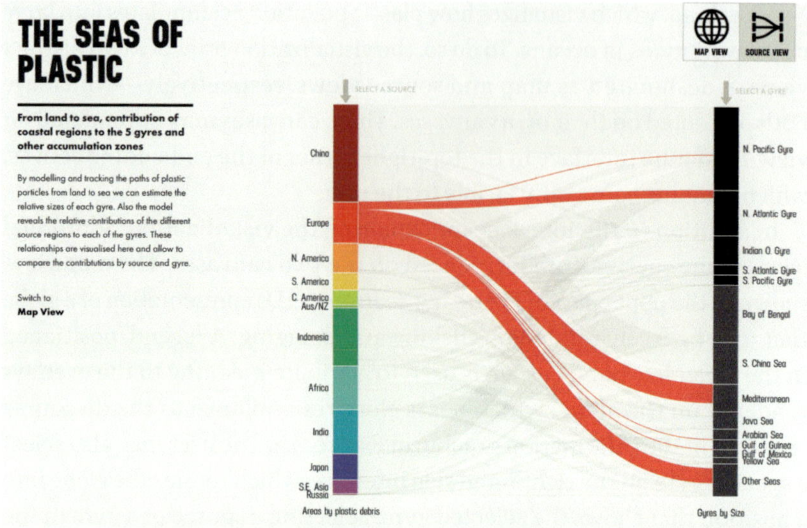

Figure 17.3. The Seas of Plastic, an interactive dynamic data visualization. Produced by Dumpark. Copyright 2018 by Dumpark. Printed with permission.

production and consumption, in order to pinpoint how meaning potential emerges. This argument is very similar to what Bateman et al. (2017, p. 221-222) propose for multimodality research in general, underlining the need to pursue analyses in an exhaustive manner, which Ledin and Machin (2018) identify as a key component of any critical inquiry as well. Such analyses should involve (1) accounting for the communicative situations involved in engaging with a data visualization, (2) identifying the canvases on which these communicative situations take place, (3) identifying the semiotic modes mobilized on these canvases, and (4) the genres that shape the semiotic modes. This does not, however, necessarily entail full-blown analyses at each stage, but can also serve as a tool for limiting the scope of investigation.

That being said, identifying the canvases and describing their properties can be proposed as a first step towards a more comprehensive analysis of production processes. Canvases inherit affordances from the materiality of the medium that carries them, and they may be manipulated in different ways for different communicative purposes. What motivates the producers to manipulate these canvases and their material affordances can be revealed using ethnographic methods (Hiippala, 2016; Zha, 2017). For the examples discussed above, the properties of the canvases are visualized in Figure 17.4, in which they are marked as being either present (+) or absent (-).

The static information graphic (1) in Figure 17.4 illustrates how certain material properties of the medium are passed down to all canvases. The

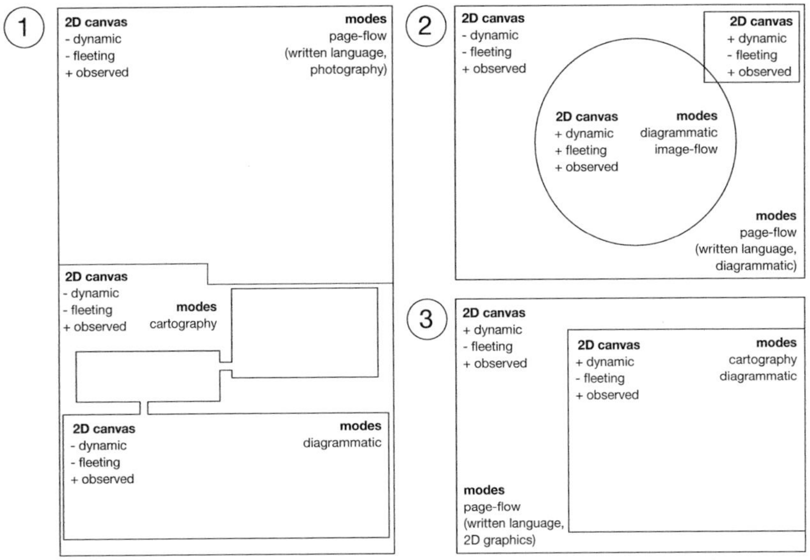

Figure 17.4 The decomposition of (1) static information graphics, (2) non-interactive data visualizations, and (3) interactive data visualizations into canvases. Illustration by T. Hiippala.

consequences are clear: a 2D medium without a material prerequisite for a temporal dimension can never be used to instantiate a dynamic canvas. This consequently rules out any semiotic modes that require this property. In contrast, the digital medium in which data visualizations (2) and (3) are realized affords controlling temporality and transience of any canvas, which also provides the foundation for interactivity. To summarize, the canvases and their properties determine which semiotic modes may appear on them, and thus their description should precede any in-depth description of the semiotic modes used and their contribution to the visualization at hand.

One contribution that emerges from mapping the canvases at play is the role of layout. Bateman (2008, p. 157) proposes the term *page-flow* for describing the semiotic mode responsible for setting up discourse relations in the layout space, which hold between contributions from distinct semiotic modes. The discourse semantics of page-flow are exemplified, for instance, by the relations that hold between the diagrammatic overlay and the underlying map in the static information graphic in Figure 17.1. The role of page-flow in organizing the spatial structure of 2D canvases is highlighted by indicating page-flow as the active semiotic mode in all visualizations in Figure 17.4. However, to what extent the discourse semantics of page-flow differ between data visualizations and entire page-based documents remains an open question for empirical research.

To sum up, the widespread use of data visualizations makes their multimodal analysis challenging, given the wealth of communicative situations in which they appear. However, only a sufficiently developed theoretical apparatus, which is able to impose control on the communicative situation in which the visualizations appear, can advance our understanding of how the visualizations work. This will undoubtedly require an extensive programme of empirical research, which must involve specialists from various fields, given the need for exhaustive analyses that cover the whole range of phenomena from production to consumption (cf. Waller, 2012; Zha, 2017).

Conclusion

In this chapter, I have attempted to highlight how much state-of-the-art theories of multimodality can reveal about data visualizations even before venturing into in-depth descriptions of semiotic modes and the discourse relations that hold between their individual contributions to the data visualization under analysis. By drawing on the notion of canvas, recently introduced in Bateman et al. (2017), I have also sought to establish a foundation for further analysis by attending closely to the underlying properties of the medium, investigating their contribution to meaning-making, as called for by Ledin and Machin (2018). Such multimodally-informed insights could provide a basis for critical insights into the use of data visualizations in society, allowing them to be strongly rooted in well-informed analyses of multimodal discourse. Supporting these critical perspectives will also require continuous refinement of multimodal theories that are applicable to data visualizations, and given the rapid spread of data visualizations into all areas of society, these theories must undoubtedly be founded on empirical research.

Acknowledgements

I would like to thank Fiona Roberts of Graphic News, Antti Lipponen of the Finnish Meteorological Institute, and Timo Franz of Dumpark for giving the permission to reproduce the data visualizations discussed in this chapter.

References

Aarseth, E. J. (1997). *Cybertext: Perspectives on ergodic literature.* Baltimore, MD: Johns Hopkins University Press.

Amit-Danhi, E. R., & Shifman, L. (2018). Digital political infographics: A rhetorical palette of an emergent genre. *New Media & Society, 20*(10), 3540-3559. https://doi.org/10.1177/1461444817750565

Bateman, J. A. (2008). *Multimodality and Genre: A foundation for the systematic analysis of multimodal documents.* London: Palgrave Macmillan.

Bateman, J. A. (2011). The decomposability of semiotic modes. In: K. L. O'Halloran & B. A. Smith (Eds.), *Multimodal studies: Multiple approaches and domains.* (pp. 17-38). London: Routledge.

Bateman, J. A. (2014a). Multimodal coherence research and its applications. In: H. Gruber & G. Redeker (Eds.), *The pragmatics of discourse coherence: Theories and applications.* (pp. 145-177). Amsterdam: Benjamins.

Bateman, J. A. (2014b). *Text and Image: A critical introduction to the visual/verbal divide.* London and New York: Routledge.

Bateman, J. A. (2017). Triangulating transmediality: A multimodal semiotic framework relating media, modes and genres. *Discourse, Context & Media, 20*, 160-174.

Bateman, J., Wildfeuer, J., & Hiippala, T. (2017). *Multimodality: Foundations, research and analysis—a problem-oriented introduction.* Berlin: De Gruyter Mouton.

Bertin, J. (1981). *Graphics and graphic information processing.* Berlin: Walter de Gruyter.

Bertin, J. (1983). *Semiology of graphics: Diagrams, networks, maps* (W. J. Berg, Trans.). Madison, WI: University of Wisconsin Press.

Bertin, J. (2001). Matrix theory of graphics. *Information Design Journal, 10*(1), 5-19.

Bounegru, L., Venturini, T., Gray, J., & Jacomy, M. (2017). Narrating networks: Exploring the affordances of networks as storytelling devices in journalism. *Digital Journalism, 5*(6), 699-730. https://doi.org/10.1080/21670811.2016.1186497

Bucher, H.-J., & Niemann, P. (2012). Visualizing science: The reception of PowerPoint presentations. *Visual Communication, 11*(3), 283-306. https://doi.org/10.1177/1470357212446409

de Haan, Y., Kruikemeier, S., Lecheler, S., Smit, G., & van der Nat, R. (2018). When does an infographic say more than a thousand words? Audience evaluations of news visualizations. *Journalism Studies, 19*(9), 1293-1312. https://doi.org/10.1080/1461670X.2016.1267592

Dick, M. (2015). Just fancy that: An analysis of infographic propaganda in The Daily Express, 1956–1959. *Journalism Studies, 16*(2), 152-174. https://doi.org/10.1080/1461670X.2013.872415

D'Ignazio, C. (2017). Creative data literacy: Bridging the gap between the data-haves and the data-have nots. *Information Design Journal, 23*(1), 6-18. https://doi.org/10.1075/idj.23.1.03dig

Elleström, L. (2010). The modalities of media: A model for understanding intermedial relations. In: L. Elleström (Ed.), *Media borders, multimodality and intermediality.* (pp. 11-48). London: Palgrave.

Engebretsen, M., & Weber, W. (2017). Graphic modes: The visual representation of data. In: C. Cotter & D. Perrin (Eds.), *The Routledge handbook of language and media.* (pp. 277-295). New York and London: Routledge.

Engelhardt, Y. (2002). *The language of graphics: A framework for the analysis of syntax and meaning in maps, charts and diagrams* (Doctoral thesis). Institute for Logic, Language and Computation, University of Amsterdam.

Halliday, M. A. K. (1994). *Introduction to functional grammar* (2nd ed.). London: Arnold.

Hiippala, T. (2016). Individual and collaborative semiotic work in document design. *Hermes: Journal of Language and Communication in Business, 55*, 45-59. https://doi.org/10.7146/hjlcb.v0i55.24288

Hiippala, T. (2017). The multimodality of digital longform journalism. *Digital Journalism, 5*(4), 420-442. https://doi.org/10.1080/21670811.2016.1169197

Holsanova, J., Holmberg, N., & Holmqvist, K. (2009). Reading information graphics: The role of spatial contiguity and dual attentional guidance. *Applied Cognitive Psychology, 23*, 1215-1226. https://doi.org/10.1002/acp.1525

Kennedy, H., Hill, R. L., Aiello, G., & Allen, W. (2016). The work that visualisation conventions do. *Information, Communication and Society, 19*(6), 715-735. https://doi.org/10.1080/1369118X.2016.1153126

Kress, G. (2014). What is mode? In: C. Jewitt, C. (Ed.), *The Routledge handbook of multimodal analysis* (2nd ed.). (pp. 60-75). London: Routledge.

Ledin, P., & Machin, D. (2018). Doing critical discourse studies with multimodality: From metafunctions to materiality. *Critical Discourse Studies.* http://doi.org/10.1080/17405904.2018.1468789

Lima, R. C., de Castro Andrade, R., Monat, A. S., & Spinillo, C. G. (2014). The relation between online and print information graphics for newspapers. In: A. Marcus (Ed.), *Design, user experience, and usability: User experience design for everyday life applications and services.* (pp. 184-194). New York: Springer.

MacEachren, A. M. (2004). *How maps work: Representation, visualization and design.* New York: Guilford.

Tan, S. (2011). Facts, opinions, and media spectacle: Exploring representations of business news on the internet. *Discourse & Communication, 5*(2), 169-194. https://doi.org/10.1177/1750481311399511

Tufte, E. R. (1983). *The visual display of quantitative information.* Chesire, CT: Graphics Press.

Tufte, E. R. (1997). *Visual explanations: Images and quantities, evidence and narrative.* Cheshire, CT: Graphics Press.

Tversky, B. (2017). Diagrams: Cognitive foundations for design. In: A. Black, P. Luna, O. Lund, & S. Walker (Eds.), *Information design: Research and practice.* (pp. 349-360). London: Routledge.

Waller, R. (2012). Graphic literacies for a digital age: The survival of layout. *The Information Society, 28*(4), 236-252. https://doi.org/10.1080/01972243.2012.689609

Ware, C. (2012). *Information visualization: Perception for design* (3rd ed.). Amsterdam: Elsevier.

Weber, W. (2017). Interactive information graphics: A framework for classifying a visual genre. In: A. Black, P. Luna, O. Lund, & S. Walker (Eds.), *Information design: Research and practice.* (pp. 243-256. London: Routledge.

Zambrano, R. N., & Engelhardt, Y. (2008). Diagrams for the masses: Raising public awareness—from Neurath to Gapminder and Google Earth. In: G. Stapleton, J. Howse, & J. Lee (Eds.) *Diagrams 2008*, volume 5223 of LNAI. (pp. 282-292). Springer, Berlin.

Zha, Y. (2017). Unpacking communication tensions in visual transmediation from print to digital papers. *Communication Design, 5*(1-2), 165-182. https://doi.org/10.1080/20557132.2017.1402501

About the author

Tuomo Hiippala is Assistant Professor of English and Digital Humanities at the University of Helsinki, Finland. His major publications include *The Structure of Multimodal Documents* (2015, Routledge) and *Multimodality: Foundations, Research and Analysis* (2017, De Gruyter, with John A. Bateman and Janina Wildfeuer).

18. Exploring narrativity in data visualization in journalism

Wibke Weber

Abstract

Many news stories are based on data visualization, and storytelling with data has become a buzzword in journalism. But what exactly does storytelling with data mean? When does a data visualization tell a story? And what are narrative constituents in data visualization? This chapter first defines the key terms in this context: story, narrative, narrativity, showing and telling. Then, it sheds light on the various forms of narrativity in data visualization and, based on a corpus analysis of 73 data visualizations, describes the basic visual elements that constitute narrativity: the instance of a narrator, sequentiality, temporal dimension, and tellability. The paper concludes that understanding how data are transformed into visual stories is key to understanding how facts are shaped and communicated in society.

Keywords: Data visualization; Journalism; Narrativity; Storytelling; Telling; Showing

Introduction

Storytelling is deeply rooted in our society. From the very beginning of time people have told stories to convey ideas and thoughts, to share experience and knowledge, to express desires and feelings, or to remember the past. They have told stories with the purpose of informing and recording, explaining and persuading, understanding and entertaining. To this day, telling stories is a pivotal activity of our everyday lives. Stories help us to make sense of the world, to create individual and cultural identity, and to evoke emotions. Because of its emotional impact and cognitive effectiveness, storytelling has become an integral part of journalism.

Engebretsen, M. and H. Kennedy (eds.), *Data Visualization in Society.*
Taylor & Francis Group, 2020
DOI 10.5117/9789463722902_CH18

A lively discourse about storytelling in journalism has developed in recent years. Triggered by new technologies and increasing media convergence, new forms and hybrid genres have emerged (e.g. audio slideshows, gamified interactives, motion graphics, or VR pieces)—often subsumed by academics and practitioners under the vague term of multimedia storytelling, long-form journalism, or online narrative journalism. These new products go far beyond the traditional text-based genres such as news, feature writing, or opinion. They cross the boundaries of images, texts and numbers, facts and fiction, distance and immersion; they conflate writing and drawing, telling and showing, narration and exploration; they combine objectivity with subjectivity, literacy with orality. Thus, they stand in the tradition of narrative journalism, also called literary journalism, which aims to find the private story behind the public story. One of these new forms that have gained tremendous momentum in the wake of data journalism is data visualization.

We are currently witnessing an increased use of data visualization in journalism, since data become only visible and publicly accessible through their visualization. Journalists and designers use not only standardized types of data visualizations like bar charts or maps, but also create new forms that are tailor-made in order to tell the story in the most understandable and engaging way. Here again, we come across the term *storytelling:* storytelling with data.

Storytelling is a buzzword in journalism, overused and with a fuzzy meaning. Studies in newsrooms have shown that when journalists use the term 'story' they often mean 'news', because both story and news refer to events (Merminod, 2016). When they talk about storytelling, they mean not only the text-linguistic practice of narrating but also describing, explaining, or arguing. The same applies to data visualization. 'The phrase "data storytelling" has been associated with many things—data visualizations, infographics, dashboards, data presentations, and so on. [...] Data storytelling is a structured approach for communicating data insights, and it involves a combination of three key elements: data, visuals, and narrative' (Dykes, 2016). Here, another term comes into play: *narrative.* In text linguistics, the narrative mode is distinguished from the text modes of description, explanation, and argumentation (Brinker, 2010). In this chapter, I focus only on the *narrative* mode in data visualizations (for argumentation see Archer & Noakes, this volume).

A term that often appears in the context of the narrative mode is *showing.* In journalism, trainers give the normative advice: 'Show, don't tell' (e.g. Mencher, 1997, p. 154). It means not describing a particular subject from the narrator's point of view (the narrator remains in the background), but allowing the reader to witness the events, to experience the emotions of the character, and to immerse him- or herself into the story through moment-by-moment actions and dialogues. 'Show, don't tell' is also the title of a workshop held at the Malofiej Summit—one of the most important events for information graphics and data visualization in journalism (see http://www.malofiejgraphics.com/). However, what does *telling* mean in data visualization, and what does *showing* mean? I will come back to this question later.

Against this backdrop, I want to explore the following terms throughout the chapter: story, narrative, narrativity, telling and showing, and what they mean in the context of data visualization. I focus primarily on journalistic pieces that are mainly based on data visualization or stand-alone graphics. The leading question is: what does narrativity mean in data visualization? Since more and more news stories are based on data, understanding the different forms of narrativity in data visualization in journalism is key to understanding and critiquing how meaning is made out of data, how this meaning is shaped by the process of visualization, and how knowledge is thus conveyed in society.

Story, narrative, narrativity, telling and showing

Story and narrativity are often used interchangeably. Both *story* and *narrative* have been defined in many ways depending on the discipline, scholarly approach, or professional field (Ryan, 2007; Bell, 1991; Genette, 1988, 1980; Lotman, 1977; Barthes & Duisit, 1975, to mention but a few). For my purpose, I regard a *narrative* as a textual, visual, or multimodal representation that presents a story. As such, a *narrative* is the semiotic product of narrating (Genette, 1988, p. 14). Every narrative is based on a story and mediated by a narrator, the person or speech position from which the story originates, or 'the individual agent who serves as the answer to Genette's question *qui parle?*' (Margolin, 2014).

What defines *story*? On a very basic level, a story is a sequence of events or happenings that are temporally structured and coherently related to each other, involving one or more characters or anthropomorphic agents or objects. According to Genette, 'as soon as there is an action or an event, even a single one, there is a story because there is a transformation, a transition from an earlier state to a later and resultant state' (1988, p. 19). For him, the sentence 'I walk' is a minimal but whole story because it implies 'a state of departure and a state of arrival' (p. 19). A foundational definition of story is given by Forster (1927, p. 130): 'The king died, and then the queen died', whereas a plot adds causality to a story: 'The king died, and then the queen died of grief'. Story is not tied to a specific genre. It works as an underlying layer in a narrative as well as in other literary works such as dramas, poems, comics, movies, or data visualizations.

While story refers to *what* is being told, the distinction between *telling* and *showing* addresses *how* the story, that is, the events, are presented in a narrative. When Genette talks about telling vs. showing, he refers to the degree to which the narrating instance is present. '"Showing" can be only a way of telling, and this way consists of both saying about it as much as one can, and saying this "much" as little as possible [...]—in other words, making one forget that it is the narrator telling' (Genette, 1980, p. 166). A 'pure narrative', the telling mode, is characterized as more distant, more mediated, and says less than the showing mode (p. 163), whereas the showing mode gives the readers the illusion that they are shown the events of a story. Guided by the explanations of Klauk and Köppe (2014), Table 18.1 displays the main features of telling vs. showing.

Table 18.1 Telling vs. showing

	Telling	Showing
Narrator and narrator's spatial position towards what is told	– presence of a narrator – mediated presentation – remote distance – only what is worth telling is presented in the narrative	– absence of a narrator – unmediated presentation – close distance, as if the events were revealed
Speed of unfolding the narrative	– fast speed, which means less detailed information – focus on summary	– slow speed, which means more detailed information – focus on scene
Dialogue	– absence of dialogues – epic	– presence of dialogues – scenic
Explicitness / implicitness	– explicitness of characters' traits, themes, meanings, or morals of the story	– implicitness of characters' traits, themes, meanings, or morals of the story
Partiality / impartiality	– partiality which includes commentary and subjective evaluation	– impartiality and objectivity
Reader's perception	– the reader gets the story told	– the reader witnesses the events of the story

In the current discourse of narratology, there is a broad range of possible meanings of telling and showing and some of the features listed are contentious. Thus 'it remains and open question whether, or to what extent, these accounts allow for unification' (Klauk & Köppe, 2014). In my context, Table 18.1 serves as a heuristic means for identifying narrative techniques in data visualization.

Narrativity is closely connected to story and narrative. Many concepts that define narrativity refer to fictional text (Abbott, 2011; Ryan, 2007). A definition of narrativity that is most suitable for the application to non-fictional genres such as data visualization is the set of conditions proposed by Ryan (2007). Ryan does not regard narrativity 'as a strictly binary feature, that is, as a property that a given text either has or doesn't have'. Instead she defines narrativity as 'a fuzzy set allowing variable degrees of membership, but centred on prototypical cases that everybody recognizes as stories' (2007, p. 28). Table 18.2 summarizes a few crucial constituents of narrativity along Ryan's set of conditions (pp. 28-31). With this sketchy framework in mind, I turn to a discussion of storytelling in data visualization.

Table 18.2 Narrative constituents

Spatial and temporal dimension	– The world is situated in time and undergoes a transformation caused by non-habitual physical events. – It is about individuated existents. – Temporal transformation excludes pure explanation, description, or argumentation.
Characters and events	– It is about characters that react emotionally to the events, which excludes weather reports and financial reports, for instance. – Some actions by the characters must be purposeful, which excludes mental events.
Sequentiality	– Sequence of events that are temporally structured and coherently related to each other, which excludes lists or a sequence of unconnected events. – The occurrence of the events must be a fact for the story world, which excludes hypotheses, instructions, or statements. – Completeness (eventfulness): the whole story is presented, which excludes fragmentary storytelling, e.g. breaking news or news about ongoing events.
Tellability	– Giving an answer to: What's the point? – Something meaningful that makes the story worth telling

Storytelling and data visualization

Several scholars have advanced the research on narrative techniques, rhetorical devices, and patterns in data visualization (e.g. Henry Riche, Hurter, Diakopoulos, & Carpendale, 2018; Weber, Kennedy, & Engebretsen, 2018;

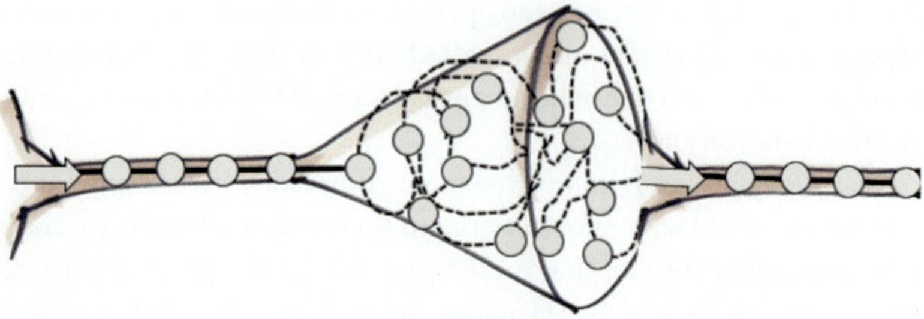

Figure 18.1. Telling, showing, telling. A modified version of the Martini-Glass structure. Illustration by W. Weber.

Brehmer, Lee, Bach, Henry Riche, & Munzner, 2016; Hullman & Diakopoulos, 2011; Segel & Heer, 2010). One concept, which is often referred to, is the author-driven and reader-driven approach introduced by Segel and Heer (2010). By author-driven, they understand a strict linear path through the visualization, heavy messaging, and no interactivity, whereas reader-driven means 'no prescribed ordering of images, no messaging, and a high degree of interactivity' (p. 1146). In case of high interactivity, the user is given maximum information to explore the data visualization.

The so-called Martini-Glass structure (Figure 18.1), a term also coined by Segel and Heer (2010), is a combination of the author-driven and reader-driven approach and often employed by data visualization practitioners (Weber, Engebretsen, & Kennedy, 2018). It comes close to what I have introduced earlier as telling vs. showing. At the beginning (the stem of the glass), the narrator, that is, the author or the production team, controls and handles the dataset from an authorial point of view (remote distance), telling the basic story found in the data in a linear way (sequentiality) by summarizing the main facts (fast speed) and emphasizing or annotating some points (partiality). Then, the data visualization opens up (at the mouth of the glass) and offers the user some room for exploration by showing the data (slow speed, close distance, impartiality), while the author steps back into the background. However, showing takes place in a limited frame predefined by the author who is in control again as soon as the user continues to click or scroll.

The author-driven/reader-driven approach, which addresses the production-reception level, can be compared to the distinction of 'explanatory' and 'exploratory', which is situated at the product level (Thudt et al., 2018, pp. 59-84; Young, Hermida, & Fulda, 2017; Kirk, 2016, pp. 77-80; Barlow, 2014). The point here is whether the message found in the data visualization is explained, or whether the visualization is presented as an analysis tool so that users can explore the datasets themselves. Strictly speaking, exploratory and explanatory are separate modes for themselves and, therefore, cannot represent the *narrative* mode as defined above. However, these types are often subsumed by scholars under the broad term data-driven storytelling (e.g. Thudt et al., 2018). What can be stated, however, is that explanatory and exploratory elements can be part of an overarching narrative frame. The following section considers whether the various techniques that constitute narrativity can be found in journalistic data visualization.

Constituents of narrativity in data visualization

In order to tell a story in a data visualization, we need techniques and stylistic devices that constitute visual narrativity. I draw on (i) findings from related studies in the field of data visualization, and (ii) insights gained from my analysis of a corpus of 73 data visualizations. The corpus, which was built for the INDVIL research project, was compiled in a very pragmatic way by looking for the latest award-winning and shortlisted data visualizations in journalism that have been selected by a jury of experts because of their qualities and standards.[1] These awards are the Malofiej Award 2018, the Data Journalism Awards 2018, and the Kantar Information is Beautiful Awards 2017. The majority of the data visualizations were produced by news organizations in Western Europe and the US. The analysis criterion relevant to my context was the mode in which the data visualization is predominantly presented: *narrative* (does it tell a story) vs. *explanatory* (does it explain something), *argumentative* (is it embedded in a text-based argumentation structure), or *exploratory* (free exploration of the data). In what follows, the techniques and stylistic devices that constitute the *narrative mode* as described in Table 18.1 and 18.2 are considered.

Presence of a narrator

While in fiction the story is presented by a narrator as a mediating instance, in journalism the story is presented by the real author or the production team, namely journalists, designers, and programmers. That means, the two different roles of narrator and real author that in narratology should be clearly

1 The research project INDVIL (Innovative Data Visualization and Visual-Numeric Literacy) is supported by the Research Council of Norway (NFR) and the Norwegian Media Authorities (RAM), www.indvil.org.

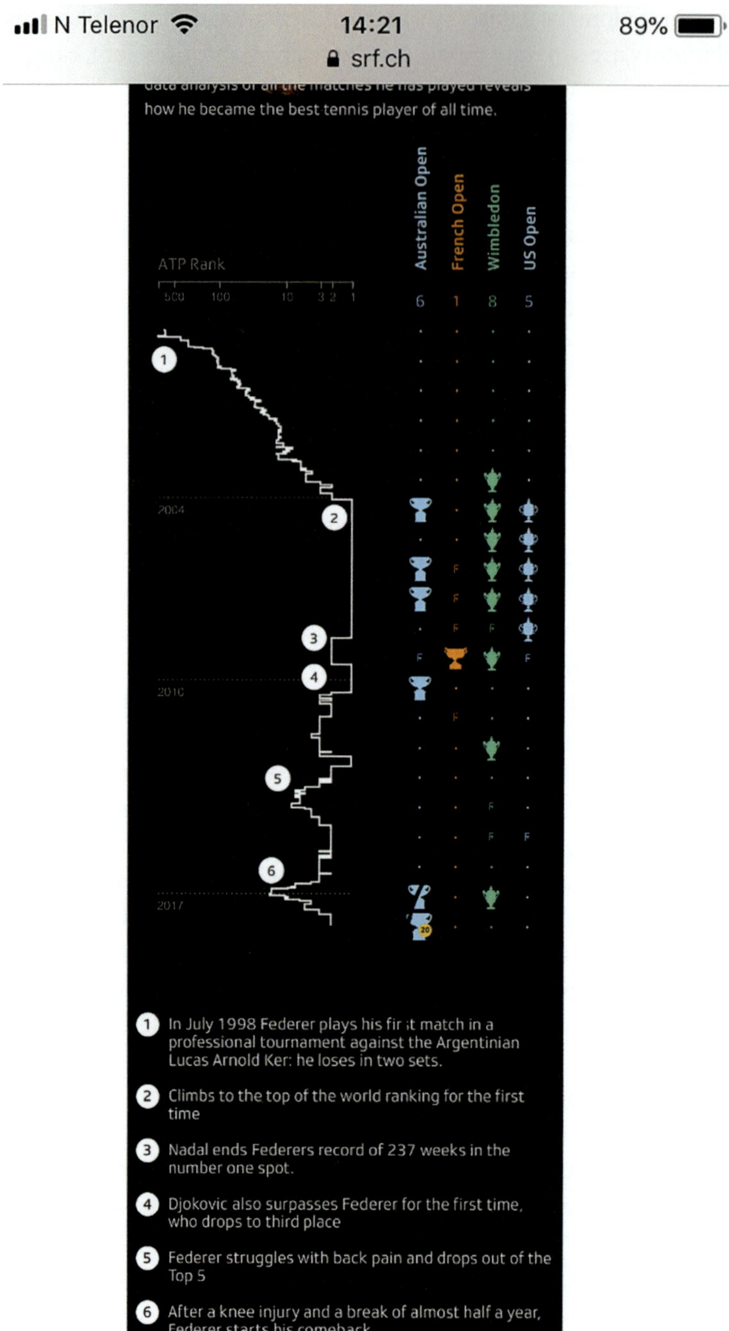

Figure 18.2. Screenshot of the intro of the data visualization '20 years, 20 titles', Mobile version. From '20 years, 20 titles', by A. Zehr et al., 2018. (https://www.srf.ch/static/srf-data/data/2018/federer/en.html#/en) Copyright SRF. Reprinted with permission.

distinguished from each other, particularly in fiction, coincide. Looking at the text elements embedded in the visualization, titles, teasers, captions, or labels can give a hint to the author's presence and the narrative mode. The author can be present in the form of personal pronouns (we) or proper nouns as for example in 'An 18-month nationwide investigation by *The Guardian* reveals, for the first time, what really happens at journey's end' [8]. The temporal specification at the end of the sentence also points to the narrative mode. Verbs of movement or change do the same job as well as adverbs of time and place: 'After hurricane Maria, Puerto Rico was in the dark for 344 days, 6 hours and 43 minutes' [7], 'How California's Most Destructive Wildfire Spread, Hour by Hour' [11], 'How a Melting Arctic Changes Everything' [9]. In contrast, adverbs of manner are more likely to indicate the explanatory mode: 'Don't waste your time at Disneyland. Here's how to avoid the lines' [4], or 'We analyzed 100,000 drawings to show how culture shapes our instincts' [5].[2]

A narrator/author can also become visible in the form of tooltips [2], which are small text boxes that pop up when the user moves the mouse cursor over an item in the graphic. Narratorship too appears through highlighting, emphasizing, or annotating certain data, making elements salient, or pointing to statistical outliers. In the line graph of the data visualization '20 years, 20 titles' [12], important milestones of Roger Federer's career are numerically labelled and annotated in a legend (Figure 18.2).

Data visualizations that are organized like a slideshow (Figure 18.3) often show all features of a story: beginning, ending, and a change in between as well as causality. In the data visualization 'Mass exodus: the scale of the Rohingya crisis' [6], the narrator/author becomes visible by commenting that the numbers rose 'dramatically each day'. This data visualization, a stand-alone graphic, displays the increase of Rohingya refugees fleeing to Bangladesh. It is based on animation, another technique that supports narrativity. The animation works like a narrating instance which decides what is presented to the reader and worth telling, summarizes the events, and controls the speed in which the events unfold. The user is given no option to intervene or stop the animation, instead the user is told the events.

Another kind of audio narration, namely sonification, is used in the data visualization 'The sound of the substantial fall of the SPD' [10]. The line graph tells the story of the substantial fall of the Social Democratic Party of Germany (SPD), based on 3,838 poll ratings from January 1998 to the end of February 2018. The data team translated the ups and downs of the visible line into music. Thus, we cannot only see the data but also

2 Emphases added. The numbers refer to the sources, see References section.

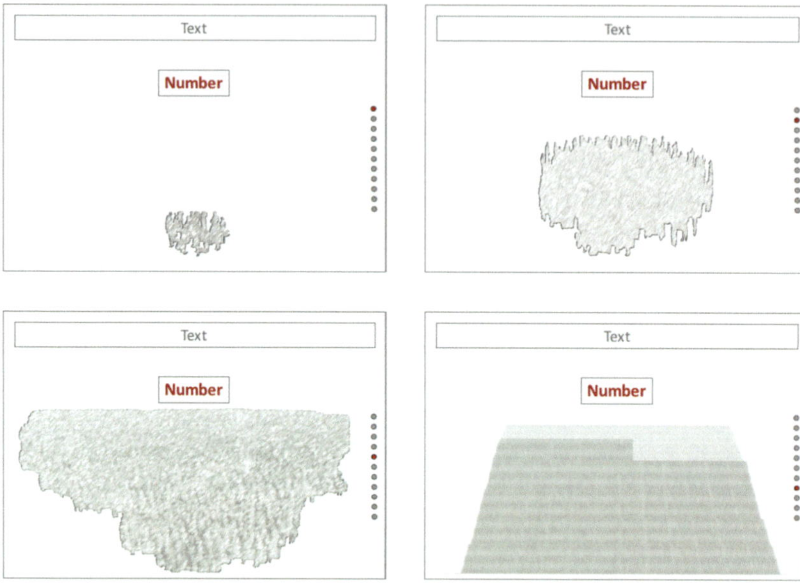

Figure 18.3. Sequential pattern with scrolling and zooming out. Drawn after the graphic 'Mass exodus: The scale of the Rohingya crisis' by C. Inton et al., 2017 (http://fingfx.thomsonreuters.com/gfx/rngs/MYANMAR-ROHINGYA/010050XD232/index.html). Reuters Graphics.

hear it. The sonification, that is, transforming data into sound, works as a narrative comment using variables such as volume, pitch, duration, tempo, and rhythm. The deep final tone sounds dramatic and seems to be the end of the party. In data videos, the voice-over narrator can be a kind of omniscient narrator who comments on actions, events, or characters and ensures coherence between the sequences.

Sequentiality

Sequentiality can be realized in several ways. One option is to use the (parallax) scrolling technique. Here, the effect is twofold: stepping from one event to the next while scrolling down or pressing the arrow keys (Stolper, Lee, Henry Riche, & Stasko, 2018, pp. 95-96), and thus, causing a transformation of the visualization. Telling stories by scrolling is called scrollytelling. The scrolling movement can be combined with zooming effects as the pattern in Figure 18.3 illustrates; it visualizes the dramatic increase of the Rohingya refugees [6]. First, the user has to scroll down, step-by-step, to see how fast the amount of refugees is growing; then, the story speeds up by replacing scrolling by animation. Each sequence is represented as a dot, which is lined up in a breadcrumb navigation placed on the right. While scrolling down is somewhat closer to the showing mode because the user is in control of the visualization and can decide whether to go forth or back, the animation represents the telling mode with fast speed and author's control.

In animation the transition from one narrative point to the next is performed smoothly and cohesively by the author; by contrast, in scrollytelling the sequences are linked through an interaction performed by the user. Next and previous buttons or the instruction 'click or press to continue', e.g., 'If you're black' [3], fulfil the same function as the horizontal or vertical scrolling. These data visualizations are called steppers because users have to click through the story step-by-step in a linear way to see how the story develops. The linking elements that combine the different sequences into a coherent whole can be for instance lines, arrows, or colour, e.g. using the same colour, while shape or size is changing. Another option for providing sequentiality is to show the different events as chapters in a navigation menu at the beginning of the story [2] or in a navigation bar [9].

Visualization types for depicting change over time

As mentioned above, the temporal dimension is crucial for storytelling. Therefore, timelines, time series graphs, flow maps (e.g. a Sankey diagram), slideshows, or data videos are well suited for storytelling. Maybe the most famous flow map is Charles Joseph Minard's map of the Russian campaign 1812-1813. 'By placing stroked lines on top of a geographic map, a flow map can depict the movement of a quantity in space and (implicitly) in time' (Heer, Bostock & Ogievetsky, 2010, p. 63). A timeline consists of a sequence of events (happenings) in chronological order, whereas a time series graph shows how several variables have changed over a specific time period. It combines temporal data (when) with numerical data (how many).

To turn other charts into narratives, we must add a temporal dimension. A pie chart, a network diagram, or a treemap alone does not represent a story, they just present facts. However, a line graph that depicts how values have changed over a time period, can tell a story. 'If you show a bar chart with a stack on top of it to indicate growth between two points in time, well, you have added a time dimension' (Kirk, 2016, p. 160). Dynamic elements such as animated points, lines, or areas often show a movement from one point to another and thus a temporal change. Examples that show a change over time are again the 'Mass exodus' graphic [6] with its animated chart of the increase of refugees and the animated map 'Thousands Cried for Help as Houston

Flooded' [1] that depicts the requests made by people who sought help during the Houston flooding.

Real-time data visualizations can be regarded as simultaneous narration. These are types of visualizations, e.g. maps, in which the data are immediately visualized after collecting them, so that the story develops simultaneously while looking at it. However, these are fragmentary stories with an open end, since neither the author nor the audience know how the story will end, and thus, do not fulfil the criterion of completeness. When Bounegru, Venturini, Gray, & Jacomy (2017) ascribe narrative potential to non-sequential exploratory data visualizations such as interactive network diagrams, they mean that network visualizations can evoke a narrative script in the mind of the recipient. The question is, however, whether the reader is able to recognize and construe these network stories in the network diagram provided by the author.

Tellability

Tellability raises the questions of what makes a story interesting and appealing to the audience, what is the point of the story. Journalists are influenced by news values when they decide which story counts as news and which does not. News values are, for instance, relevance and impact of an event, negativity, proximity (geographical and cultural nearness), superlativeness, novelty (new and unexpected aspects), eliteness of individuals, and personalization which is the human face of an event (Caple & Bednarek, 2016, p. 439).

Most of the data stories I considered meet the criterion of tellability since they focus on something that is novel to the audience, unexpected, or surprising. The topics that are covered in the data visualizations analysed deal with relevance and impact (racial discrimination [3]), superlativeness (Rohingya refugee crisis [6]), personalization (people seeking help during the Houston flood [1]), eliteness (celebrities and their life [12]), negativity (story of the melting Arctic [9]), novelty and personalization (what really happens to homeless people [8]).

Conclusion

The overarching question of this chapter was: What does narrativity mean in data visualization? Through my analysis of data visualizations in journalism and drawing on related studies on data-driven storytelling, I identified various techniques that constitute narrativity in data visualization. These are:

- The instance of a narrator that becomes present in the form of micro text elements (e.g. title, teaser, captions, and tooltips), through design elements and stylistic devices (e.g. making some data more salient through colour or size), or through aural interpretation.
- Sequentiality, which can be expressed through scrolling, animation, and dynamic transition effects (e.g. zoom, dissolve, wipe, fade).
- The temporal dimension, which can be displayed best through data visualization types such as timelines, time series graphs, flow maps, slideshows, data videos, and other charts that show a change over time.
- Tellability, which addresses the journalistic question of what makes a story worth telling.

When it comes to the distinction between telling and showing, both can be found in narrative data visualizations. In the telling mode the message that is found in the data is communicated through a mediated instance, that is, the reader gets the message told. The showing mode comes into play as soon as the visualization prompts the reader to interact within a given set of options such as zooming, filtering, or selecting objects, but without leaving the narrative frame. This limited frame of interaction where the reader is given more control and the narrator remains in the background can be seen as a dialogue-like communication process. The extent to which this showing mode can be characterized as impartial and objective remains to be discussed since data visualizations are always subject to a design process and thus have subjective traits, even though they may appear objective and impersonal (Kennedy, Hill, Aiello, & Allen, 2016).

Throughout this chapter, it has become clear that the journalistic advice 'Show, don't tell' does not fit well in the context of data visualization. Instead, 'Show and tell' would be more appropriate. The analysis of the corpus has also shown that it is a common practice to employ both modes in one single data story to attract and keep the reader's attention. At the same time, it has become obvious that the nature of journalistic storytelling is changing enormously, and data visualization is shaping this change. This change will affect how we shape facts, communicate news, and share knowledge in society in future.

References

Abbott, H. P. (2014). Narrativity. In: P. Hühn, J. C. Meister, J. Pier, W. Schmid (Eds.), *The living handbook of narratology*. Hamburg: Hamburg University. Retrieved August 24, 2018 from http://www.lhn.uni-hamburg.de/article/narrativity

Barlow, M. (2014). *Data visualization: A new language for storytelling*. Sebastopol, CA: O'Reilly Media.

Barthes, R., & Duisit, L. (1975). An introduction to the structural analysis of narrative. *New Literary History,* 6(2), 237-272. http://doi.org/10.2307/468419

Bell, A. (1991). *The language of news media.* Oxford: Blackwell.

Brehmer, M., Lee, B., Bach, B., Henry Riche, N., & Munzner, T. (2016). Timelines revisited: A design space and considerations for expressive storytelling. *IEEE Transactions on Visualization and Computer Graphics, 23*(9), 2151-2164. http://doi.org/10.1109/TVCG.2016.2614803

Brinker, K. (2010). *Linguistische Textanalyse. Eine Einführung in Grundbegriffe und Methoden.* Berlin: Erich Schmid.

Bounegru, L., Venturini, T., Gray, J., & Jacomy, M. (2017). Narrating networks: Exploring the affordances of networks as storytelling devices in journalism. *Digital Journalism, 5*(6), 699-730. https://doi.org/10.1080/21670811.2016.1186497

Caple, H., & Bednarek, M. (2016). Rethinking news values: What a discursive approach can tell us about the construction of news discourse and news photography. *Journalism: Theory, Practice and Criticism, 17*(4), 435-455. https://doi.org/10.1177/1464884914568078

Dykes, B. (2016, March 31). Data storytelling: The essential data science skill everyone needs. *Forbes.* Retrieved August 26, 2018 from https://www.forbes.com/sites/brentdykes/2016/03/31/data-storytelling-the-essential-data-science-skill-everyone-needs/#6126f53e52ad

Forster, E. M. (1927). *Aspects of the novel.* New York: Harcourt, Brace and Company.

Genette, G. (1980). *Narrative discourse: An essay in method* (J. E. Lewin, Trans.). Ithaca, NY: Cornell University Press.

Genette, G. (1988). *Narrative discourse revisited* (J. E. Lewin, Trans.). Ithaca, NY: Cornell University Press.

Heer, J., Bostock, M., & Ogievetsky, V. (2010). A tour through the visualization zoo. *Communications of the ACM, 53*(6), 59-67.

Henry Riche, N., Hurter, C., Diakopoulos, N., & Carpendale, S. (Eds.). (2018). *Data-driven storytellling.* Boca Raton, FL: A K Peters/CRC Press.

Hullman, J., & Diakopoulos, N. (2011). Visualization rhetoric: Framing effects in narrative visualization. *IEEE Transactions on Visualization and Computer Graphics, 17*(12), 2231-2240.

Kennedy, H., Hill, R. L., Aiello, G., & Allen, W. (2016). The work that visualisation conventions do. *Information, Communication and Society, 19*(6), 715-735. https://doi.org/10.1080/1369118X.2016.1153126

Kirk, A. (2016). *Data visualisation: A handbook for data driven design.* London: Sage.

Klauk, T., & Köppe, T. (2014). Telling vs. Showing. In: P. Hühn, J. C. Meister, J. Pier, W. Schmid (Eds.), *The living handbook of narratology.* Hamburg: Hamburg University. Retrieved August 24, 2018 from http://www.lhn.uni-hamburg.de/article/telling-vs-showing

Lotman, J. (1977). *The structure of the artistic text.* Ann Arbor, MI: University of Michigan Press.

Margolin, U. (2014). Narrator. In: P. Hühn, J. C. Meister, J. Pier, W. Schmid (Eds.), *The living handbook of narratology.* Hamburg: Hamburg University. Retrieved August 24, 2018 from http://www.lhn.uni-hamburg.de/article/narrator

Mencher, M. (1997). News reporting and writing (7th ed.). London and Chicago: Brown & Benchmark.

Merminod, G. (2016). *Telling stories from the newsroom: A linguistic ethnographic account of dramatization in the broadcast news.* Working Papers in Urban Language & Literacies. Paper 197. Retrieved August 26, 2018 from http://www.academia.edu/28219120/WP197_Merminod_2016._Telling_stories_from_the_newsroom_A_linguistic_ethnographic_account_of_dramatisation_in_the_broadcast_news

Ryan, M.-L. (2007). Toward a Definition of Narrative. In: D. Herman (Ed.), *The Cambridge companion to narrative.* (pp. 36-62). Cambridge: Cambridge University Press.

Segel, E., & Heer, J. (2010). Narrative visualization: Telling stories with data. *IEEE Transactions on Visualization and Computer Graphics, 16*(6), 1139-1148. http://doi.org/10.1109/TVCG.2010.179.

Stolper, C. D., Lee, B., Henry Riche, N., & Stasko, J. (2018). Data-driven storytelling techniques: Analysis of a curated collection of visual stories. In: N. Henry Riche, C. Hurter, N. Diakopoulos, & S. Carpendale (Eds.), *Data-driven storytellling.* (pp. 85-105). Boca Raton, FL: A K Peters/CRC Press.

Thudt, A., Walny, J., Gschwandtner, T., Dykes, J., & Stasko, J. (2018). Exploration and explanation in data-driven storytelling. In: N. Henry Riche, C. Hurter, N. Diakopoulos, & S. Carpendale (Eds.), *Data-driven storytellling.* (pp. 59-84). Boca Raton, FL: A K Peters/CRC Press.

Young, M. L., Hermida, A., & Fulda, J. (2018). What makes for great data journalism? A content analysis of data journalism awards finalists 2012–2015. *Journalism Practice, 12*(1), 115-135. https://doi.org/10.1080/17512786.2016.1270171

Weber, W., Engebretsen, M., & Kennedy, H. (2018). Data stories: Rethinking journalistic storytelling in the context of data journalism. *Studies in Communication Sciences, 18*(1), 191-206. https://doi.org/10.24434/j.scoms.2018.01.013

Data visualizations: Examples taken from the corpus analysed

[1] Aisch, G., Almukhtar, S., Ashkenas, J., Bloch, M., Carlsen, A., Delreal, J. A., Fessenden, F., ... Yourish, K. (2017, August 30). Thousands Cried for Help as Houston Flooded. *The New York Times.* Retrieved from https://www.nytimes.com/interactive/2017/08/30/us/houston-flood-rescue-cries-for-help.html

[2] Aisch, G., Almukhtar, S., Ashkenas, J., Bloch, M., Burgess, J., Carlsen, A., Fessenden, F., ... Williams, J. (2017, August 31). Maps: Tracking Harvey's Destructive Path Through Texas and Louisiana. *The New York Times*. Retrieved from https://www.nytimes.com/interactive/2017/08/24/us/hurricane-harvey-texas.html#rainfall

[3] Bedi, N., & Humburg, C. (2017, April 4). If you're black. *Tampa Bay Times*. Retrieved from http://www.tampabay.com/projects/2017/investigations/florida-police-shootings/if-youre-black/

[4] Fox, J. Krishnakumar, P., & Schleuss, J. (2017, July 12). Don't waste your time at Disneyland. Here's how to avoid the lines. *Los Angeles Times*. Retrieved from http://www.latimes.com/projects/la-fi-disneyland-ride-wait-time/#ride-times

[5] Ha, T.-H., & Sonnad, N. (2017, June 15). How do you draw a circle? *Quartz*. Retrieved from https://qz.com/994486/the-way-you-draw-circles-says-a-lot-about-you/

[6] Inton, C., Wu, J., Scarr, S., Cai, W., & Singh, K. (2017, September 21). Mass exodus: the scale of the Rohingya crisis Mass exodus. *Reuters Graphics*. Retrieved from http://fingfx.thomsonreuters.com/gfx/rngs/MYANMAR-ROHINGYA/010050XD232/index.html

[7] Lu, D., & Alcantara, C. (2018, April 4 updated). After Hurricane Maria, Puerto Rico was in the dark for 344 days, 6 hours and 43 minutes. *The Washington Post*. Retrieved from https://www.washingtonpost.com/graphics/2017/national/puerto-rico-hurricane-recovery/?utm_term=.5d71b968db1a

[8] Outside in America team (2017, December 20). Bussed out: How America moves its homeless. *The Guardian*. Retrieved from https://www.theguardian.com/us-news/ng-interactive/2017/dec/20/bussed-out-america-moves-homeless-people-country-study

[9] Roston, E., & Migliozzi, B. (2017, April 19). How a Melting Arctic Changes Everything. *Bloomberg*. Retrieved from https://www.bloomberg.com/graphics/2017-arctic/

[10] Tröger, J., Pätzold, A., Timcke, M.-L., Klack, M., & Wendler, D. (2018, March 1). Der Sound zum tiefen Fall der SPD (The substantial fall of the SPD). *Berliner Morgenpost*. Retrieved from https://interaktiv.morgenpost.de/spd-absturz-sound/

[11] Watkins, D., Griggs, T., Lee, C., Park, H. Singhvi, A. Wallace, T. Ward, J. (2017, October 21). How California's Most Destructive Wildfire Spread, Hour by Hour. *The New York Times*. Retrieved from https://www.nytimes.com/interactive/2017/10/21/us/california-fire-damage-map.html

[12] Zehr, A., Schmidli, J. Nguyen, D.-Q., Boa, T., & Guillemot, L. (2018, January 28). 20 years, 20 titles. *SRF*. Retrieved from https://www.srf.ch/static/srf-data/data/2018/federer/en.html#/en

About the author

Wibke Weber is Professor of Media Linguistics at ZHAW Zurich University of Applied Sciences. Her areas of research are data visualization, information graphics, visual storytelling, visual semiotics, multimodality, and augmented and virtual reality. From 2001 to 2013 she worked as Professor of Information Design at Stuttgart Media University (HdM, Germany).

19. The data epic: Visualization practices for narrating life and death at a distance

Jonathan Gray

Abstract

This chapter proposes the notion of the 'data epic', which is examined through two works of 'cinematic data visualization': *The Fallen of World War II* and *The Shadow Peace: The Nuclear Threat*. These pieces mobilize an aesthetics of distance to narrate life and death at scale, in past and possible global conflicts. While previous studies of quantification emphasize the function of distance in relation to aspirations of objectivity, this chapter explores other narrative and affective capacities of distance in the context of 'public data culture'. The data epic can thus enrich understanding of how data are rendered meaningful for various publics, as well as the entanglement of data aesthetics and data politics involved in visualization practices for picturing collective life.

Keywords: Data politics; Data aesthetics; Data practices; Sociology of quantification; Distance; Scale

Introduction

'In Visual Education we should think of the onlooker's emotional habits, but that does not imply that we have to make charts and their captions emotional. People like to get an opportunity to judge for themselves and to reach their decisions without feeling themselves bullied by "visual dictators" who take care of the public's "visual food"'. (Neurath, 1944, p. 65)

'In my work, I try to find ways to make statistical information less boring and intimidating. I believe it's often appropriate to express numbers with emotion and cinematic drama, particularly when there is a humanitarian

Engebretsen, M. and H. Kennedy (eds.), *Data Visualization in Society*.
Taylor & Francis Group, 2020
DOI 10.5117/9789463722902_CH19

> component. I'm hoping that new forms of data-driven storytelling can help us as an informed democracy close the troubling gap between expert and public opinion'. (Halloran, cited in Dvorsky, 2017)

> '[...] it's important to slow down and take in the data part by part. That required a lot of shifting of gears these days, because my screen was a veritable anthology of narratives, and in many different genres. I had to shift between haiku and epics, personal essays and mathematical equations, Bildungsroman and Götterdämmerung, statistics and gossip [...]. The temporalities in these genres ranged from the nanoseconds of high-frequency trading to the geological epochs of sea level rise, chopped into intervals of seconds, hours, days, weeks, months, quarters, and years. [...] The economic sublime!' (Robinson, 2017)

How do data visualizations enable different ways of making sense of life and death 'at a distance'? How can they articulate not just 'ways of knowing' but also 'ways of feeling' with data? This chapter examines two projects by software developer, data analyst, and filmmaker Neil Halloran: *The Fallen of World War II* (2015) on deaths during the war; and *The Shadow Peace: The Nuclear Threat* (2017) on nuclear weapons, nuclear war scenarios, and peacekeeping efforts (both projects are available online at: http://www.fallen.io/). These have been variously described by their creator as 'cinematic data visualization', 'interactive documentary' and 'animated data-driven documentary about war and peace' (Halloran, 2015, 2017; and see http://www.neilhalloran.com/).

The pieces are said to exemplify a novel way of doing data visualization. Regarding the first, writer and researcher Steven Pinker asks, 'Who would have thought that bar graphs (admittedly, with the help of haunting music) could overflow with human pathos?'. He goes on to comment that 'data graphics has become a major new medium of intellectual exposition and artistic expression' and that the 'war death data' are 'stunning' and 'emotionally ravaging' (Pinker, 2015). Both pieces were also critically acclaimed amongst practitioner communities, including winning prizes in the 'Information is Beautiful' data visualization awards for the 'Motion Infographic' (2015) and 'Humanitarian' (2017) categories. They were both crowdfunded and freely distributed online, with the former piece winning a 'Best of 2015' award from video-sharing platform Vimeo.

Both of these works by Halloran use data visualizations to narrate war deaths and the potential effects of nuclear weapons at the level of populations, over periods of decades and centuries. As such, a defining feature of both is mobilizing data to tell stories 'at a distance' and to comprehend the societal consequences of war and nuclear weapons across space and time, beyond the level of individual incidents. These works explore the narrative and affective capacities of distance, in a way which can be distinguished from (but which may also serve to reinforce and potentially reify) its role as a methodological ideal in the production of knowledge and objectivity. The data epic can thus enrich understanding of how data are rendered meaningful for various publics, as well as how data aesthetics and data politics are entangled in visualization practices for picturing collective life.

Public data cultures and aesthetics of distance

Practices of quantification and datafication are often used to understand and narrate phenomena at scale in order to identify and articulate patterns, trends, and dynamics across different cases and settings. As Porter puts it, 'quantification is a technology of distance', which aims to support objectivity by attempting to produce 'knowledge independent of the particular people who make it' (1996, p. ix). A similar argument is made by Daston and Galison, who argue that 'emotional distance' was considered a condition of objectivity, citing nineteenth-century statistician Karl Pearson's call for citizens to set aside their 'own feelings and emotions' in order to be impartial and impersonal (2010, pp. 29, 380, 196). Social and historical studies of quantification have explored how numbers can be put to work in the service of institutions of objectivity in science, management, and governance (see e.g. Porter, 1986, 1996; Desrosières, 2002; Rottenburg, Merry, Park, & Mugler, 2015).

Such data practices have significance outside of these institutional settings, including as part of what I propose to call 'public data cultures' through which various publics are invited to participate in making sense with data. For example, the work of Marie and Otto Neurath and the 'visual education' activities of their Isotype Institute sought to create a common pictorial language of 'isotypes'—pictorial representations of data—used in public exhibitions, pamphlets, and other materials about a wide range of issues and areas of life such as demography, economics, work, health, agriculture, industry, and politics (Neurath & Kinross, 2009). They thus aimed to use data not just to advance science or management, but also in the service of advocacy, journalism, and democratic participation (Rayward, 2017). The subtraction of emotion, decoration, and detail in such projects can be understood in relation to modernist ideals of an 'unaesthetic aesthetics' (Galison, 1990).

The use of data to produce distant perspectives is not limited to impartial, 'facty' aesthetics. As Kennedy and Hill note, data visualizations can have 'emotional and affective, not just cognitive and rational' capacities (2017, p. 10). Halloran's work explores the aesthetic and affective dimensions of the distance that data performs. Heterogeneous data about population, mortality, war, and peace are assembled, visualized, animated, and sequenced in the context of animated graphics and interactive features in order to tell stories about the fate of human creatures at sweeping scales across time and space. These 'thin descriptions', as Porter calls them (2012), are mobilized to facilitate different kinds of experiences and ways of making sense of collective life at a distance.

Such aesthetics of distance may be understood in relation to recent work on the 'data sublime' (Liu, 2004; Davies, 2015; Stallabrass, 2007). The data sublime is said to arise from Stallabrass's contention that 'the impression and spectacle of a chaotically complex and immensely large configuration of data' can play a role similar to 'mountain scenes and stormy seas' for nineteenth-century viewers (2007, p. 82.). As Davies suggests in relation to big data technologies, we may find an aesthetic of awe which 'functions beyond empiricism' through a 'sheer quantitative magnitude' which is 'as disturbing as exciting' (2015). These notions of the data sublime draw on traditions of thinking about the aesthetics of the sublime that rose to prominence in the eighteenth century—such as Burke's notion that the sublime produces 'the strongest emotion which the mind is capable of feeling' associated with a 'great extreme of dimension' (1998, pp. 36, 66). Kant says that the sublime is the 'absolutely great' and shifts emphasis from 'objects of the sense'—which can elicit but never embody such greatness—to a 'faculty of mind transcending every standard of the senses' (2007, pp. 78, 81). This includes what he calls the 'mathematically sublime', which is not simply 'greatness of number' but 'the fact that in our onward advance we always arrive at proportionately greater units' (p. 87).

Halloran's work can be viewed as an emerging style of data practice which I characterize as the 'data epic'. The projects have moments of tilting towards a data sublime, as well as other aesthetic and affective dimensions which I shall explore in the following sections. They affirm the affective and narrative capacities of data visualizations, which operate in combination with other images, film, music, audio effects, textual annotations, overlays, and voice-over. The two pieces can be compared with other forms of 'conventionalized representations' (Becker, 2007) about war, demography, and mortality—from documentary film and war memorials, to isotypes and mythological constellations—as well as giving rise to their own conventions and techniques for making sense with data visualization that are shared across the two projects. The data epic may be viewed as an emerging area for research and practice, contributing to recent work on the emotional capacities of data (Kennedy & Hill, 2017).

The Fallen of World War II

The Fallen of World War II is an 18-minute, interactive documentary about the 'human cost of the second World War'. Rather than focusing on 'individual war stories', its main characters are isotype-style representations of 'tens of millions of people whose lives were cut short by the war' (Halloran, 2015). While the repeated pictorial figures strongly resemble the visual style of the Isotype Institute, they are mobilized in a way which departs from the latter's unadorned and unemotional aspirations.

The setting for the piece starts with a blackboard, and transitions to an abstracted endless canvas whereby historical events are re-staged with data. Blackboard figures sketched onto a virtual blackboard are transformed into polished, computer-generated markers and timelines. The limits of the screen are used as a device to emphasize scale, such that timelines and graphs spill beyond its limits, and the screen pans or zooms out in order to account for the large numbers of deaths. A soundtrack of bass tones and reverb effects contributes to an atmosphere of endless space and timelessness.

The narrative starts with a chalk line for the 'average lifespan', the statistical equivalent of 'everyman' or the ordinary person. This is used as a device to transition from an individual perspective to the collective scale of 'lives cut short', such that numerous white lines enter from the left of the screen and terminate in a red block representing WWII (Figure 19.1).

While emphasizing the scale of death, *The Fallen* also uses visual strategies to connect visual representations of large numbers with individual lives. For example, we are shown a mass of white silhouettes with more individual details and features pouring out of a single isotype figure (Figure 19.2). This is accompanied by a rushing sound like a rainstick that suggests a sound for each individual as well as for the collective. This collective is then subsumed back into the isotype.

These isotype figures are then stacked up into charts to represent the scale of war deaths by nationality, by region of conflict, and by battle. Again, sound design plays a crucial role in the experience of distance and scale: a click is added for each figure, until the click becomes a whir. To convey the vast numbers of Soviet deaths, the screen slowly pans up a red column and the music fades out until there is nothing but a single note, the sound of high winds

Figure 19.1. The white timelines of individual lives ending in the red block of WWII. From *The Fallen of WWII*. Retrieved from http://www.fallen.io/ww2/. Copyright 2015 by N. Halloran. Reprinted with permission.

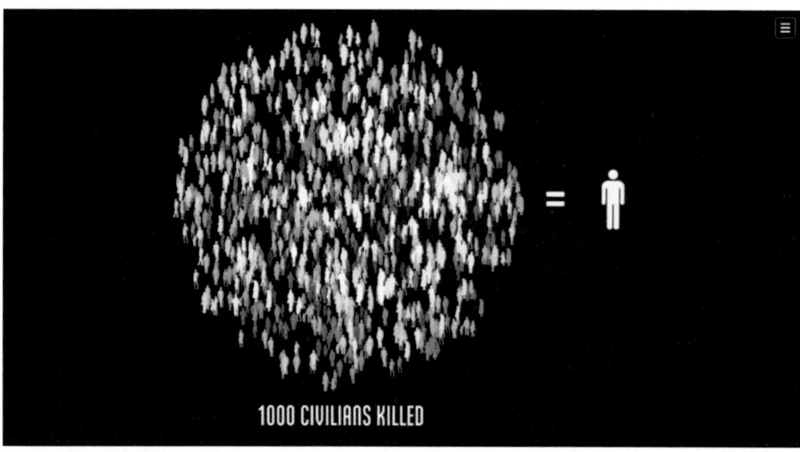

Figure 19.2. Group of silhouettes rendered equivalent to an isotype figure. From *The Fallen of WWII*. Retrieved from http://www.fallen.io/ww2/. Copyright 2015 by N. Halloran. Reprinted with permission.

and the whirring sound of the isotypes being added (Figure 19.3). Eventually the count stops and the figure of '8.7 million' appears. A similar approach is taken with people killed in the Holocaust: the screen zooms out from a huge block of isotypes representing millions of people killed during the Holocaust, with sound effects contributing to a sense of the vastness of the scene.

Other techniques are used to allude to individual lives within the vast multitudes of deaths. This includes combining charts with depictions of scenes at a human scale. Military drums accompany the transition from

Figure 19.3. Panning up a long column of Soviet deaths. From *The Fallen of WWII*. Retrieved from http://www.fallen.io/ww2/. Copyright 2015 by N. Halloran. Reprinted with permission.

isotypes to photographs of battles. A photograph showing a soldier about to shoot a woman and child transitions into a silhouette outline, which is then shown as one amongst many outlines contained within a single isotype, which itself is one amongst many isotypes representing deaths in concentration camps. These techniques are not intended to undo or negate the distance which is articulated through the visualizations, but rather to modify how they are meaningful by connecting vast scenes to recognizable and relatable ones. It is precisely by alternating between scales that such an epic narrative of life and death at a distance is enabled.

After exploring deaths across the world, the piece zooms out to show 70 million deaths for the war, depending on 'who is counting and what civilian deaths get included'. The total estimated deaths in WWII are then compared with a chart showing deaths in other wars and atrocities. The empty space above the bars in the chart is filled with light bars that extend beyond the screen, as the narrator comments: 'peace is a difficult thing to measure: it is a bit like counting the people who didn't die and wars that never happened'. The quantification of death is thus contrasted with the difficulty of quantifying peace, to which viewers are nevertheless encouraged to attend.

The Shadow Peace: The Nuclear Threat

The theme of accounting for peace is taken up in a sequel, *The Shadow Peace: The Nuclear Threat*, which examines scenarios of nuclear war and

Figure 19.4. Comparing population of total living with total dead. From *The Shadow of Peace: The Nuclear Threat*. Retrieved from http://www.fallen.io/shadow-peace/1/. Copyright 2017 by N. Halloran. Reprinted with permission.

peacekeeping efforts to avoid it. The setting for this piece is also a vast and apparently limitless canvas. Just as *The Fallen* begins with the demographic average, so *The Shadow Peace* commences with a visualization of demographic entities: falling cubes representing 4.6 people born into the world every second, each with its own accompanying sound. The narrative shifts from this fathomable number to the aggregation of '140 million births a year', accompanied by deep synth and reverb sounds suggestive of a large space. The screen pans out to show falling cubes as a barely visible trickle onto an enormous pyramid representing the entire living population. Another trickle comes down from the living towards another pyramid of the dead. The two pyramids are depicted as part of a vast hourglass, against a background of what appears to be clouds or distant nebulae (Figure 19.4). Thus data visualizations are mobilized to show us the entirety of human life and death at a distance in the form of a cosmic symbol representing the passage from life to death. This cosmic setting is later reinforced with constellations of stars which transform into weapons and graphs.

Just as isotypes represent a thousand people in *The Fallen*, pyramids are broken down into blocks representing one million people in *The Shadow Peace*, before being rearranged onto a timeline showing the rise and fall of earth's population over time. The scale of deaths is illustrated through a combination of isotype charts, photographs, and annotated maps narrating events such as nuclear strikes on Japan during WWII.

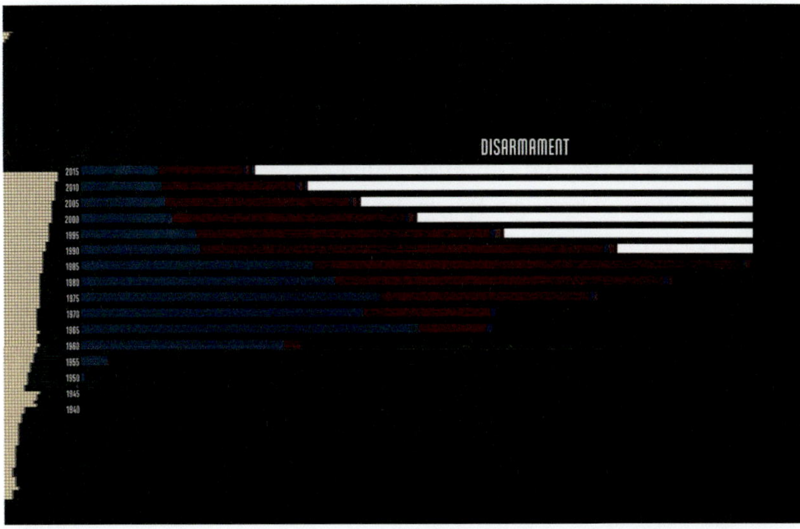

Figure 19.5. Visualizing nuclear disarmament alongside proliferation. From *The Shadow of Peace: The Nuclear Threat*. Retrieved from http://www.fallen.io/shadow-peace/1/. Copyright 2017 by N. Halloran. Reprinted with permission.

To examine the possible catastrophic consequences of nuclear war, a scenario of '2,000 strikes' shows a map of the earth with red dots indicating nuclear strikes appearing rapidly across the US along with the names of cities—'New York', 'Cleveland', 'Yonkers'—appearing and fading from view, before panning to show strikes in Western Europe, Eastern Europe, and Asia. A timeline showing deaths per decade depicts the effects of this nuclear war scenario as equivalent to 'almost ten World War IIs in three weeks', accompanied by discussions of nuclear winter and nuclear famine.

The focus then shifts to nuclear weapons—showing peaks in the mid-1980s, and the effects of disarmament highlighted by white bars showing the negative space above bars representing numbers of weapons (Figure 19.5), echoing the visual approach to highlighting peace in *The Fallen*. The narrative then turns to the proliferation and non-proliferation of nuclear weapons, different approaches to prevent nuclear war, accompanied by outlines of weapons presented as constellations of stars under the heading 'instruments of war' and a list of prevention strategies under 'instruments of peace'. The piece concludes with a discussion of 'what works in preventing war', reviewing the activities and effects of UN Peacekeepers, and contrasting belief in the technologies underpinning nuclear weapons with scepticism about 'statistics and the science of less tangible things, such as peace'.

Aesthetics and politics of the data epic

The Fallen and *The Shadow Peace* use similar aesthetic approaches for using data visualizations to make sense of life and death at a distance. They both have moments of the 'data sublime', such as the portrayal of historical events in settings of limitless space; the use of music and sound effects to indicate vast numbers of lives and deaths; the use of panning, zooming, and animated effects to dramatize the revelation of large numbers; and the use of thematic motifs alluding to visual cultures of death, spirituality, and the cosmos, such as war memorials and mythological constellations (Table 19.1).

Table 19.1 Comparison of features contributing to aesthetics of data sublime in Halloran 2015 and 2018

	The Fallen of World War II	**The Shadow Peace: The Nuclear Threat**
Implied setting	– Endless canvas.	– Space.
Sound	– Lack of voice-over contributes to dramatization of key scenes; – Sound effects imply vast space; – Sonification to accompany data visualizations—from individual clicks to whirring to indicate scale; – Wind noises emphasize height of graphs; – Emotional piano music.	– Lack of voice-over contributes to dramatization of key scenes; – Sound effects imply vast space; – Sonification to accompany data visualizations—from individual clicks to whirring to indicate scale; – Submarine-like sounds and deep synths give sense of space; – Melancholy music for strike scenes.
Panning and zooming effects	– Slow upward pan to show scale of Russian military deaths.	– Slow downward pan and outward zoom to show pyramid representing populations; – Slow pans across globe to show projected nuclear strikes.
Thematic motifs	– Equidistant layout of many figures alludes to war memorial practices used to indicate scale (e.g. in scenes with US soldiers accompanied by falling flag, figures representing deaths in Holocaust, and in image used for video still).	– Cosmic-scale hourglass representing total living and dead populations; – Earth from space; – Constellations to show weapons and charts, invoking mythological sense of the eternal.

In addition to these aspects of the data sublime, other strategies are used not just to produce an aesthetic of distance but also to connect these vast scales to the lives and deaths of individuals. This includes the use of graph features (such as segmentation, annotation, and visual cues); the use of photographs and other media to highlight individual people and events; zooming and panning effects; and the use of sonic textures to provide a sense of individuals within collectives, such as clicks to represent individual deaths (Table 19.2).

Table 19.2 Comparison of features contributing to connection between scale and individual in Halloran 2015 and 2018

	The Fallen of World War II	The Shadow Peace: The Nuclear Threat
Graphs features	– Segmentation and annotation to indicate different battles, events, countries, and causes of death; – Showing silhouettes alongside isotype figure to emphasize number of individuals represented by each unit.	– Segmentation and annotation to indicate different events, countries, and causes of death; – Showing trickle of blocks depicting birth and death rates to emphasize individuals within population pyramids.
Media formats	– Transitions into photographs to highlight individual people, battles.	– Transitions into photographs and video clips to highlight individual people and events.
Panning and zooming effects	– Zooming into timeline and highlighting groups of isotypes in order to emphasize individual events.	– Slow pan across globe showing nuclear strikes, with points being gradually added to emphasize each incident.
Sound	– Sound effects accompanying data visualizations to emphasize click of each isotype being added as well as whirring to indicate aggregates; – Military drums to associate graphs with battle scenes depicted in photographs.	– Sound effects accompanying data visualizations to emphasize tone of each birth, death, and nuclear strike as well as whirring to indicate aggregates.

Such strategies serve both to highlight people and situations at a more familiar and relatable scale, perhaps to counter what Boltanski calls the 'massification' of suffering associated with a politics of pity as opposed to a politics of justice (1999, p. 13). At the same time, we are shown the relationship between individual actors and scenes on the one hand, and aggregates, trends, and patterns at a larger scale on the other, such that the former are enlisted to rhetorically support and validate the graphs, charts, timelines, and other data visualizations. Halloran characterizes this in terms of the addition of 'weight':

> Bar charts are great for showing relative scale, but they can feel disconnected from what it is they represent. By building the bars out of stacks of figures, each representing 1,000 people who died, I tried to make the bars feel bigger and weightier. (Neil Halloran, creator of Fallen.io, cited in Emory, 2015)

His visual techniques to traverse from the aggregated numbers of the graph (portraying death at a distance, sometimes inflected with a sense of the sublime) to the granularity and comparative intimacy of photographs and video clips can be read in the context of what Latour calls the 'zoom effect' (2014, p. 121), fabricating a smooth trajectory between different and discontinuous knowledge-making and cultural practices. Numbers and statistics about war deaths may be seen as 'thin descriptions' (Porter, 2012), subtracting detail, decontextualizing people and events, articulating and attending to only certain aspects of situations through social practices of commensuration (making comparable through common metrics) and quantification (making numbers) which thereby enable styles of reasoning and sense-making across situations, space, and time (Espeland & Stevens, 1998, 2008; Verran, 2010, 2015). Halloran's strategies to recontextualize and thicken data by visually associating graphical aggregates to images of individuals could be read as a form of what Desrosières calls 'proof in use realism', whereby data are treated as 'self-sufficient' and 'without footnotes to interfere with the message' (2001, p. 346). While Halloran intersperses his narrative with verbal caveats about controversies of estimation, quantification, probability, and inference, the data visualization practices of the two pieces stabilize and solidify numbers, and portray them as 'given', at least provisionally, for the purposes of narrating past and future events from afar, including through the anchoring of averages, estimates, and provisional counts with audiovisual 'weight' and zoom effects.

Such visual practices may thus risk implying a misleading sense of immediacy, foregrounding the direct representational capacities of numbers in a way which leads attention away from social, cultural, and political processes involved in their making—including practices of counting, classifying, estimating, averaging, contesting, and publishing them. Even operations as apparently straightforward as counting can be contentious and dependent on fragile networks, infrastructures, and practices, such as determining group membership, interpreting scientific images, or estimating crowd sizes (Martin & Lynch, 2009). The settings of the infinite canvas and the cosmic surround may be taken as a rhetorical embodiment of what Haraway calls 'disembodied vision' and the 'god trick of seeing everything from nowhere' which she associates with the visualizing technologies of science, management, and the military (1988, pp. 581, 590). As Lauren Klein recently puts it, 'the view from a distance, is, of course, as much of a view from a particular place as a view from up close' (2018).

How might we surface the particular situated perspectives that underpin these data epics? As well as looking at the design process, one might also examine the making, selection, and translation of data which shape how life and death are rendered intelligible and experienceable through data visualizations (Gray, Bounegru, Milan, & Ciuccarelli, 2016; Gray, 2018). Both pieces involve gathering and animating different types of numbers from different information sources and data infrastructures (Gray, Gerlitz, & Bounegru, 2018), which are listed in a 'data sources' section, including both original and processed datasets. A Google Fusion table file linked to *The Fallen* gives a list of 602 sources, most of which are from Wikipedia (81 percent), with the remaining 19 percent from a combination of academic, military, governmental, and hobbyist websites. These sources have been transformed through a combination of normalizing, adjusting, averaging, and interpolating in order to render numbers commensurable and visualizable. As well as the work to integrate and harmonize different sources of information, the apparent continuity between them is further enabled through the use of common visual formats, styles, sounds, colour, motifs (such as isotypes), and other effects. These provide an aesthetic vocabulary for dramatizing quantified collectives in narratives at an epic scale.

The data epic may thus be considered as another emerging area of practice wherein one might study how 'ideology, power and politics are at work in data visualisation' (Kennedy, Hill, Aiello, & Allen, 2016, p. 732), including in relation to the making of public data, and the particular ways in which data are made public. The broad narratives of war, peace, stability, and violence which are performed through data are not without baggage. War deaths are considered in terms of battles between nation states, rather than the colonial projects in which they were involved. Viewers are invited into particular subject positions in witnessing these narratives. The question of who has agency and who can do what to whom in situations of life and death, peace and war, may be considered in terms of what Mbembe calls the 'necropolitical', or 'contemporary forms of subjugation of life to the power of death' (2003, p. 39). Halloran cites Steven Pinker's argument about the decline of violence after WWII as the inspiration for the end of *The*

Fallen. But the selection and interpretation of events in order to make this argument are not uncontested, with some suggesting that 'war and violence are not declining, but they are being transformed', in particular through the replacement of interstate conflict with violence in poorer countries armed and supported by richer ones (Mann, 2018, p. 37).

The techniques and politics of producing aesthetics of distance exemplified in Halloran's data epics—oscillating between vast, sublime perspectives across space and time and the texture of particular situations at a relatable scale—are an emerging area of public data culture for both researchers and practitioners to attend to. They may be relevant in relation to narrating and making sense of other complex and transnational issues such as climate change, migration, and inequality, where it may be desirable to attend to dynamics of injustice with different registers, scales, and temporalities, as well as scrutinizing the means through which such perspectives are produced.

References

Becker, H. S. (2007). *Telling about society*. Chicago: University of Chicago Press.

Boltanski, L. (1999). *Distant suffering: Morality, media and politics*. Cambridge & New York: Cambridge University Press.

Burke, E. (1998 [1757]). *A philosophical enquiry into the origin of our ideas of the sublime and beautiful*. Oxford: Oxford University Press.

Daston, L., & Galison, P. (2010). *Objectivity*. New York: Zone Books.

Davies, W. (2015, January 12). The Data Sublime. *The New Inquiry*. Retrieved from https://thenewinquiry.com/the-data-sublime/

Desrosières, A. (2001). How real are statistics? Four possible attitudes. *Social Research*, *68*(2), 339-355.

Desrosières, A. (2002). *The politics of large numbers: A history of statistical reasoning* (C. Naish, Trans.). Cambridge, MA: Harvard University Press.

Dvorsky, G. (2017, November 8). Disturbing data visualization shows just how many people would die in a nuclear war. *Gizmodo*. Retrieved from https://gizmodo.com/disturbing-data-visualization-shows-just-how-many-peopl-1797748588

Emory, S. (2015, June 28). Data visualizing WWII's devastating death toll. *VICE*. Retrieved from https://creators.vice.com/en_uk/article/4xqq73/inside-look-data-visualizing-wwiis-death-toll

Espeland, W. N., & Stevens, M. L. (1998). Commensuration as a social process. *Annual Review of Sociology*, *24*(1), 313-343. https://doi.org/10.1146/annurev.soc.24.1.313

Espeland, W. N., & Stevens, M. L. (2008). A sociology of quantification. *European Journal of Sociology / Archives Européennes de Sociologie, 49*(3), 401-436. https://doi.org/10.1017/S0003975609000150

Galison, P. (1990). Aufbau/Bauhaus: Logical positivism and architectural modernism. *Critical Inquiry, 16*(4), 709-752.

Gray, J. (2018). Three aspects of data worlds. *Krisis: Journal for Contemporary Philosophy, 1.*

Gray, J., Bounegru, L., Milan, S., & Ciuccarelli, P. (2016). Ways of seeing data: Toward a critical literacy for data visualizations as research objects and research devices. In: S. Kubitschko & A. Kaun (Eds.), *Innovative methods in media and communication research.* (pp. 227-251). Cham: Palgrave Macmillan.

Gray, J., Gerlitz, C., & Bounegru, L. (2018). Data infrastructure literacy. *Big Data & Society, 5*(2), 1-13. https://doi.org/10.1177/2053951718786316

Halloran, N. (2015). *The Fallen of WWII* [Video file]. Retrieved May 1, 2018 from http://www.fallen.io/ww2/

Halloran, N. (2017). *The Shadow Peace: The Nuclear Threat* [Video file]. Retrieved May 1, 2018 from http://www.fallen.io/shadow-peace/1/

Haraway, D. (1988). Situated knowledges: The science question in feminism and the privilege of partial perspective. *Feminist Studies, 14*(3), 575-599. https://doi.org/10.2307/3178066

Kant, I. (2007). *Critique of judgement* (J. C. Meredith & N. Walker, Trans.). Oxford: Oxford University Press.

Kennedy, H., Hill, R. L., Aiello, G., & Allen, W. (2016). The work that visualisation conventions do. *Information, Communication and Society, 19*(6), 715-735. https://doi.org/10.1080/1369118X.2016.1153126

Kennedy, H., & Hill, R. L. (2017). The feeling of numbers: Emotions in everyday engagements with data and their visualisation. *Sociology, 52*(4), 830-848. https://doi.org/10.1177/0038038516674675

Klein, L. (2018). *Distant reading after Moretti.* Paper presented at the MLA2018, New York. Retrieved from http://lklein.com/2018/01/distant-reading-after-moretti/

Latour, B. (2014). Anti-Zoom. In S. Pagé, L. Bossé, H. U. Obrist, & C. Staebler (Eds.), *Olafur Eliasson: Contact.* Paris: Flammarion. Retrieved May 1, 2018 from http://www.bruno-latour.fr/node/609 Liu, A. (2004). Transcendental data: Toward a cultural history and aesthetics of the new encoded discourse. *Critical Inquiry, 31*(1), 49-84.

Mann, M. (2018). Have wars and violence declined? *Theory and Society, 47*(1), 37-60.

Martin, A., & Lynch, M. (2009). Counting things and people: The practices and politics of counting. *Social Problems, 56*(2), 243-266. https://doi.org/10.1525/sp.2009.56.2.243

Mbembe, A. (2003). Necropolitics. *Public Culture, 15*(1), 11-40. https://doi.org/10.1215/08992363-15-1-11

Neurath, O. (1944). The human approach to visual education. *Health Education Journal, 2*(2), 61-66. https://doi.org/10.1177/001789694400200204

Neurath, M., & Kinross, R. (2009). *The transformer: Principles of making isotype charts*. London: Hyphen Press.

Pinker, S. (2015, August 23). On my radar: Steven Pinker's cultural highlights. *The Guardian*. Retrieved from https://www.theguardian.com/culture/2015/aug/23/on-my-radar-steven-pinker-psychologist-author

Porter, T. M. (1986). *The rise of statistical thinking, 1820-1900*. Princeton, NJ: Princeton University Press.

Porter, T. M. (1996). *Trust in numbers: The pursuit of objectivity in science and public life*. Princeton, NJ: Princeton University Press.

Rayward, W. B. (Ed.). (2017). *European modernism and the information society: Informing the present, understanding the past*. London: Routledge.

Robinson, K. S. (2017). *New York 2140*. New York: Orbit.

Rottenburg, R., Merry, S. E., Park, S.-J., & Mugler, J. (Eds.). (2015). *The world of indicators: The making of governmental knowledge through quantification*. Cambridge: Cambridge University Press.

Stallabrass, J. (2007). What's in a face? Blankness and significance in contemporary art photography. *October, 122*, 71-90.

Verran, H. (2010). Number as an inventive frontier in knowing and working Australia's water resources. *Anthropological Theory, 10*(1-2), 171-178. https://doi.org/10.1177/1463499610365383

Verran, H. (2015). Enumerated entities in public policy and governance. In: E. Davis & P. J. Davis (Eds.), *Mathematics, substance and surmise*. (pp. 365-379). New York: Springer, Cham.

About the author

Jonathan Gray is Lecturer at the Department of Digital Humanities, King's College London; Cofounder of the Public Data Lab; and Research Associate at the Digital Methods Initiative (University of Amsterdam) and the médialab (Sciences Po, Paris). More about him can be found at jonathangray.org and he tweets at @jwyg.

20. What a line can say: Investigating the semiotic potential of the connecting line in data visualizations

Verena Elisabeth Lechner

Abstract

The line is a graphical element widely used in data visualizations, its purpose often being to signal a connection between other visual elements. Based on social semiotic theory, this article investigates what semiotic functions connecting lines can have and how these functions can be related to variations in form. The results show that, in addition to the basic function of connecting elements, such lines can also indicate the level of certainty, direct the viewer to read the information either as a narrative or a conceptual claim, indicate patterns of cohesion, and regulate the viewer's position. These findings allow for further empirical research on the formation of visual conventions.

Keywords: Visual variables; Relation; Link; Metafunction; Modality; Arrow

Introduction

New digital forms of data visualization, as they appear for example on online newspaper pages and the webpages of organizations, companies, and private persons, offer the possibility to make data accessible for specialists as well as the broad public. The particular ways in which such graphical forms make meaning to the readers contribute to their social power in society, as Krippendorff states: 'We do not react to the physical properties of things, but act on what they mean to us' (1998, pp. 01_8).

A central task of various types of data visualizations (such as network visualizations, route maps, and others) is to show how different visual

Engebretsen, M. and H. Kennedy (eds.), *Data Visualization in Society.*
Taylor & Francis Group, 2020
DOI 10.5117/9789463722902_CH20

elements are connected. Such connections are often represented by lines, which are in the focus of this chapter.[1] This chapter deals with the meaning potential of this basic element of the language of graphics, and it is thus concerned with the detail level of data visualizations. Although several examples will be given, this chapter stays on the theoretical level, whilst also opening up for practical investigations.

The use of graphical lines to represent connections between elements is an old technique (Bertin, 2011; Brinton, 1939) but still ubiquitous in current data visualizations. Nevertheless, their potential for making meaning has changed in the course of time, as the options for visual representation have increased, especially through the advent of digital production techniques and output devices. Today, the fact that connections are often represented by lines in data visualizations can be observed not only in the number of published data visualizations that have this characteristic (see Figure 20.1 as an example). It can also be observed within several tools available for digital creation of data visualizations (e.g. D3.js, Tableau, R). However, what functions these connecting lines have, and what effects different visual appearances of the lines have on their meaning potential, are issues which have not been widely researched.[2]

This chapter asks: What semiotic functions can a connecting line in a data visualization have, in addition to the basic function of indicating a connection between two visual elements?

Raising this question is necessary, not only for the scientific community, in order to generate more knowledge about visual language, but also for practitioners on the production side, in order to raise their awareness on how to communicate as nuanced and clearly as possible with their readers. Thus, a central aim for the chapter is to offer a language for discussing the functions of smaller elements within data visualizations. This is fundamental, because the meaning potential of visual elements informs important decisions in the design process.[3]

1 Besides the connective function, lines may have many other functions, e.g. to define contours, to separate elements, to lead the eye, or to function as a base element for textures and patterns (Poulin, 2012, p. 29).

2 It is possible to find literature about meaning potentials for different visual appearances of lines in general (cf. Habermann, 2015, p. 649; Horn, 1998, pp. 147-148; Ware, 2013, p. 225). However, the suggested meaning potentials are not directly transferable to connecting lines within the context of digital data visualizations.

3 It has to be noted here that other reasons than the desired function (e.g. the circumstances in the data visualization developer tools or aesthetical reasons) can influence the designer's decisions.

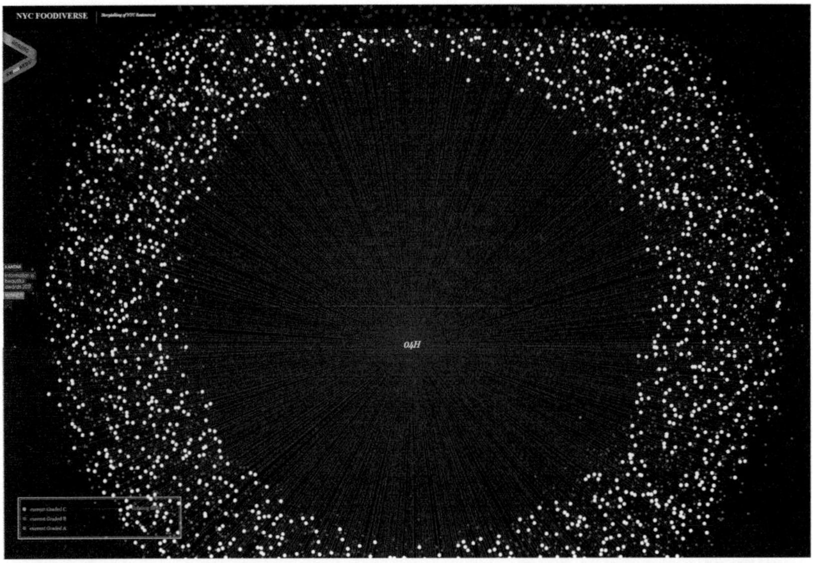

Figure 20.1. Example of a data visualization using lines to represent the connections between sanitary problems (central group of purple letter and number codes) and the restaurants in Manhattan, NYC they occurred in, represented as dots in the outer circle. From 'NYC FOODIVERSE' by W. Su, 2017 (http://nycfoodiverse.com). Copyright 2017 by W. Su. Reprinted with permission.

The approach chosen to answer the question will be presented in three steps. I will start with the central element, the connecting line itself, and describe especially *how* the line connects. In the second step, the elements that are being connected are discussed, in other words, *what* the line connects. In the last part, I describe how the connecting line is *integrated into the whole* data visualization and thus contributes to the creation of larger structures of information. However, before these steps can be taken, I need to make some terminological clarifications and outline the theoretical framework of the discussion.

Terminological considerations

The two terms that identify the object of study, namely *line* and *connection*, are used in many contexts and with a number of different meanings. Within the field of graphics, the French cartographer Jacques Bertin identified the line as one of three basic elements in the language of graphics, together with the point, and the area (2011, p. 271). Decades earlier, the Russian painter Wassily Kandinsky named the point and the line as the two elements that 'constitute the conclusive material for an independent kind of painting—graphic' (1947, p. 20). Concerning the formal characteristics of the line, he stated that the line is a product of the moving point (1947, p. 57). This movement is what provides the line with its main formal characteristics. Wucius Wong points out that the breadth of a line is 'extremely narrow' and 'its length is quite prominent' (1993, p. 45). However, as the gestalt laws state, single elements grouped in a certain way, or incomplete lines, can also be perceived as lines (Lauesen, 2005, pp. 68-69). Thus, in this chapter the term *line* refers to all kinds of visible lines, including incomplete lines and arrangements of visual elements that can be perceived as lines.

Returning to the central term *connecting line*, this refers to lines which have the basic function of establishing or indicating a connection. This means that it must be possible to identify the two parts of a connection: the single *components* and the *relationship*, as Bertin describes it (2011, p. 271), or the *nodes* and the *connector*, to mention the terms used by Engelhardt (2002, p. 40). The terminology used to describe the phenomenon of connections varies, since many synonyms exist, such as *relation*, *relationship*, *link*, and *tie* (Fergusson, 1992, p. 88). The terms are used slightly differently in various corners of the field (Brinton, 1939, pp. 43-72; Engelhardt, 2002, p. 40; Richards, 1984, p. 3/21; Ware, 2013, pp. 221-226). However, Kress and van Leeuwen, who look at graphics from a linguistic perspective similar to this chapter, use the compound term *connecting line* (2006, p. 59), which I have chosen to adopt.

Theoretical framework

Data visualizations, like other types of semiotic material, offer a specific way to communicate meaning. In order to make meaning out of a data visualization, the reader has to apply certain rules, which help him or her to decode what the producer of the data visualization wanted to communicate. On the other side, the producer of the data visualization most likely also had similar rules in mind when deciding on this specific form of visual representation. The meaning potential carried by the visual forms through the application of such shared rules defines the social function of the forms.

In the understanding of how these rules evolve, traditional semiotics and social semiotics differ on certain central aspects. In the former case, rules are seen as predefined and more or less consistent, and the communicating persons have to learn these rules before they are able to apply them, either in production or in interpretation (Hodge & Kress, 1988, p. 12). In contrast to that, in social semiotics, as van Leeuwen (2005, pp. 47-48) describes, people actively participating in social activities are seen as the ones who generate these rules—on the basis of certain culturally shared codes. He further explains that semiosis is an ongoing process, where the sign users themselves have the power to influence and change the rules. This again implies that those rules are seen to be rather unstable and to a high degree dependent on the social situation.

Returning to the case of data visualizations, which are often produced for a large and diverse target group, we can assume that some rules exist, connecting forms to meanings. But they might be somewhat different from how they were years ago and might also be dependent on the social context. M. A. K. Halliday laid the theoretical basis for seeing text as 'a sociological event, a semiotic encounter through which the meanings that constitute the social system are exchanged' (1978, p. 139, emphasis deleted). However, in the centre of his research stands verbal language. In their seminal work *Reading Images: The Grammar of Visual Design* (2006, p. 2), Theo van Leeuwen and Gunther Kress state that visual structures, just like linguistic structures, invite a particular interpretation, that is formed by experience and social interaction.[4]

From these theoretical abstractions, we can conclude that the process of meaning-making in contexts involving data visualization is a process where certain culturally formed, relatively stable codes and conventions interplay with a set of more unstable, situated rules concerning the exact meaning of the visual forms displayed. This interplay also defines the meaning of connecting lines, and calls for empirical research to investigate which semiotic functions are conventionalized and which are not.

Halliday defined three universal functions in verbal language, also known as 'metafunction[s]' (2004, p. 30), understood as different aspects of the meaning potentials of a clause. Any clause, any verbal utterance, carries all three metafunctions simultaneously: the 'ideational' (what is said about the world), the 'interpersonal' (how social relations between the participants are constructed), and the 'textual metafunction' (how the parts construct a coherent whole) (pp. 30-31). As Kress and van Leeuwen adapted Halliday's concept of social semiotics to other semiotic modes, they also applied the concept of these metafunctions to the analysis of visual expressions (2006, p. 13), and during the last decades, their work has been adapted by many other researchers. Yet, for every new social

4 It should be underlined that they also reflected critically on this comparison between visual and verbal structures and pointed out that this similarity easily can be overemphasized (see also Kress & van Leeuwen, 2006, p. 76).

semiotic study, the systems of meaning making have to be defined again in order to make a systematic analysis possible. This is the case because different types of visual material offer different semiotic choices and need to be interpreted in different ways. Thus, it is also necessary to define the system of choices activated in the kind of visual material investigated in this book. To develop the basis for that, focusing on connecting lines, is the contribution of this chapter.

Towards an analytical procedure

Functions related to the connecting line itself

Having given a brief insight into terminology and the theoretical framework for the study, I will now present a proposed method for analysing types and functions of connecting lines. Starting with the central element, the connecting line itself, its main function first has to be pointed out. This function is already implied in the word *connecting,* and therefore works as a basic selection criterion for the kind of lines that are to be investigated. As Clive James Richards noted, a line can have a verb-like function, and in the verbal translation of a figure showing two letters with a line in between, he states: 'A is connected to B' (1984, p. 3/21). In the words of Halliday, this corresponds to the ideational meaning, which says something about a process, or 'goings-on' (2004, p. 170). More precisely, the line represents the process itself. Secondly, the connecting line might also say something about the associated circumstances, e.g. whether the connection is strong or weak. The third component of a process—the participants—is determined at the ends of the line, showing what is connected.

Following the proposed analytical structure, we can summarize that a line *connects* certain objects in a *certain manner.* It represents a connection in a specific way, and it can, among other things, also point to the certainty of this connection.[5] In verbal language we have several alternatives to express the certainty of a piece of information, building up the modality system of the language in question (Halliday, 2004, p. 147). This system offers means to express the level of certainty that the speaker wants to give a certain claim—e.g. choosing between *This is probably true* and *This is certainly true.*

5 In his visual grammar of relationship representations Colin Ware named, for example, the strength of a connection as a characteristic that could possibly be expressed by different line weights (2013, p. 225).

Hodge and Kress assumed that modality markers can also appear in other kinds of media, although they considered them to be not so clearly articulated as the ones in verbal language (1988, pp. 121-122). Kress and van Leeuwen further investigated modality in terms of visual communication, including examples like illustrations, photographs, and pieces of art (2006, pp. 159-180). Because modality markers are closely related to the social participants in the communication process and are used to build shared truths, they consider modality as a phenomenon to be categorized as part of the interpersonal rather than the ideational metafunction (2006, pp. 159-160).

Van Leeuwen notes that different coding systems can have different kinds of *coding orientations*—like the 'naturalistic', 'technological', 'sensory', and 'abstract' orientations (van Leeuwen, 2005, pp. 168-170). As he understands it, these different orientations mean that the scales of modality may have different types of markers, or criteria for what is regarded true and realistic, and what is not. A line graph may show little details of the background, compared to a photograph, but that does not indicate that what is represented in the diagram is not true (p. 167). Within the abstract coding system, 'visual truth is abstract truth' (p. 168). 'The more an image [...] represents the general pattern underlying superficially different specific instances, the higher its modality from the point of view of the abstract truth. This is expressed by reduced articulation', he explains further (p. 168). This means that if a data visualization is seen as being part of this coding system, an abstract way of visualizing data conveys an impression of truth.

Visualizations of past and future paths of hurricanes can serve as an example of data visualizations that often contain a degree of uncertainty, such as the data visualization *Irma is following a well-worn path* (Dottle, King, & Koeze, 2017).[6] Here, the future, uncertain path of Hurricane Irma is shown as a dashed line, within a shape surrounded by another dashed line. The interruption of the lines therefore serves as the modality marker within this example. Here, the lack of sufficient data is signalled visually in the data visualization. In other cases, the data available may be precise and sufficient, but for certain reasons (like privacy protection) the visualizations are intentionally made to look imprecise, through the application of uncertainty markers (Dasgupta, Chen, & Kosara, 2012, p. 1022). Yet, although it might appear clear in the example of Hurricane Irma's path, there does not exist any general and recognized description of how certainty is expressed in graphical material through different forms of connecting lines. Sometimes, like in *Musicmap* (Crauwels, 2016, see https://musicmap.

6 See https://fivethirtyeight.com/features/what-lies-in-irmas-path/.

info/), dashed lines are used for purely compositional reasons, in order to separate them from other lines. The same can be said about other visual variables shaping the physical appearance of the line (like colour, shape, etc.). How these visual variables indicate specific types of connections, whether through convention or explicit explanation, is an issue that calls for extensive empirical investigation.

Functions related *to what* the line connects

As stated earlier, to recognize a connecting line as such, it must be possible for the reader to identify not only the line itself, but also the connected components. When looking at different types of data visualizations, it becomes apparent that sometimes lines are used to connect two different elements (as in network diagrams, see Figure 20.1), whereas in other types the lines connect two different states of the same element (as in route maps). In either case, the function type in question belongs to the ideational type, saying something about states in the world.

In order to trace the graphical lines to the natural, non-digital world, a relevant source is Tim Ingold (2007, pp. 41-43), who writes about lines from an ethnological viewpoint. He divides lines into five groups, two of which are called 'threads' and 'traces' (p. 41).[7] In his account, *threads* (such as a washing line, an electrical circuit, a tightrope, or a skein of yarn) seem to correspond to the former group of lines, connecting two different elements. *Traces* (such as a scratched line or the slime trail of a snail), on the other hand, relate to the connecting lines of the second type, connecting different states of the same element.

The way that connecting lines relate the connected components to each other forms their *representational structure*, a concept investigated by Kress and van Leeuwen (2006) within many different visual media. The two main categories into which they divide their investigated material are 'narrative structures' and 'conceptual structures' (p. 79). What is represented in narrative structures are 'unfolding actions and events, processes of change, transitory spatial arrangements' (p. 79). They contain 'vectors' (p. 59) which show a direction. In data visualizations, connecting lines can work as vectors when the direction is made explicit, e.g. by an arrowhead or a tapering body, as in Figure 20.3. Conceptual structures, on the other hand, have no vectors, and represent 'participants in terms of their more generalized

7 The three other groups he calls 'cuts, cracks and creases' (2007, p. 44), 'ghostly lines' (p. 47), and 'lines that don't fit' (p. 50).

and more or less stable and timeless essence, in terms of class, or structure or meaning' (p. 79). In other words, narrative structures always contain a form of action, whereas conceptual structures describe a phenomenon in a certain state. Although originating in different disciplines, it seems obvious that Ingold's traces to a certain degree correspond to the concept of narrative structures, while threads correspond more closely to the concept of conceptual structures.[8] Based on both sources, I suggest that connecting lines have the function of directing the viewer to read the information either as a narrative or as a conceptual claim, and that the way they do this through their visual appearance in current data visualization design is an issue that calls for both theoretical and empirical investigation.

Functions related to the line as part of larger text units

After having started this investigation on the micro level focusing on the connecting line itself, and then extending it to the connected units, it is now time to have a look at the surrounding context, that is, the wholeness of the data visualization. Some data visualization types, like network visualizations or tree diagrams, traditionally contain many connecting lines, often even lines interconnected with each other. Other data visualization types, like flow maps, might either show only one or a few lines, which are not necessarily interconnected (although they may cross each other). Such examples make it obvious that connecting lines contribute to, and are integrated in a bigger whole, a composition of semiotic elements. This observation is a starting point for analyses that focus on the textual, also called the compositional metafunction. At this level, *cohesion* is a core concept.

Linguists working on the discourse level have a long tradition of describing connectedness in verbal texts (Sanders & Pander Maat, 2006, p. 591). For the English language for example, Halliday and Hasan published their pioneering book *Cohesion in English* already in 1976, of which some main ideas shall be explained in the following (1994, pp. 1-4). According to them, what makes a text be regarded as such, is that it forms a recognizable, coherent unit of meaning. For that, it needs to have meaning relations that combine the single text units. These are called 'cohesive properties' (1994, p. 4). These properties come into action when 'the interpretation of some element in the discourse is dependent on that of another' (1994, p. 4, emphasis deleted). In other words,

8 Both Ingold and Kress and van Leeuwen emphasize that it is not always possible to distinguish their research material exhaustively with their categories. They rather see them as a tool for describing what is represented (Ingold, 2007, p. 50; Kress & van Leeuwen, 2006, p. 86).

a cohesive text contains connections between the single elements of the text, which help the reader to understand the meaning of the entire text.

However, the phenomenon of cohesion is manifest not only in verbal texts. Theo van Leeuwen has investigated forms of cohesion in the field of multimodal texts (2005, pp. 179-268), the principles of which will be shortly introduced here. He lists four ways of constructing cohesion, namely: 'composition', 'rhythm', 'information linking', and 'dialogue' (p. 179). *Composition,* he explains, works with the placement of elements in space. For van Leeuwen, whether an element is placed at the bottom or the top of a page, to the left or to the right, in the centre or in the margin, has an impact on its meaning potential. This impact is often based on metaphors from the physical world. Composition, as he continues, is the spatial equivalent to *rhythm.* This in turn is formed by a transition between two opposing states repeated in the dimension of time, such as soft and loud, fast and slow, big and small, and so on. *Information linking* has to do with the ways that one piece of information can be related to another piece. A *dialogic* structure, as the fourth form of cohesion he suggests, appears when more than one voice is perceived either simultaneously or sequentially, like in a spoken dialogue, or in a film track, where the flow of images and the music track may establish a dialogue (pp. 179-268).

When investigating the cohesive functions of connecting lines in digital data visualization, I propose to focus on *composition* and *information linking,* for a number of reasons.

Composition is built up by the spatial arrangement of the constituting elements in the visual object. In the case of data visualizations containing several connecting lines, the conscious placement of such connections, which include the connecting line as well as the connected elements, can be used to imply a certain meaning potential. On a macro level, the composition of these elements can build up specific types of data visualizations and help to define the roles of the connected elements. The configuration of connecting lines can indicate, for example, sequences, hierarchies, or networks, offering very different roles for each of the involved visual elements (see Figure 20.2).

In some types of interactive data visualization, the user is enabled to change the placement of the nodes and its connections manually. The visualizations in the report *Panama papers—the power players* (ICIJ, 2017), for instance, offer this affordance. (Figure 20.3 is a static screenshot of one of them.)[9]

Information linking, as another way to construct cohesion (van Leeuwen, 2005, pp. 219-247) shall here be discussed in further detail. Borrowing the

9 See https://panamapapers.icij.org/the_power_players/

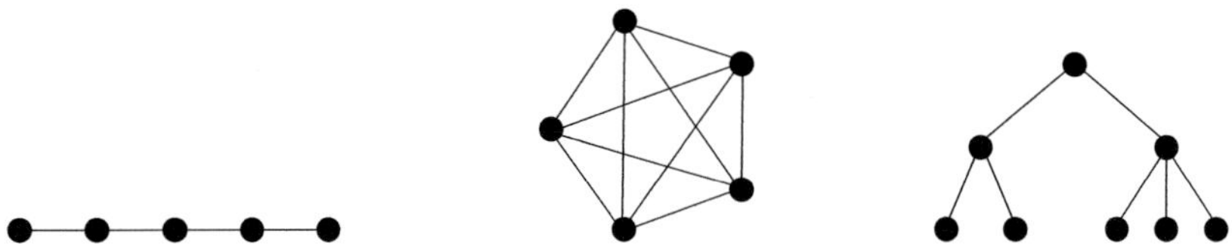

Figure 20.2. Three exemplary compositions of connecting lines and their connected elements. Illustration by V. E. Lechner.

linguistic concept of conjunctions from cohesion within verbal texts (as described in Halliday & Hasan, 1994, pp. 336-338) van Leeuwen states that links are 'temporal, logical or additive' (2005, p. 222). He further explains that if a temporal link occurs, this points to the fact that the two single pieces of information happen either at different points in time or in parallel. A logical link highlights that one of the information pieces 'gives a reason for, a condition of, or a comparison with the information in the first item' (p. 223). If it is not a temporal or logical link, yet the one item adds information to that given by the other, he concludes that additive linking takes place.[10]

Whereas in verbal language, the different linking types can be determined because of explicitly used words like conjunctions, in data visualizations, such 'cohesive tie[s]' (1994, p. 329), as Halliday and Hasan call them, might not always be so obvious. But as van Leeuwen shows with examples of multimodal, non-linear texts (2005, pp. 226-247), they do exist and have to be found by the reader to form the storyline. For this process, the surrounding context plays an important role, as it might influence which linking type might be the most relevant in a specific data visualization.

In Figure 20.3, we see a visualization where the connecting lines indicate a combination of additive and logical linking. The connections around Sigmundur Davíð Gunnlaugsson (former prime minister of Iceland), could be verbally translated to: *Gunnlaugsson is registered in address X and is a shareholder of company Y. Similarly, his wife is also registered at the same address* and *is also a shareholder of the same company* (*which was registered by a consulting firm*). The cohesion markers *and* and *similarly* point to additive respectively logical linking. The example shows that translating the data visualization into text may help to detect the ways in which lines are used as cohesion markers.

As the previous examples show, *composition* and *information linking* are relevant when studying cohesion formed by graphical lines in data

10 Examples for cohesion markers within verbal text: *then, next, meanwhile* (temporal linking); *because, for that reason, otherwise* (logical linking); *and, or* (additive linking). These examples can be found in van Leeuwen (2005, pp. 222-224) as well as in Halliday & Hasan (1994, pp. 336-338).

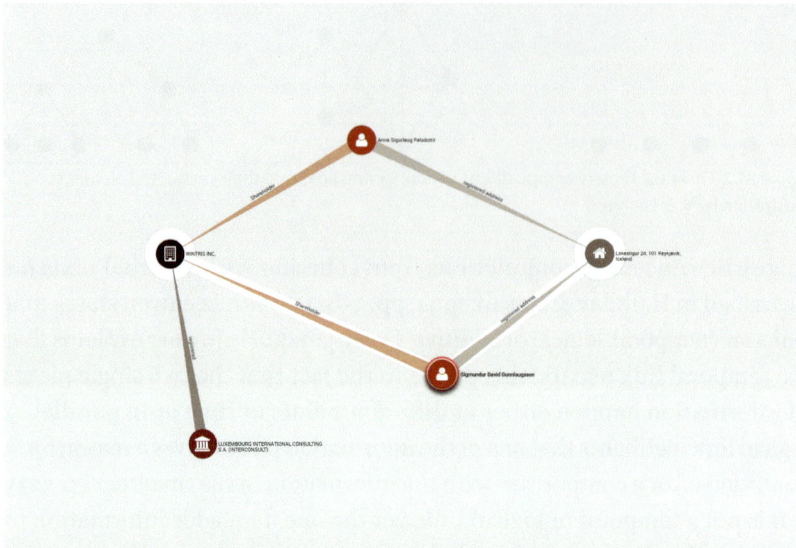

Figure 20.3. Visualization of the connections related to Sigmundur Davíð Gunnlaugsson. From 'Panama Papers—The Power Players' by The International Consortium of Investigative Journalists, 2017 (https://www.icij.org/investigations/panama-papers/the-power-players/). Copyright 2017 by ICIJ. Reprinted with permission.

visualizations. However, *rhythm* and *dialogue* can also form cohesion in such forms of textual expressions.

Rhythm can be established through visual representations of processes occurring over time, analogue to the rhythmic structures perceived in music (van Leeuwen, 2005, p. 182). In data visualizations, this form of cohesive structure can be perceived through observations of visual repetitions and certain patterns of such. It can either be shown by an animation or by presenting these repetitions in a linear sequence in a static presentation. Connecting lines can play a role in such a rhythmic organization of a data visualization. A visualization in the news site of *The Washington Post* shows flight patterns after the Brussels attacks on March 22, 2016 (Muyskens, 2016)[11]. In the early morning, all planes are flying directly to Brussels airport and build up a regular pattern of moving lines. The lines connect the planes to geographical points. But suddenly their flight routes change and the lines representing them develop an irregularity because the planes turn back before entering the airport. This example points to the fact that it is not always the coherent pattern itself that forms the most interesting feature

11 https://www.washingtonpost.com/news/wonk/wp/2016/03/23/watch-what-happened-to-flight-patterns-in-the-moments-after-brussels-attacks/

of a (visual) text, but rather the instances of violation of the pattern. The rhythm is disrupted, and attention is attracted.

In many publicly available data visualizations, a *dialogue* between different semiotic modes appears, between the visual forms, verbal elements, numbers, and sometimes dynamic modes like music or speech. How this dialogue is organized in time and space is interesting to analyse, but not so relevant for the investigation of connecting lines.

Exploring the functions of the line in relation to the total data visualization, we may also look at two aspects of interpersonal meaning, namely the ways in which the reader's position is regulated through *frame size* and *perspective*. According to Kress and van Leeuwen (2006, pp. 124-129) social distance between a human represented on an image (e.g. a photo) and the viewer of the image is managed by different frame sizes, 'close distance', 'middle distance', and 'long distance' (pp. 125-126). They further suggest that a similar set of relations is possible between the viewer and depicted non-human elements. In data visualizations, the connecting lines can be presented from a very far distant position (showing much of the surrounding context) or from very near, just as if the viewer could touch them. In interactive data visualizations, where the user is able to zoom, this could even be changed manually (as in *Musicmap* (Crauwels, 2016)).[12] Such interactive mechanisms offer the reader a position as an active participant in the communication, being able to choose the frame size and therefore also the position from where the data visualization is observed.

Besides frame size, the chosen perspective also influences the relation between the viewer and the represented objects. Data visualizations are often presented in a direct frontal or a top-down angle, which adds to their aura of objectivity, whereas other angles rather indicate subjectivity (Kress & van Leeuwen, 2006, pp. 135-151). Placing the connecting line in a two- or three-dimensional space makes it possible for the connecting line to indicate perspective. In data visualizations, a top-down angle is e.g. often used for route maps, where the movement of certain objects is shown on a map. One such example is the first visualization in the news article *Bussed out: How America moves its homeless* (Outside in America team, Bremer, & Wu, 2017), showing the route of a homeless person relocating in the US.[13] In another visualization from the same article, the same kind of geographical movements are presented as curved lines viewed from a frontal perspective

12 see http://www.musicmap.info

13 see http://www.theguardian.com/us-news/ng-interactive/2017/dec/20/bussed-out-america-moves-homeless-people-country-study

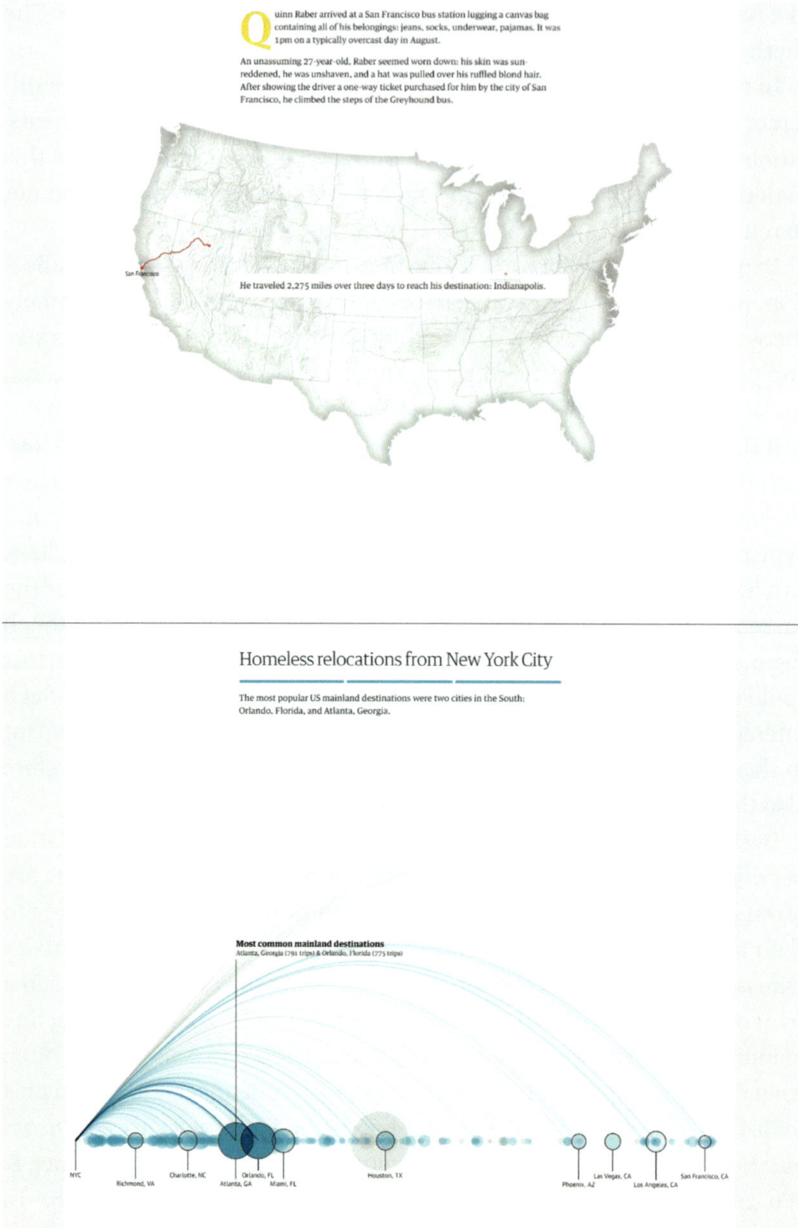

Figure 20.4. Two visualizations of spatial movement, using a top-down angle in a route map (upper picture) and a frontal perspective in an arc diagram (lower picture). From 'Bussed out: How America moves its homeless' by Outside in America team, N. Bremer and S. Wu, 2017 (http://www.theguardian.com/us-news/ng-interactive/2017/dec/20/bussed-out-america-moves-homeless-people-country-study). Copyright 2017 by *The Guardian*. Reprinted with permission.

(see Figure 20.4). If we apply Kress and van Leeuwen's principle, both perspectives indicate an objective representation of reality. Similar to frame size, some interactive data visualizations, such as Kim Albrecht's *Cosmic Web* (n.d.) also include possibilities to manually change the perspective.[14]

Conclusion

Beyond the basic function of connecting, four semiotic functions of connecting lines in the context of data visualizations have been identified and described in this chapter.[15] Connecting lines can potentially be used:

1) To indicate the level of certainty of a specific connection as a modality marker.
   Both the visual scales of the modality markers as well as what the exact values indicate in a certain context need to be investigated further on a corpus of data visualizations before they can be used for analysing single data visualization examples.
2) To direct the viewer to read the information either as a narrative or as a conceptual claim—how things *develop* or how things *are*.
   It should be possible to identify the two types of representational structures with the help of the surrounding context of the data visualization. What kinds of sub-categorization are possible and reasonable can only be discovered with the help of a corpus of data visualizations.
3) To indicate patterns of cohesion in a data visualization, and to indicate the role of particular objects in the context of the whole.
   Composition and information linking are especially relevant when investigating a data visualization as a cohesive textual unit.
4) To regulate the reader's position, by regulating the physical relation between the viewer and the connecting line(s).
   The concepts of frame size and perspective, such as proposed by Kress and van Leeuwen (2006), are directly applicable. However, the effects on the reception side call for further research.

In order to see if and how these potential functions are realized in current data visualization design, empirical research on larger corpora of data visualizations is demanded. Such studies would also offer insights in the

14 See http://cosmicweb.kimalbrecht.com/viz/#1

15 This list is not meant to be exhaustive, but it may serve as a starting point for further investigations.

evolving process of conventionalization—the forming of rules for making meaning through data visualizations, widely shared on both the production side and the viewer side. The stronger the conventions become, the stronger data visualizations' role will be in society, because they will afford a more nuanced communication. The relevance of these fields of knowledge can be demonstrated with the social impact of the following two examples. If a data visualization showing the potential future path of a hurricane (which might be indicated by connecting lines as in *Irma is following a well-worn path* (Dottle, King, & Koeze, 2017)) is interpreted incorrectly, this might have an impact on whether or not people decide to leave their homes. Similarly, if a data visualization about problems of a disadvantaged group in society (such as *Bussed out: How America moves its homeless* (Outside in America team et al., 2017), where connecting lines indicate the journey of homeless people taking part in a relocation programme) provokes a long social distance instead of compassion, social awareness might not be developed, and action might not be taken.

References

Albrecht, K. (n.d.). *Cosmic web* [Project page]. Retrieved March 31, 2018 from http://cosmicweb.kimalbrecht.com/viz/#1

Bertin, J. (2011). *Semiology of graphics: Diagrams, networks, maps* (W. J. Berg, Trans.). Redlands, CA: ESRI Press.

Brinton, W. C. (1939). *Graphic presentation*. New York: Brinton Associates.

Crauwels, K. (2016). *Musicmap: The genealogy and history of popular music genres*. Version 1.0.1. Retrieved April 21, 2018 from http://www.musicmap.info

Dasgupta, A., Chen, M., & Kosara, R. (2012). Conceptualizing visual uncertainty in parallel coordinates. *Computer Graphics Forum, 31*(3), 1015-1024). https://doi.org/10.1111/j.1467-8659.2012.03094.x

Dottle, R., King, R., & Koeze, E. (2017, September 8). What lies in Irma's path [Blog post]. Retrieved from https://fivethirtyeight.com/features/what-lies-in-irmas-path/

Engelhardt, Y. (2002). *The language of graphics. A framework for the analysis of syntax and meaning in maps, charts and diagrams* (Doctoral thesis, University of Amsterdam). Retrieved from http://dare.uva.nl/search?arno.record.id=105970

Fergusson, R. (Ed.). (1992). *The Penguin dictionary of English synonyms and antonyms* (revised). London: Penguin Books.

Habermann, H. (2015). *Grundlagen der Gestaltung: Industrie-Design / Kommunikations-Design*. Göttingen, Germany: Cuvillier Verlag.

Halliday, M. A. K. (2004). *An introduction to functional grammar* (3rd ed). London: Arnold.
Halliday, M. A. K. (1978). *Language as social semiotic: the social interpretation of language and meaning.* London: Arnold.
Halliday, M. A. K., & Hasan, R. (1994). *Cohesion in English* (13th impression). New York: Longman Group Limited.
Hodge, R. I. V., & Kress, G. (1988). *Social semiotics.* New York: Cornell University Press.
Horn, R. E. (1998). *Visual language: Global communication for the 21st century.* Bainbridge Island, WA: MacroVU.
ICIJ. (2017). Panama Papers—The Power Players [Investigation project page]. Retrieved July 30, 2018 from https://panamapapers.icij.org/the_power_players/
Ingold, T. (2007). *Lines: A Brief History.* Abingdon: Routledge.
Kandinsky, W. (1947 [1926]). *Point and line to plane: Contribution to the analysis of the pictorial elements.* (H. Dearstyne & H. Rebay, Trans.). New York: Solomon R. Guggenheim Foundation. Retrieved from http://archive.org/details/pointlinetoplaneookand
Kress, G., & van Leeuwen, T. (2006). *Reading images: The grammar of visual design* (2nd ed.). London: Routledge.
Krippendorff, K. (1998, December). *Design discourse: a way to redesign design.* Keynote speech presented at the Society for Science of Design Studies, Tokyo, Japan. Retrieved from https://repository.upenn.edu/asc_papers/227
Lauesen, S. (2005). *User interface design: A software engineering perspective.* Harlow: Pearson Education.
Muyskens, J. (2016, March 23). Watch what happened to flight patterns in the moments after Brussels attacks. *The Washington Post.* Retrieved from https://www.washingtonpost.com/news/wonk/wp/2016/03/23/watch-what-happened-to-flight-patterns-in-the-moments-after-brussels-attacks/
Outside in America team, Bremer, N., & Wu, S. (2017, December 20). Bussed out: how America moves its homeless. *The Guardian.* Retrieved from http://www.theguardian.com/us-news/ng-interactive/2017/dec/20/bussed-out-america-moves-homeless-people-country-study
Poulin, R. (2012). *The language of graphic design: An illustrated handbook for understanding fundamental design principles.* Beverly, MA: Rockport.
Richards, C. J. (1984). *Diagrammatics: An investigation aimed at providing a theoretical framework for studying diagrams and for establishing a taxonomy of their fundamental modes of graphic organization* (Doctoral thesis). Royal College of Art, London, United Kingdom.
Sanders, T., & Pander Maat, H. (2006). Cohesion and coherence: Linguistic approaches. In: *Encyclopedia of Language & Linguistics* (2nd ed.). (pp. 591-595). https://doi.org/10.1016/B0-08-044854-2/00497-1

Su, W. (2017a). NYC FOODIVERSE. Retrieved July 31, 2018 from https://jiahao01121.github.io/Restaurant_data_viz/second_draft_mar_25/index.html

Su, W. (2017b). NYC FOODIVERSE—Storytelling of NYC Restaurant. Retrieved April 25, 2018 from http://nycfoodiverse.com/

van Leeuwen, T. (2005). *Introducing social semiotics*. London & New York: Routledge.

Ware, C. (2013). *Information visualization: Perception for design* (3rd ed.). Waltham, MA: Morgan Kaufmann.

Wong, W. (1993). *Principles of form and design*. New York: John Wiley & Sons.

About the author

Verena Elisabeth Lechner is currently a PhD student at the University of Agder, Norway. She holds an MA in InterMedia from the University of Applied Sciences Vorarlberg, Austria and is especially interested in the visual appearances and meaning potential of graphic elements within data visualizations.

21. Humanizing data through 'data comics': An introduction to graphic medicine and graphic social science

Aria Alamalhodaei, Alexandra P. Alberda, and Anna Feigenbaum

Abstract
In recent years data visualization scholars and practitioners have drawn attention to the need for data to be humanized. In addition to making complex information more coherent, visualizations can work to incorporate empathy and help audiences connect to information. Addressing this call for humanizing data visualization, this chapter considers the emergent area of 'data comics', looking at how the new fields of graphic medicine and graphic social science deal with numeric data. We examine recent data comics from graphic medicine and graphic social science that exemplify the complexities and potential of presenting data in humanizing ways. Our discussion is framed around what we call the EMA framework, considering the Epistemic (knowledge and perspective), Methodological (ways of working), and Aesthetic (practices of representation).

Keywords: Humanizing data; Data comics; Graphic medicine; Graphic social science

The call for humanizing data visualization

In recent years data visualization scholars and practitioners have drawn attention to the need for data to be humanized (D'Ignazio & Klein, 2016; Gray, Bounegru, Milan, & Ciuccarelli, 2016; Kennedy, Hill, Aiello, & Allen, 2016; Lupton, 2017). Looking at possibilities for haptic and visceral data manifestations, Lupton (2017, p. 15) has called for the 'generation of

Engebretsen, M. and H. Kennedy (eds.), *Data Visualization in Society.*
Taylor & Francis Group, 2020
DOI 10.5117/9789463722902_CH21

alternative or counter perspectives and greater opportunities for people to "feel" their data in ways which make sense in the context of their own lives'. In a piece that circulated around social media, data visualizer Giorgia Lupi provocatively asked, 'Can a data visualization evoke empathy and activate us also at an emotional level, and not only at a cognitive one? Can looking at a data visualization make you feel part of a story of a human's life?' (2017). Resonating across these calls for humanizing data is an acknowledgement that while data visualizations often do a good job of clearly presenting information visually, more emphasis can be placed on creating empathy and connecting data to audiences.

In this chapter we introduce 'data comics', looking at the emergent fields of graphic medicine and graphic social science in relation to humanizing data visualization. Recent work by Bach, Wang, Farinella, Murray-Rust, & Riche (2018) on data comics explains the potential of the medium in communicating data-driven stories and provides practical information and theoretical research on how this potential might be achieved. Graphic medicine is an umbrella term used to bring together a growing number of comics that engage with illness, disability, and the healthcare system (Green & Myers, 2010). Graphic social science refers to the use and potential uses of comics in public communication about social science (Carrigan, 2017). By looking at examples from graphic medicine and graphic social science that explicitly engage emotive and empathetic narratives, we explore what Bach et al. (2018) describe as the potential for comics to humanize data. We do so by considering the Epistemic (knowledge and perspective), Methodological (ways of working), and Aesthetic (practices of representation) dimensions of these exemplary data comics.

Both graphic medicine and graphic social science mobilize the graphic medium of comics to engage with data communication in ways that aim to be approachable, accessible, and relatable (McCloud, 1993; Green & Myers, 2010; Williams, 2014; Czerwiec et al., 2015; McNicol, 2016). By *approachable* we mean that the comics medium is familiar to people (McCloud, 1993). They are *accessible* because the information is presented to the readers using iconography that is familiar to targeted cultural audiences (Williams, 2014; Czerwiec et al., 2015; Bach et al., 2018). Finally, data comics rely on elements of storytelling and visual narrative in order to make information *relatable,* using personal experiences as a basis for the interpretation process (Bates, 2012; Bowman, 2017; McNicol, 2016).

In addition, research into graphic visualization has shown that the effective use of text and image can enhance understanding of complex information, especially in low literacy and vulnerable audiences (Green & Myer, 2010; Ahmed-Husain & Dunsmuir, 2014; Al-Jawad & Frost, 2014;

Kassai et al., 2016). For example, beyond patient-doctor understanding, which was the original target relationship of graphic medicine (Czerwiec et al., 2015), comics have been examined as an effective communication medium to enhance social behaviours in young people with autism spectrum disorder (Ahmed-Husain & Dunsmuir, 2014), and pre-surgery education for paediatric patients in order to improve post-surgery recovery (Kassai et al., 2016; current study by Thomas & Schirrmeister, 2018).

Introduction to the EMA framework

To understand how graphic medicine and graphic social science can humanize data visualization through their use of data comics, we created a framework which we call 'EMA' that is structured around three pillars. These pillars are intended to guide data visualization evaluation and the future production of data comics. The pillars are: Epistemic (knowledge and perspective), Methodological (ways of working), and Aesthetic (practices of representation). These three pillars capture the varied formal conventions and underlying theoretical premises of graphic medicine and graphic social science. The EMA framework is a preliminary attempt to attend to the potential benefits for humanizing data that graphic medicine and graphic social science offer through their use of data comics.

The Epistemology pillar of our framework draws on trajectories of feminist knowledge production, specifically the work of Standpoint theorists Patricia Hill Collins (1990) and Sandra Harding (1986), in order to critically interrogate what counts as a data point, a dataset, or a data visualization. Is it only the scatter plot or pie chart that can effectively frame data? Or can comics, graphic novels, textiles, and other media also count as data visualization? Further, our framework acknowledges the *partiality* of knowledge. As Caroline Ramazanoğlu and Janet Holland point out, 'knowledge is partial both in the sense of being "not-total" and in the sense of being "not impartial"' (2002, p. 66). This does not mean that one's knowledge of the world is always *insufficient* to make claims about it; rather, it acknowledges that truth claims are always already tied up in the political reality from which they are formed.

The second pillar of our EMA framework, Methodology, attends to the material reality out of which graphic social science and graphic medicine arise, that is, the conditions of production from the point of data collection to the ethics of distribution and remediation. Although some practitioners in graphic medicine and graphic social science join the research project after the data have been collected, many artists are also involved in constructing the research model and collecting data. In these instances, we feel that the following questions are vital: how are the data collected, scraped, and analysed? How are the research 'subjects' included in the design framework, if at all? In what format are the graphics made available to the public?

The EMA framework also rejects the binary between numbers and stories in social science research. While categories such as 'qualitative' and 'quantitative' are useful for broadly categorizing types of research, this binary can also function as 'an obstacle to communication and methodological advancement as it reifies false distinctions; for example, between words and numbers, constructivist and positivist inquiry, and subjectivity and objectivity' (Sandelowski et al., 2013). Rather, we echo other recent scholars, artists, and practitioners, who employ a mixed-methods approach to provide a rich and complex examination of their subject, such as Kate Mclean's (2017) sensory maps and the Data for Black Lives group (Data for Black Lives, n.d.).

The final pillar of our EMA framework considers aesthetics as 'forms that inform'. Aesthetic decisions relating to colour, iconography, and graphic choice can powerfully shape audience perception. Further, they are an integral part of data communication and accurately conveying the empirical results of data. With the introduction of computerized modelling and user-friendly software such as Tableau, there is already a rapidly solidifying 'aesthetic' of data visualization, identified primarily by symmetry, clean lines, and preset colour palettes. These aesthetic principles can be a powerful and elegant way to clearly translate data. But as the aesthetics of data visualizations becomes more established, this limits what we think of as a data visualization. Alternatively, comics and graphic novels operate in very different aesthetic registers. For example, the majority of comics are still hand-drawn, which leaves a palpable 'imprint' of the artist's hand that isn't present in a digitally-produced image. We consider how these factors can shape the way we understand data and, critically, how these factors can work to humanize data.

The remainder of this chapter uses the EMA framework to consider how data comics deal with numeric data, offering a humanizing approach to the visual communication of information. Before turning to specific examples drawn from graphic medicine and graphic social science, we offer a brief history of these two emergent fields in turn.

History of graphic medicine

In 2007, the term 'graphic medicine' was coined by Dr Williams, a physician, writer, and comics artist (Green & Myers, 2010). As a broad, growing field, graphic medicine addresses healthcare, illness, disability, patient education, treatment and experiences, and practitioner experiences. The phrase provides an umbrella term to bring together a growing number of comics that engage with these issues. Today, 'graphic medicine' is also a critically acclaimed organization of the same name (see www.graphicmedicine.org). Works classified as graphic medicine cross a variety of comics genres, including webcomics, graphic pathographies (health and illness memoirs), informational comics, comics strips, single panels, and video/audio installations. In 2015, graphic medicine scholars, artists, and practitioners published the seminal text *Graphic Medicine Manifesto,* an interdisciplinary collection of comics and essays that laid out how the comic, as a medium, serves as a way of communicating knowledge and experience to medical practitioners and students. The manifesto looks at the shifting iconography of illness and the power of self-representation. Advocates of graphic medicine see the potential of enhancing effective communication through the direct, collaborative involvement of patients, practitioners, and artists.

Including quantitative data in graphic medicine is a way of juxtaposing such data with a patient's lived encounter of an illness or disability. Graphic medicine works often reclaim the human side of health experiences from the clinical lexicon upheld by healthcare systems (Charon, 2006; Farthing & Priego, 2016a and b; Priego, 2016; Czerwiec et al., 2015). In particular, graphic pathographies—first person-centred illness narratives in the comics medium—bring out the humanizing aspects involved in the process of making them. As graphic pathographies involve acts of personal storytelling, they are well suited to the task of humanizing data around the lived experiences of illness or disability, offering ill people an opportunity to recover their voice, as people beyond medicalized patients (Frank, 2013; Green & Myers, 2010).

Christina Maria Koch (2016, p. 29) argues that 'the visual-verbal medium of comics is particularly apt in showing how intricately mental states are bound up with lived bodily experience and an embodied sense of self'. In the comics medium, the somatic and psychological experience of one's changing health identity is found in hypervisualized graphic embodiment that allows for a humanizing representation that shows how a person experiences part of the medical process—for example, a diagnosis or proposal for treatment—allowing access to some of the inner world of emotion that is difficult to represent in other visual forms. This can be as simple as a thought bubble or split panel that adds layers to the narrative of an interpersonal interaction or event. In this way the comics medium allows for more complex human experiences to be made visible to readers. For the purpose of understanding how we might humanize data visualizations, the next section examines an example of graphic medicine through our EMA Framework.

Graphic medicine and data visualization in *Taking Turns*

Taking Turns (2017) is an account of nurse and comics artist M. K. Czerwiec's experience working in an HIV/AIDS care unit during the AIDS epidemic from 1994-1999. Czerwiec's hand-drawn line graph, which spans from 1981-2000, uses data produced by amfAR (The Foundation for AIDS Research; Figure 21.1; see https://amfar.org/). The artist includes small drawings alongside the graphical line to mark important historical moments in the AIDS crisis. By doing so, Czerwiec creates an emotional narrative out of the data visualization's timeline (x-axis) and the known deaths of AIDS victims in the US (y-axis). Czerwiec adds three illustrations and caption boxes to further contextualize the HIV/AIDS epidemic. However, it is in the images that Czerwiec links numerical data to emotive narratives for an emotional impact.

Epistemological. The hand-drawn style of this data visualization, discussed in greater detail in Jill Simpson's chapter (this volume) in this collection, serves as a reminder of the human who produced it. Czerwiec's epistemological decision to use data in conversation with her own life experience enhanced the emotive narrative behind the numerical information presented in the visualization through her aesthetic choices. She used existing data collected and refined by amfAR. By embedding these data into a personal narrative, readers get to know the larger dataset in a smaller scale. the placement of Czerwiec's hand-drawn graph at the end of her emotive narrative shows readers the larger context of the HIV/AIDS epidemic. Simultaneously, embedding the national statistics within her story contextualizes the numerical data through a small number of people's life stories.

Methodological. The rise and fall of the deaths on the graph are similar to the structure of a basic narrative. The three images included are that of (1) the introduced antagonist, (2) a nurse MK's conflicts, and (3) the hope embedded in a dénouement. The inclusion of comics elements offers a shift in the existing ways amfAR data are produced and distributed. Readers can interact with the graph and larger illness narrative physically by turning back to the moments in the story to which the data refer. This does not occur in traditional interactions or encounters with data visualizations and is a methodological affordance of the comics medium.

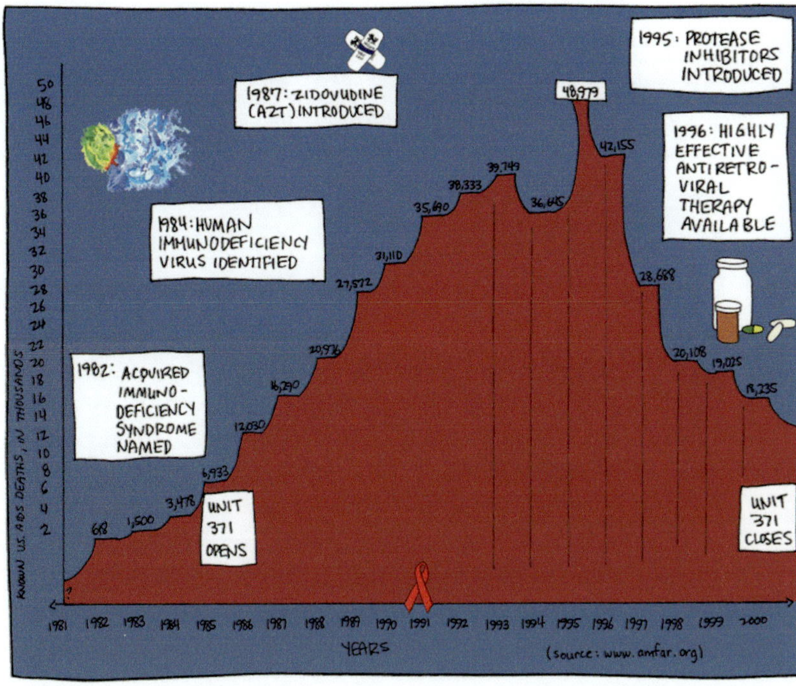

Figure 21.1. Hand-drawn amfAR line graph. Reprinted from *Taking Turns* (n.p.), by M. K. Czerwiec, 2017, University Park: The Pennsylvania State University Press. Copyright 2017 by M. K. Czerwiec. Reprinted with permission.

According to Bach et al. (2018, p. 2), a key principle of 'space-oriented' genres of narrative visualization, which they list as infographics, charts, and posters, is that a reader can relatively quickly and effortlessly interpret the information. In this encounter, the communicability of the data visualization will have an impact on how long the reader will engage with the information, unlike 'time-oriented' genres, like videos or animations, where the time is predetermined (p. 2). Bach et al. (2018, p. 2) classify the data comics genre as being both spatially and temporally oriented, which allows readers to choose their own pace and also has a narrative structure that research has shown 'is intrinsically easier to remember and facilitate readers['] engagement and persuasion'. The time spent with the emotive narrative adds depth to the line graph even with the inclusion of three images discussed next for their aesthetic contribution to the amfAR data. Temporal and spatial scales create affordances in the comics medium that, combined with the hand-drawn characteristic of Czerwiec's line graph, can create experiences with a data visualization that are akin to the principles of data storytelling. It creates longer and more intimate interactions with emotive narratives than traditional interactions with graphs sometimes provide.

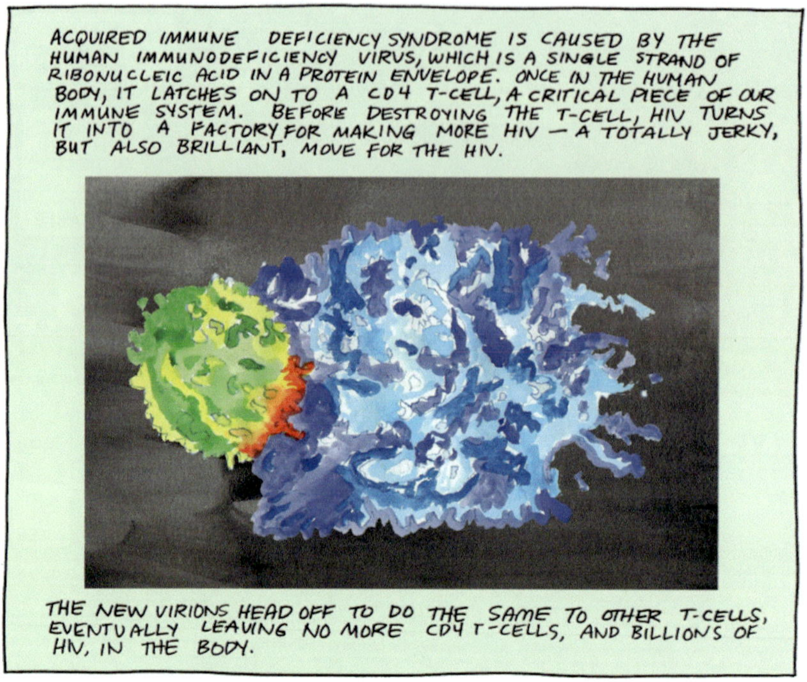

Figure 21.2. HIV virus cell. Reprinted from *Taking Turns* (p. 6), by M. K. Czerwiec, 2017, University Park: The Pennsylvania State University Press. Copyright 2017 by M. K. Czerwiec. Reprinted with permission.

Aesthetics. The line graph is included at the end of Czerwiec's text, so we can assume that readers are encountering it after they have read the longer emotive narrative and are able to link the images back to their seminal place in the story. Readers are assumed to recognize the HIV virus cell (the green-yellow and blue abstract image) from the beginning of the story when Czerwiec describes and illustrates HIV and AIDS for her readers (Figure 21.2). This image appears when the virus is referred to in the story. The comics medium allows the creator to visualize the virus, thus transforming it into the antagonist of the story, rather than a sick body for one individual. During the HIV/AIDS epidemic, victims of this disease were stereotyped and HIV/AIDS became synonymous with the gay male body. By separating the virus from its carrier, Czerwiec challenges the way that social stigma affects marginalized populations.

The second image in the line graph is an image of AZT pills. This is a reminder of when MK accidently stabbed herself with a needle after treating a patient in her unit (Figure 21.3). AZT is prescribed to her in order to attempt to fight any transmission that may have occurred. The inclusion of the pill in

Figure 21.3. ATZ pills prescription. Reprinted from *Taking Turns* (p. 83), by M. K. Czerwiec, 2017, University Park: The Pennsylvania State University Press. Copyright 2017 by M. K. Czerwiec. Reprinted with permission.

the graph reminds readers of the fear, anxiety, and unknowns surrounding the HIV/AIDS epidemic as seen through nurse MK's personal experience. The experience of reading this section of the narrative could have mirrored to some degree these emotions: does nurse MK now carry the virus?

The third image that Czerwiec includes in the graph is of medication bottles and pills. This image, as its accompanying caption states, is the HAART (Highly Active Anti-Retroviral Therapy) medications that became available in 1996. The bottles and pills come from a single full-page panel with the caption, 'And then hope arrived' (Figure 21.4). Returning to the line graph, we see that the ATZ pills are placed above the highest reported deaths in 1995-1996, thereby aligning fear and death; whereas the location of the HAART medications, with the decreasing reported deaths, provides a feeling of hope to the visualization.

Using our EMA framework we are able to examine how Czerwiec's use of data visualizations and the comics medium in a graphic pathography not only brings clinical evidence to these stories, but also contextualizes data by embedding it in an emotive narrative.

Figure 21.4. HAART medication introduction. Reprinted from *Taking Turns* (p. 146), by M. K. Czerwiec, 2017, University Park: The Pennsylvania State University Press. Copyright 2017 by M. K. Czerwiec. Reprinted with permission.

Graphic social sciences

Just as graphic medicine has offered a new way of thinking about and relating to medical research, the burgeoning field of graphic social science seeks to establish itself within the social sciences (Alamahodaei, Alberda, Feigenbaum, 2017). In June 2017, the Graphic Social Science Research Network was established to provide a forum for scholars, artists, and publishers to formally consider the practical and theoretical implications for the integration of graphics into social science. While some efforts have been made to adapt research articles and theses into the comics medium (Priego, 2016), many affiliated with the network are interested in embedding data visualizations to communicate research findings to stakeholders impactfully, through graphic, emotive narratives. Just as graphic medicine highlights the socially embedded and psychologically contextualized nature of illness, the work explored in this section uses personal experience to extrapolate larger claims about social and political realities, and the ways that these realities in turn shape everyday life.

Funny Weather and graphic social science

British comics artist, graphic novelist, and zinester Kate Evans produces work that significantly parallels graphic medicine and offers a social sciences example of data comics. Like Czerwiec's, Evans's comics stretch the comics medium to include biographies (*Red Rosa,* 2015), comics journalism (*Threads: From the Refugee Experience,* 2017), and educational guides on breastfeeding. Although Evans is not an academically trained social scientist, her works can be classified as 'graphic social science' for the ways that they draw on social science research to graphically represent complex numerical data in the comics form.

Epistemological. Funny Weather We're Having at the Moment: Everything you Didn't Want to Know About Climate Change but Probably Should Find Out (2006), or simply *Funny Weather,* is Evans's take on the topic of climate change. The comic covers an impressive amount of data, visualizing complex meteorological processes that account for rising sea levels and average global temperatures. She constructs three characters that lead the viewer through the narrative. One of the characters is a young boy, and he reflects the viewer: we learn alongside him about the realities of the carbon supermarket, and his youthfulness is used to represent naivety or youthful idealism and the will to change society for the better. The second character is a nameless man in a suit, a cigar poking out of his paunchy mouth—a 'fat cat' who represents elites who contribute to the manufacturing of dangerous emissions. Throughout the comic, the suited man is constantly pushing back against the young boy's questions. He dismisses the boy's suggestions that countries develop alternative energy sources, rationalizes the phenomenon of rising temperatures, and offers straw man objections to climate data. In one panel, he's depicted towering over the young boy, his face contoured in rage (Figure 21.5). The text beside his image reads, 'Who says climate change is even happening anyway? I'm not convinced! We need more proof!' In this sense, he represents broadly antagonistic social attitudes towards human-driven climate change.

While these characters may seem over-the-top, they are hyperbolic manifestations of two opposing subject positions vis-a-vis the larger issue of human-driven climate change. We have the banker, who has a financial interest in 'business as usual'; and the boy, who represents the inheritors of today's political decisions. It is the standpoint of the characters, rather than objectivity, that most concerns Evans. In an interview, she states:

> In your so-called objectivity you're missing out a layer of political information that people need to make sense of the world. I don't attempt to be

Figure 21.5. Explanation of methane gas. Reprinted from *Funny Weather We're Having at the Moment: Everything you Didn't Want to Know About Climate Change but Probably Should Find Out* (n.p.), by K. Evans, 2006, Oxford: Myriad Editions. Copyright 2006 by Kate Evans. Reprinted with permission.

> objective in the representations I make. What I do is I make a representation of events that's consistent with the facts, but I make it as emotionally engaging as possible to the reader. (K. Evans, personal communication, January 26, 2018)

Note how 'facts' are not opposed to subjectivity; rather, Evans's comments acknowledge the ways in which epistemic knowledge is always situated and made legible by one's embeddedness in social and political systems.

Methodological. Evans's methodology reflects her interest in creating empowering educational tools to guide social change. In preparing a new comic, she spends substantial time reviewing the source data in order to translate them into an easy-to-understand format for the lay reader. From the perspective of an activist and comics artist, it is the scientific report or

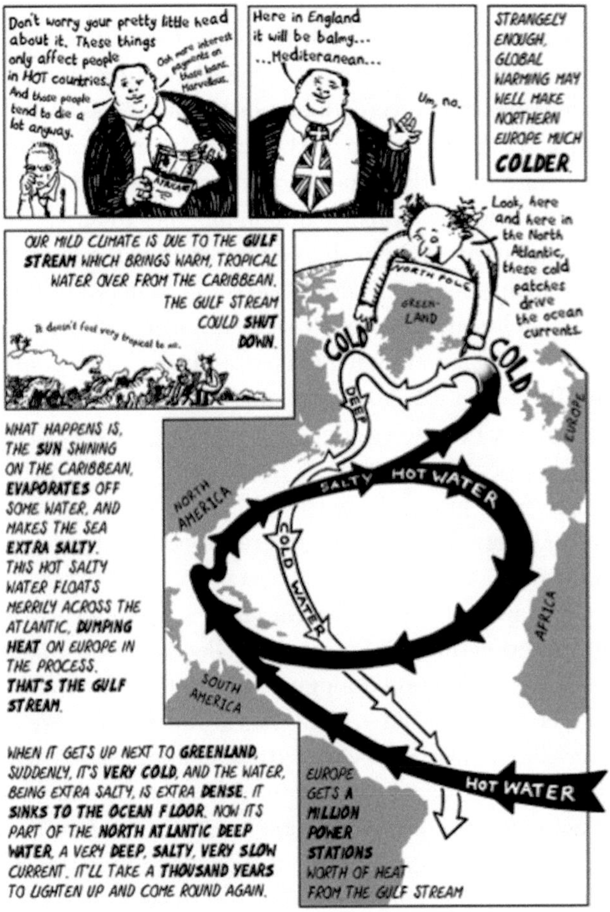

Figure 21.6. Explanation of the Gulf Stream. Reprinted from *Funny Weather We're Having at the Moment: Everything you Didn't Want to Know About Climate Change but Probably Should Find Out* (n.p.), by K. Evans, 2006, Oxford: Myriad Editions. Copyright 2006 by Kate Evans. Reprinted with permission.

the data table that obscures, rather than highlights, the truth. The reality of climate change that *Funny Weather* addresses, as reflected in meteorological and geological information, is made illegible by 'science-speak'. The comic becomes a vehicle to both demystify and translate these complex data, ultimately with the goal of spurring her reader to pressing action. For example, in Figure 21.6 she includes an illustration of the Gulf Stream in order to debunk the idea that climate change will only affect people living in hot climates. The illustration is accompanied by a narrative explanation of the phenomenon, which allows the reader to more easily understand how the Gulf Stream will be affected by rising global temperatures.

Aesthetics. Like other works from graphic medicine, *Funny Weather* employs a hand-drawn, unsophisticated aesthetic. This simple style recalls the feminist and queer zine subcultures of the 1980s and 1990s, in which Evans produced many of her earlier works. Common to many zines from this period is an emphasis on DIY production methods: writing and drawing all content by hand, cutting and pasting images to form collages, and stapling or sewing the zine's binding. It was also common for zines to be Xeroxed for distribution, giving many of them their signature black-and-white, shadowed, and irregular appearance. These visual components were highly aligned with an ethical political framework (that is, anti-capitalist, feminist, or DIY). Zines were proto-blogs, allowing makers and particularly young women an opportunity to create, share stories, and engage in political and social issues (Deibert, 2014; Piepmeier, 2009). But recalling these aesthetic elements is not simply an homage; it is directly invoking the same principles that were common to zine culture. Here we see how aesthetic elements intersect with methodology that encourages grassroots involvement in social and political issues.

We encounter the graphs alongside the characters, looking with and through them. Encountering the data in this way collectivizes the process of understanding, as the characters dialogue with each other to clarify graphical meaning. As a social activist, Evans has always been interested in questions of accessibility. In an interview with *Scientific America*, when asked why comics are a good teaching tool for difficult science, Evans relates the power of humour to demystify complex statistics. 'People are having fun', she says. 'When you create that, it's very easy to get the message across'. In Figure 21.7, for example, the figure of the scientist is seen dancing in a grass skirt and flip-flops next to a graph depicting rising Earth surface temperatures. Despite the seriousness of the graph's content, by using laughter, silliness, and absurdity, Evans is able to tackle the topic of climate change in a disarming format.

Lessons from data comics

Comics and related graphic forms allow for inclusion of the affective and personalized (Czerwiec et al., 2015; Williams, 2014). In each of the case studies we explored, practices of graphic storytelling are used to expand the realms of possibility for the visual depiction of numerical data. These artists' works humanize data by incorporating visual elements of the comics medium to engage with some of the broader issues that these data both

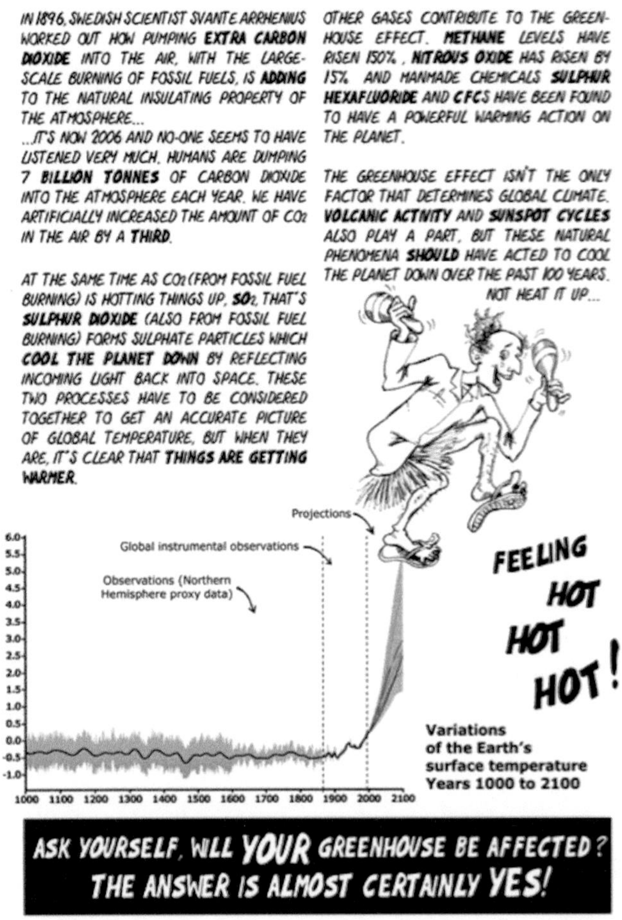

Figure 21.7. Graph of the Earth's surface temperature from year 1000-2100. Reprinted from *Funny Weather We're Having at the Moment: Everything you Didn't Want to Know About Climate Change but Probably Should Find Out* (n.p.), by K. Evans, 2006, Oxford: Myriad Editions. Copyright 2006 by Kate Evans. Reprinted with permission.

represent and produce in society. By applying our EMA Framework, we can see how data comics represent an approachable, accessible and relatable aesthetic form that can allow for new ways of knowing and understanding data, strengthening the connections between lived experience and numeric information.

Reading data comics from the emergent fields of graphic medicine and graphic social science in relation to data visualization enables us to see how comic forms can be engaged to visualize human reactions and encounters with data, how data come to be known or experienced, and what data do in terms of shaping our lives and the lives of others. We argue that by using the

EMA framework, we can explore the epistemic, methodological, and aesthetic possibilities for expanding what data visualization can do. Graphic medicine and graphic social science could have an impactful role to play in humanizing data visualization. These graphic works allow us to expand representations of personhood beyond traditional statistical ways of symbolizing people in data visualizations. Engaging with data comics to visualize information can humanize the personal narratives behind the numbers.

References

Ahmed-Husain, S., & Dunsmuir, S. (2014). An evaluation of the effectiveness of Comic Strip Conversations in promoting the inclusion of young people with autism spectrum disorder in secondary schools. *International Journal of Developmental Disabilities, 60*(2), 89-108. http://doi.org/10.1179/2047387713Y.0000000025

Al-Jawad, M., & Frost, L. (2014). Creating and analysing practitioner comics to develop a meaningful ward manifesto for a new dementia care unit. *International Practice and Development Journal, 4*(2). Retrieved September 19, 2019 from https://www.fons.org/Resources/Documents/Journal/Vol4No2/IPDJ_0402_06.pdf

Alamahodaei, A., Alberda, A., & Feigenbaum, A. (2016, November). *Giving graphic narratives to numbers.* Paper presented at An Agenda for Graphic Social Science, London.

Appadurai, A. (1988). *The social life of things: Commodities in cultural perspectives.* Cambridge: Cambridge University Press.

Bach, B., Wang, Z., Farinella, M., Murray-Rust, D., & Riche, N. H. (2018). Design patterns for data comics. *Proceedings of the 2018 CHI Conference on Human Factors in Computing Systems—CHI 18.* http://doi.org/10.1145/3173574.3173612

Bates, V. (2012). 'Misery loves company': Sexual trauma, psychoanalysis and the market for misery. *The Journal of Medical Humanities, 33*(2), 61-81. http://doi.org/10.1007/s10912-012-9172-x

Bordo, S. (1987). *The flight to objectivity: Essays on Cartesianism and culture.* Albany, NY: State University of New York Press.

Bowman, D. (2017). The moral of the tale: Stories, trust, and public engagement with clinical ethics via radio and theatre. *Journal of Bioethical Inquiry, 14*(1), 43-52. http://doi.org/10.1007/s11673-016-9766-5

Boy, J., Pandey, A. V., Emerson, J., Satterthwaite, M., Nov, O., & Bertini, E. (2017). Showing people behind data. *Proceedings of the 2017 CHI Conference on Human Factors in Computing Systems—CHI 17.* http://doi.org/10.1145/3025453.3025512

Carrigan, M. (2017, June 21). What is graphic social science? [Blog post]. Retrieved from https://markcarrigan.net/2017/06/21/what-is-graphic-social-science/

Charon, R. (2006). *Narrative medicine: Honoring the stories of illness*. New York: Oxford University Press.

Collins, P. H. (1990). *Black feminist thought: Knowledge, consciousness, and the politics of empowerment*. Boston: Unwin Hyman.

Czerwiec, M., Williams, I., Squier, S. M., Green, M. J., Myers, K. R., & Smith, S. T. (2015). *Graphic medicine manifesto*. University Park, PA: The Pennsylvania State University Press.

Czerwiec, M. K. (2017). *Taking turns: Stories from HIV/AIDS Care Unit 371*. University Park, PA: The Pennsylvania State University Press.

Data for Black Lives. (n.d.). *Data for black lives*. Retrieved November 9, 2018 from http://d4bl.org/

Deibert, R. (2014). *DIY citizenship: Critical making and social media*. Cambridge, MA: MIT Press.

D'Ignazio, C. (2015, December 1.) What would feminist data visualization look like [Blog post]. Retrieved from https://civic.mit.edu/2015/12/01/feminist-data-visualization/

D'Ignazio, C., & Klein, L. (2016, October). *Feminist Data Visualization*. Paper presented at and published in the workshop proceedings from the Workshop on Visualization for the Digital Humanities at IEEE VIS Conference, Baltimore, Maryland.

Evans, K. (2006). *Funny weather we're having at the moment: Everything you didn't want to know about climate change but probably should find out*. Oxford: Myriad Editions.

Farthing, A., & Priego, E. (2016a). Data from 'graphic medicine' as a mental health information resource: Insights from comics producers. *Open Health Data, 4*(1), e3. http://doi.org/10.5334/ohd.25

Farthing, A., & Priego, E. (2016b). 'Graphic medicine' as a mental health information resource: Insights from comics producers. *The Comics Grid: Journal of Comics Scholarship, 6*, 3. http://doi.org/10.16995/cg.74

Frank, A. W. (2013). *The wounded storyteller: Body, illness, and ethics*. Chicago & London: University of Chicago Press.

Fyfe, G., & Law, J. (Eds.). (1988). *Picturing power: Visual depiction and social relations*. London & New York: Routledge.

Gordon, A. F. (2008). *Ghostly matters: Haunting and the sociological imagination*. Minneapolis: University of Minnesota Press.

Gray, J., Bounegru, L., Milan, S., & Ciuccarelli, P. (2016). Ways of seeing data: Toward a critical literacy for data visualizations as research objects and research devices. In: S. Kubitschko & A. Kaun (Eds.), *Innovative methods in media and communication research*. (pp. 227-251). Cham: Palgrave Macmillan.

Green, M. J., & Myers, K. R. (2010). Graphic medicine: Use of comics in medical education and patient care. *BMJ* 340(c863). https://doi.org/10.1136/bmj.c863

Haraway, D. (1988). Situated knowledges: The science question in feminism and the privilege of partial perspective. *Feminist Studies, 14*(3), 575-599. http://doi.org/10.2307/3178066.

Harding, S. (1986). *The science question in feminism*. Cornell: Cornell University Press.

Kassai, B., Rabilloud, M., Dantony, E., Grousson, S., Revol, O., Malik, S., Ginhoux, T., Touil, N., Chassard, D., Pereira de Souza Neto, E. (2016). Introduction of a paediatric anaesthesia comic information leaflet reduced preoperative anxiety in children. *British Journal of Anaesthesia, 117*(1), 95-102. http://doi.org/10.1093/bja/aew154

Kennedy, H., Hill, R. L., Aiello, G., & Allen, W. (2016). The work that visualisation conventions do. *Information, Communication and Society, 19*(6), 715-735. https://doi.org/10.1080/1369118X.2016.1153126

Koch, C. M. (2016). 'When you have no voice, you don't exist'? Envisioning disability in David Small's Stitches. In: C. Foss, J. W. Gray, & Z. Whalen (Eds.), *Disability in comic books and graphic narratives.* (pp. 29-43). London: Palgrave Macmillan. https://doi.org/10.1057/9781137501110_3

Lupi, G. (2017, January 30). Data humanism: The revolutionary future of data visualization. *Print Magazine.* Retrieved from http://www.printmag.com/information-design/data-humanism-future-of-data-visualization.

Lupton, D. (2017). Feeling your data: Touch and making sense of personal digital data. *New Media & Society, 19*(10), 1599-1614. https://doi.org/10.1177%2F1461444817717515

McCloud, S. (1993). *Understanding comics: The invisible art.* New York: HarperCollins Publishers.

McLean, K. (2017). Mapping the invisible and the ephemeral. In: A. Kent and P. Vujakovic (Eds.), *Routledge handbook of mapping and cartography.* (pp. 500-515). New York: Routledge.

McNicol, S. (2016). The potential of educational comics as a health information medium. *Health Information & Libraries Journal, 34*, 20-31. http://doi.org/10.1111/hir.12145.

Piepmeier, A. (2009). *Girl zines: Making media, doing feminism.* New York: New York University Press.

Priego, E. (2016). Comics as research, comics for impact: The case of *Higher Fees, Higher Debts. The Comics Grid: Journal of Comics Scholarship* 6, 16. http://doi.org/10.16995/cg.101

Ramazanoğlu, C., & Holland, J. (2002). *Feminist methodology: Challenges and choices.* London: Sage Publications.

Refaie, E. (2012). *Autobiographical comics: Life writing in pictures.* Jackson: University Press of Mississippi.

Sandelowski, M., Voils, C. I., Crandell, J. L., & Leeman, J. (2013). Synthesizing qualitative and quantitative research findings. In: C. Tatano Beck (Ed.), *Routledge international handbook of qualitative nursing research.* (pp. 347-356). New York: Routledge.

Thomas, C., & Schirrmeister, A. (2018, May 28). Don't be afraid of surgery! Comic monkey takes pediatric patients by the hand [Blog post]. Retrieved from https://www.graphicmedicine.org/dont-be-afraid-of-surgery-comic-monkey-takes-pediatric-patients-by-the-hand/

Williams, I. (2014). Graphic medicine: The portrayal of illness in underground and autobiographical comics. In: V. Bates, A. Bleakley, & S. Goodman (Eds.), *Medicine, health and the Arts: Approaches to the Medical Humanities.* (pp. 64-84). London: Routledge.

About the authors

Aria Alamalhodaei is an independent writer and researcher. She received her MA in Art History from the Courtauld Institute of Art. She is co-author of *The Data Storytelling Workbook* (Routledge, forthcoming 2020) with Dr. Anna Feigenbaum, creative studio Minute Works and comics illustrator Alexandra Alberda.

Alexandra P. Alberda is a PhD researcher in the Department of Journalism, English, and Communication at Bournemouth University, where she is researching public engagement with comics in medical humanities. Her research interests are curatorial practices, public engagement with graphic medicine, comics studies, medical humanities, and the politics and ethics of representation.

Anna Feigenbaum is a Principal Academic in Digital Storytelling at Bournemouth University. Her research and practice focus on communication and social justice. She is co-author of *The Data Storytelling Workbook* (Routledge, forthcoming 2020) with Aria Alamalhodaei, creative studio Minute Works and comics illustrator Alexandra Alberda.

Section V

Data visualization and inequalities

22. Visualizing diversity: Data deficiencies and semiotic strategies

John P. Wihbey, Sarah J. Jackson, Pedro M. Cruz, and Brooke Foucault Welles

Abstract

This chapter explores the complicated dynamics that are inherent to the practice of data visualization involving issues of race and identity. We focus on data from the US Census and the profound questions that are raised as visual forms purport to represent groups. After reviewing historical context and related limitations and controversies, we present a project that explores a novel approach to visualizing US immigration patterns, an approach that relies on visual metaphors and algorithmic construction of visualization patterns based on massive sampling of Census microdata. The chapter suggests that the use of innovative expressive techniques to convey insights through poetic, and thus less literal, and limiting, forms is a way of grappling with underlying deficiencies in administrative population data.

Keywords: Data visualization; Immigration; Race; Diversity; Computational design; Data art

Introduction

The vocabulary of diversity, pluralism, multiculturalism, and the proverbial 'melting pot' are often invoked in contemporary discourse to characterize the complex, highly fraught, and extraordinarily multilayered history of immigration, race, and cultural identity in the United States. Ideas about American identity—who people truly 'are' at some essential level, where they come from, how they choose to be identified, and how majority cultures may identify them—continue to evolve over time. Through this discursive

Engebretsen, M. and H. Kennedy (eds.), *Data Visualization in Society*.
Taylor & Francis Group, 2020
DOI 10.5117/9789463722902_CH22

space, groups work to access forms of cultural and political recognition and resources, all the while potentially excluding and/or including others as the boundaries of identity are asserted, renegotiated, and contested.

These dynamics echo throughout American history, and manifest in a series of vexing questions: Who is to be counted as a citizen, with full associated rights? Who is included or excluded from a wide variety of identifying categories, such as 'Indian/Native American', 'Asian', 'Latino', or white? What is 'blackness' or 'whiteness'? Who should use hyphenated identities based on unique descent and ancestry, and why? How are multiracial persons, a growing portion of the population, to be identified? While cultural debate has been, and likely always will be, sprawling and unsettled around such questions, the formal locus of this debate is the decennial US Census, mandated by the Constitution to count persons.

Media representations of many kinds—novels, films, songs, paintings, journalism—have been used to explore the changing nature of the country, bringing to light, for example, how enslaved and indigenous peoples, and their descendants, have struggled to gain equality and how waves of immigrants have entered the country and challenged dominant power structures maintained by white Protestants of European descent. Such media representations have played a vital role in reconceptualizing notions of what it is to be 'American' and in surfacing important experiences that may have otherwise been culturally marginal.

As a relatively newer form of media particular to the digital era, interactive data visualization provides novel affordances that open up new possibilities for exploring evolving notions of human identity. In this chapter, we present an example from our own work which attempts to push the boundaries of discourse about labelling and identity. This unique project leverages administrative data from the US Census to tell the sweeping story of immigration history and cultural identity in America. The project, which draws upon Census recordings of persons' countries of origin primarily over the period 1830 to 2015, deploys visual metaphor and computational techniques to expand the expressive meanings and possibilities around themes of diversity. We see the project as a particular form of discourse that both grapples with the challenges of reductionism and inclusivity/exclusivity, and that semiotically projects complex ideational and compositional meanings that speak deeply to a general theme of cultural diversity. Because deficiencies in data are a problem for all sorts of reasons, including visualization challenges, we sought to address these by experimenting with a form of visualization that works with limited/deficient data.

To contextualize the experimental case study we produce, we first situate this visualization work in the intellectual history relating to the underlying census data and the limitations embedded in it. We explore how any picture of 'diversity' based on these administrative data necessarily, and tragically, excludes certain types of persons, with African Americans and Native Americans being two categories of persons whose origins in this country do not fit into narratives about diversity through immigration.

With these caveats in mind, our project nevertheless focuses on the census 'country of origin' information, using statistical estimates, to render a picture of American diversity that evolves, grows, and complicates understanding over time. There are, of course, many ways of portraying diversity, and immigration is one of those: it is a subset of diversity. As will be explained, given how unrepresentative race data in the census are, we focused on immigration specifically, as extracting the immigration data is a much more accurate task that provides a reliable basis for visualization. We chose to deploy the visual metaphor of tree rings to evoke the complexity and interdependence of a biological ecosystem. Historical immigration patterns are shown as a set of tree rings, which are encoded by processing millions of samples of US Census microdata, from a pool of nearly 2 billion individual records. As time advances, the tree grows, forming rings of immigration. Each ring corresponds to a decade. Cells are deposited in layers, with each cell corresponding to 100 immigrants.

Our efforts focused on a central research question: *Given the known constraints, what would a dynamic picture of US diversity, as a function of immigration, look like?* Further, how might artistic, design, and poetic strategies work to enhance knowledge and interest in the diversity of the country, signifying truths and conveying important insights that may transcend the limitations of underlying, literal data? Interventions around such a question, of course, bear crucially on urgent political questions and current discourses about cultural diversity and public policy proposals, and we take up these questions with this background in mind.

Visualizing migration and identity: A brief US history

During the second half of the nineteenth century, visualizations of Census-based numbers first began appearing in government statistical abstracts; some of these figures began examining the distribution of different ethnic groups throughout the country as a function of immigration (US Census Bureau (n.d.). Statistical abstracts of the United States). Immigration as a

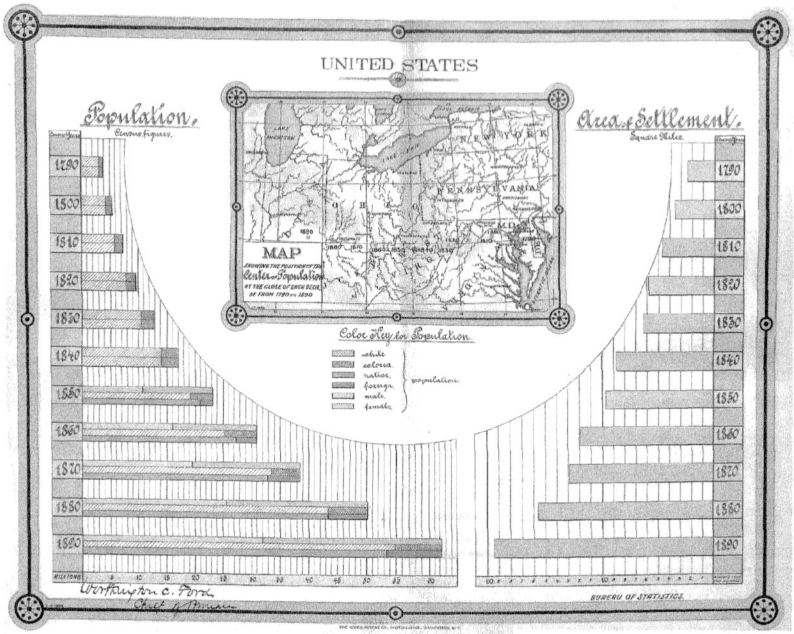

Figure 22.1. Map/chart included in 1896 US Census documents, showing growth of racial and demographic groups and territorial expansion. From US Census Bureau (1896). Statistical abstract of the United States 1897—Part 2. (https://www.census.gov/library/publications/1898/compendia/statab/20ed.html). Public domain.

phenomenon also became known through non-quantitative representations such as drawings, posters, paintings, and other hand-drawn and printed media forms. The question of place of birth was added in 1850, following the beginning of a dramatic increase in immigration (Gibson & Jung, 2006).

Of course, migration and the movement of peoples are network-driven processes, lending themselves readily to visualization (Portes & Rumbaut, 1990). Immigration records are limited, though, and certain ethnic groups can only be traced back so far; thus most representations are constrained by the available data (Daniels, 1989). Full-scale histories that attempt to recover the nuances of European and American migration, for example, have rarely been attempted (Nugent, 1995). In any case, the United States began keeping records of persons entering the country at ports in 1820 and, although prone to inaccuracy, this gave way to an idea of change in population volume due to external flows and eventually to visualizations of these numbers (Handlin, 1959).

The history of data visualization relating to US immigration is not well documented, and to our knowledge, there is no extant comprehensive history. We performed an environmental scan of the literature/relevant materials

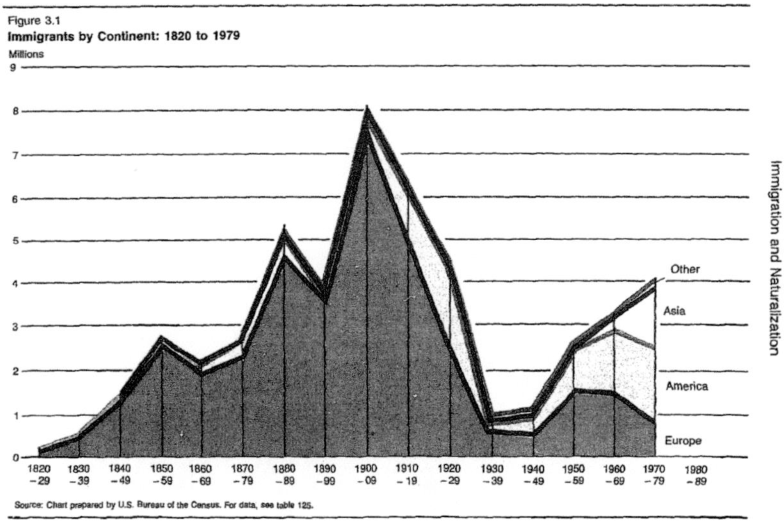

Figure 22.2. Chart included in 1983 US Census Bureau documents, showing the relative contribution of various continents to immigration totals in the United States. From US Census Bureau (1983). Statistical abstract of the United States: 1984—Section 1 Population. (https://www.census.gov/library/publications/1983/compendia/statab/104ed.html). Public domain.

and found a number of visualizations in the US Library of Congress's virtual trove of historical documents; US Census Bureau materials, especially the statistical abstracts; statistical atlases of censuses; books about immigration; and on digital news sites and data blogs. While far from a comprehensive search, we examined numerous visualizations spanning from 1828 to 2018. Representative examples included a 1984 map (see Figure 22.1) from the US Census that depicts immigrants by origin from 1820 to 1979 (US Census Bureau, 1983), as well as data visualizations from contemporary media outlets such as Vox that show 200 years' worth of data trends (Chang, 2017).

Traditionally, most visualizations of immigration to the United States have involved some sort of map, including land plot, county, density, and flowchart maps. An early example printed by the Census Bureau (1896) illustrates how the intersection of identity and geography were being represented and imagined in the nineteenth century, with categories of 'white', 'coloured', 'native', and 'foreign' delineated (see Figure 22.2).

As will be discussed, certain classes of people are wholly excluded from any such maps. Glaringly, the precise African countries of origin of slaves and their ancestors are not included in this historical narrative, nor are the indigenous nations from which Native Americans came, even as they became US citizens through subjugation. That said, it is a point that bears further

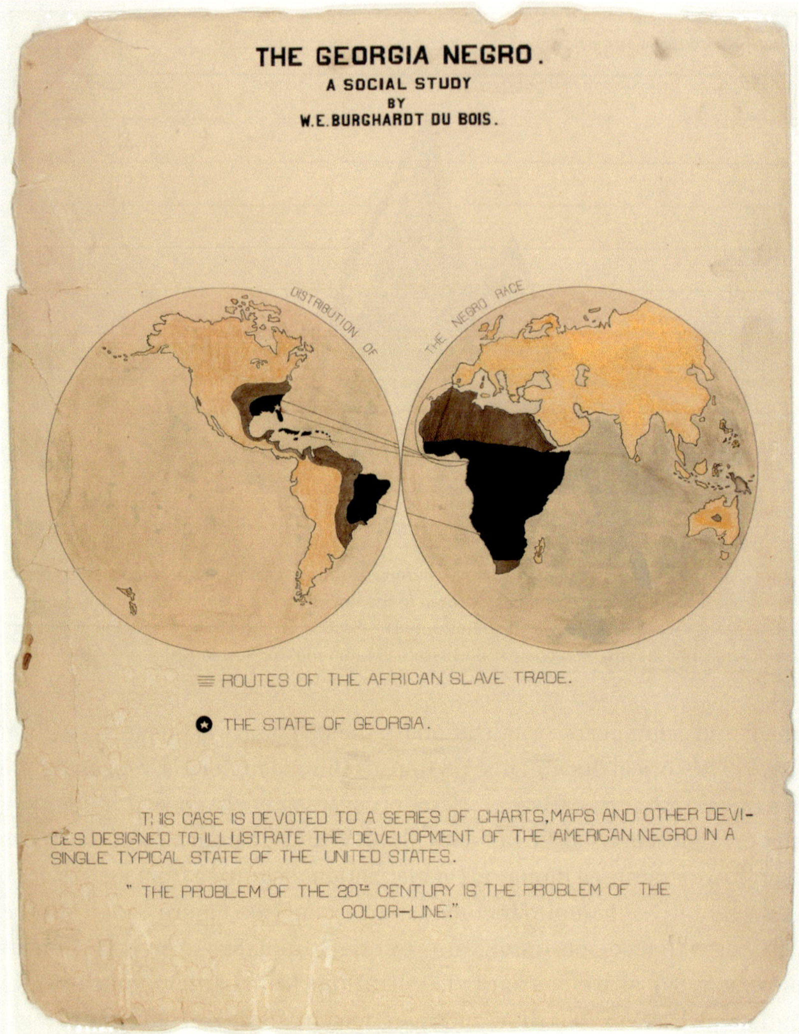

Figure 22.3. Historical rendering by sociologist W.E.B. Du Bois of trajectory of African slave trade to the Americas. From Du Bois, W.E.B. (1900). *The Georgia Negro: A social study.* [Map] Library of Congress Prints and Photographs Division, Washington, D.C. Public domain.

research that there are early examples of both African-American scholars and folk artists and Native Americans tracing their own history through visualizations. These would include, for example, a 1900 map created by the pioneering social scientist W. E. B. Du Bois about the trajectory of the slave trade from regions of the African continent to the Americas (see Figure 22.3).

For Native Americans, there are examples of data-related artefacts such as 'winter count' calendar-pictorial cloths and skins (Lakota, 1902), as well

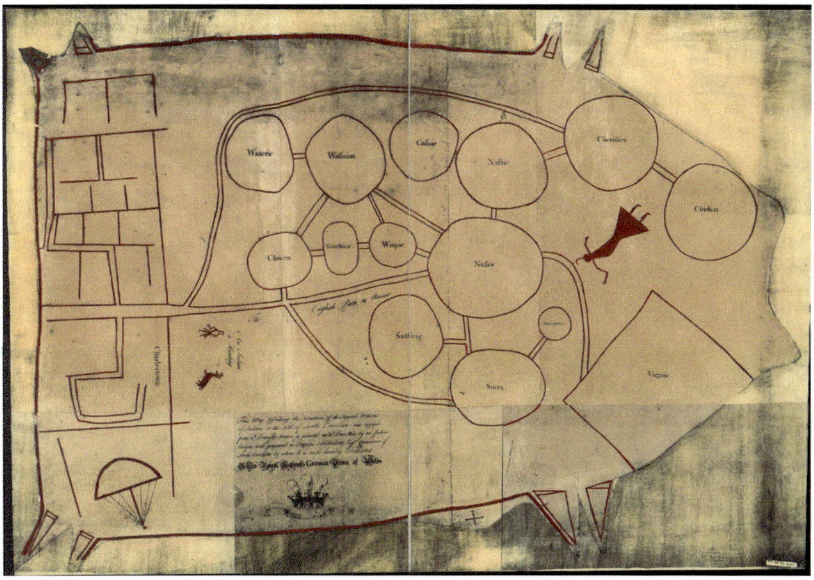

Figure 22.4. Native American map rendering on deerskin of tribal information and location. From Nicholson, F. (1724/1900). *Map of the several nations of Indians to the northwest of South Carolina.* [S.l.: s.n.] [Map]. Retrieved from the Library of Congress. (https://www.loc.gov/item/2005625337/). Library of Congress, Geography and Map Division. Public domain.

as examples of local and regional maps with population features done by indigenous persons (see Figure 22.4).

Exclusion in Census data

The US Census is intimately tied not only to questions of identity but to power and inequality, and the racial categories encoding many generations of non-white persons remain highly problematic as data sources. The categories included in the first Census in 1790 speak to this: 'free white' men and women were counted with specifics about their age, dates of birth or death, profession, and familial role. Next to these categories appear only two others: 'non-taxed Indian' and 'Slave'. Together these categories reflect the power and entrenched racial ideologies central to the nation's inception: white men and women mattered and were counted in detail because their numbers (and economic and social capital) mattered to questions of political representation and policy; Native Americans—who include over 500 tribes with distinct cultural practices and languages—are lumped into the category of 'non-taxed Indian', a term that reflects their non-citizen status and non-inclusion in apportionment counts in determining political representation; and finally, 'slaves', a term that, to those in government, stood in for people of African descent in the Americas so clearly that they did not bother to specify what we now think of as a racial category because 'slave' and Black were presumed to be synonymous (Zinn, 2015).

Rendered invisible for decades following the first census were Asian Americans and Latinxs—groups that have long been a part of the American fabric but whose early numbers were considered too low to matter enough to count, who were not yet 'raced' in the American imagination, or who primarily resided in parts of the country yet to become politically consequential. For example, in the 1860 census, the brand new state of California included 'Chinese' as a category in the census—a reflection of the presence of Chinese labourers in the West—but this category was not included in any other state (Hart, 2009).

Likewise, much of the Southwest between Census years 1790 and 1860 was either not yet a part of the United States (rather controlled variously by Spanish or Mexican governments) or relatively new territories and states without much population or political representation. Thus, the need to count those who would now be considered Latinx—and in fact even a federally recognized racial or ethnic category to describe their various origins—simply did not exist. It was well over 100 years after the 1860 census, in 1970, before the federal government would make the first attempt to count 'Hispanics' as an ethnicity (Cohn, 2010; US Census Bureau (n.d.). Measuring Race and Ethnicity Across the Decades: 1790-2010).

The evolution of census categories is a clear example of how racial categories—while socially constructed and ever changing in response to sociocultural context—are central to questions of inequality and belonging in the United States. As various groups have sought to maintain and gain power throughout American history, and as socio-political contexts have been shifted by war, labour demands, economic upheaval, migration, and activism, 'race' as an identity category worth counting has shifted, as well. Over time, the US Census Bureau, entrenched in the original exclusionary ways of thinking about identity visible in the 1790 census, has responded, sometimes slowly and under pressure and sometimes rapidly when groups are deemed a threat, to these shifts. Generally, changes in census categories are spurred by new understandings of who in America should be counted—who matters. Importantly, however, mattering and being counted are not always a positive thing: take, for example, the case of 'slaves' who mattered to their owners for the purposes of economic gain and political representation; or the Chinese whose counting lead to the passage of the draconian 1882 Chinese Exclusion Act.

The creative data visualization project we discuss below begins in the nineteenth century; with its focus on immigration, the project's substantial visual forms really begin to take shape with the 1870 census—the first after the Civil War. As important context, it is worth noting that this census reflects how deeply important questions of white racial purity became to those in power in the context of reconstruction and the attempted social gains of African Americans. The category of 'slave' as a stand-in for African American is removed. In its place new categories arise alongside 'White' and 'Indian': 'Black', 'mulatto', 'quadroon', and 'octroon'. These categories reflect the racial anxieties of whites in power during reconstruction who embraced racial pseudo-science based on mythologies of 'Black blood' to justify their fear that increased gains by African Americans would lead to mixed-race children who would sully the purity of the 'white race' and throw the existing racial order into chaos. Suffice it to say, census workers carrying out counts in the nineteenth century—and well into the twentieth—were given detailed instructions that would both offend and appal relative to contemporary standards, about how to assess and record racial distinctions and determine the cultural identity of many different kinds of people.

It was in 1890 that Asian Americans began to be counted in the national census—largely as a result of increased immigration of Japanese and Chinese men who worked first as labourers in agriculture and railroads and whose increasing numbers were perceived as an economic and cultural threat. This is reflected in the fact that *only* the categories 'Chinese' and 'Japanese' are added to the census at the time despite the lesser presence of Korean, Filipino, and other Asian labourers in the same industries (Takaki, 2012). Over time, the counting of Asian Americans by the census became more inclusive, sometimes in response to perceived threats, and other times as a result of political activism and lobbying by Asian American groups who sought to challenge the perception of Asianness as an always-unamerican-Other category. Among the most shameful examples of how questions of power and oppression are tied to the census is that the United States government used records from the 1940 census to find and intern Japanese families during World War II—illustrating that a demographic survey in the context of xenophobic ideology is anything but a simple count (Aranti, 2018).

Between 1900 and 1940 the mulatto, quadroon, and octroon categories were dropped from the census as racist blood quantum science was debunked and it became clear that the mere existence of mixed-race African Americans would not, in the context of entrenched American anti-blackness, dissolve the conditions of the black/white racial binary. During this period also, the 'race' category of 'Mexican' came and went from the 1930 census, and the categories 'Hindu' and 'Korean' were added in response to increasingly visible populations of people with South Asian and Korean origin. To be clear, none of these categories are a race—Mexican and Korean are nationalities within the Latinx and Asian ethnic and racial groups—further examples of how the federal government itself has contributed to mischaracterizations and misunderstandings about race, national origin, and ethnicity.

Likewise, the appearance of 'Indian' and 'Hindu' on the 1940 census as race categories is almost amusing in retrospect given that neither is an accurate term for the people they are supposed to describe, and if used as they were then now would cause great confusion. In 1940 'Indian' was still inaccurately being used to describe Native Americans and 'Hindu' still being used to describe immigrants from India, Pakistan, and Bangladesh who, notably, were not all religiously Hindu but include Christians, Muslims, Buddhists, Hindus, and other religions. This conflation of race and nationality with a religion—the most visible to outsiders in India—again shows how imperfect census categories can be, especially as defined by those with racial and political power who often misunderstand enormously large and diverse ethnic and racial groups. It was not until 1950 that the Census Bureau changed 'Indian' to 'American Indian' in the census and 1980 until both distinctions among Native American groups and South Asian groups began to be disaggregated.

The census categories, of course, cannot tell us specifics regarding experiences of racialization in the United States, as questions of identity weigh heavily and uniquely on communities because of other forms of de jure and de facto policy and tradition. Even after the 2000 census allowed responders to acknowledge the very American experience of being descended from multiple groups by checking more than one box, some worried this was a blow to the power of collective identity politics even as others felt seen for the first time. Among Native Americans, for example, the possibility of checking more than one box falls within weighty debates and policies about blood quantum and tribal membership laws, 'real Indians', and federal recognition of tribal status. Two Americans who check 'American Indian' in the census may, for example, have radically different understandings of the political, social, and cultural weight of that identity depending on their phenotypical experience, the families and communities to which they formally and informally belong, and federal tribal recognition policy (Jarvis, 2017; Schmidt, 2011). Likewise, because 'Hispanic or Latino' designates *ethnicity* as opposed to race and can apply to anyone from Latin America and other countries colonized by Spain, Portugal, and France, a black-skinned Haitian American, white-skinned Chilean American, indigenous Mexican American, and Asian Filipino American might all check the category (and one or more others), but understand these categories—and themselves—in radically different ways (Amaro & Zambrana, 2000).

As the United States approaches the 2020 census, new debates and concerns about the visibility and counting of identity have arisen. In particular, the Trump Administration has introduced a question asking respondents to specify if they are, or are not, American citizens which has raised concerns among human rights and immigrant rights groups. These groups fear that at the least the citizenship question might dissuade people from responding, leading to inaccurate counts in particular of immigrants of colour who seek to have an increased voice in American politics, and at worst might be used, as has been the case in the past, to target immigrant communities.

Visualizing Immigration and Identity

To address the exclusions discussed above and attempt to produce a visualization of available data, we endeavoured to create a project about US immigration that would explore novel expressive forms. We chose to sample from nearly 2 billion instances of microdata in order to get the finest granularity in terms of location of origin that we could, per state, displayed in decennial increments, and dating to as far back as 1790 when available. Census summary tables frequently lack all of this information, necessitating a sampling method. Furthermore, using the finest granularity is the most accurate way of extracting immigration counts and accounting for subtle differences in place of origin. In the face of inherent data problems, we explore new visualization forms, specifically tailored to the dataset and its context. Knowing the profoundly problematic nature of racial designations, we chose instead to focus on country of origin reports in the census data in order to gesture broadly at the diversity of the country and show its layers of complexity.

As mentioned, our case study explores historical immigration patterns (1830-2015), which are shown as a set of tree rings, drawing on millions of samples of US Census microdata, from a pool of nearly 2 billion individual records. As time advances, the tree grows, forming rings of immigration. Each ring corresponds to a decade. Cells are deposited in layers, and each cell corresponds to 100 immigrants.

The underlying dataset consists of samples of questionnaires from the US Census made available through IPUMS, a repository for statistical agencies that is maintained by the University of Minnesota (Ruggles et al., 2017). We queried the US state of residence, age, and place of origin of each person since 1790. (It should be noted that a large amount of territories were only incorporated as states after 1790, meaning that data for these states were only available after a certain year.) The places of origin originally had 571 denominations. Using these data, we calculated estimates for the number of native-born persons and the number of immigrants who arrived in each decade. After reviewing the data, these places of origin were grouped into seven cultural-geographical groups: Canada, Europe, Latin America, Asia, Oceania, Africa, and the Middle East. Colours were assigned accordingly, creating a swirling spiral of various hues.

The precise evolution of the result is detailed further below, but first we present here the general pattern:

We employ visual metaphor for a variety of reasons. First, metaphor is useful to suggest other ways of thinking about the data, generating meanings that a bar chart, for example, cannot. Metaphor is also useful to embed meaning by the authors—in this case, and among others, inclusiveness. Metaphors can be used to convey figurative meanings that are recognizable and familiar, contributing to memorability (Cox, 2006); and figurative approaches allow for expressiveness and uniqueness, which contribute to stickiness (Borkin et al., 2013).

Lakoff and Johnson (1980) pioneered the view of metaphors from a cognitive perspective, framing a theory on 'conceptual metaphors', which map structural properties between a source domain and a target domain and represent a cross-domain mapping process. With this, one can understand one domain in terms of another. The metaphorical expression to convey such processes is just a linguistic expression, a surface realization of such cross-domain mapping (Lakoff, 1993; Lakoff & Johnsen, 1980; Chandler, 2017).

The meaning of the visualization is intertwined with the aesthetic qualities of the artefact, as it attempts to connote notions of wonder and to play with ideas of transformation, recurring growth, and evolution. (The study of tree rings is called 'dendrochronology', a term we have used and playfully co-opted in exhibiting the data to audiences; likewise, a studio exhibit at our home institution of Northeastern University that shows prints of the tree rings was entitled 'Naturalizing Immigration'.) The rather dry, clinical, and exacting qualities of traditional data visualization forms (Tufte, 1983)—bar charts, pie charts, line graphs—are eschewed in favour of a ludic, curiosity-evoking, and, we hope, more sublime figurative style that attempts to match thematically the country's diversity itself, while avoiding claims of finality and starkly direct quantitative comparisons among groups, whose essential nature are highly problematic (Cruz, 2015). We attempt to solve the very real problems of data integrity by, in effect, moving to a poetic and expressive level.

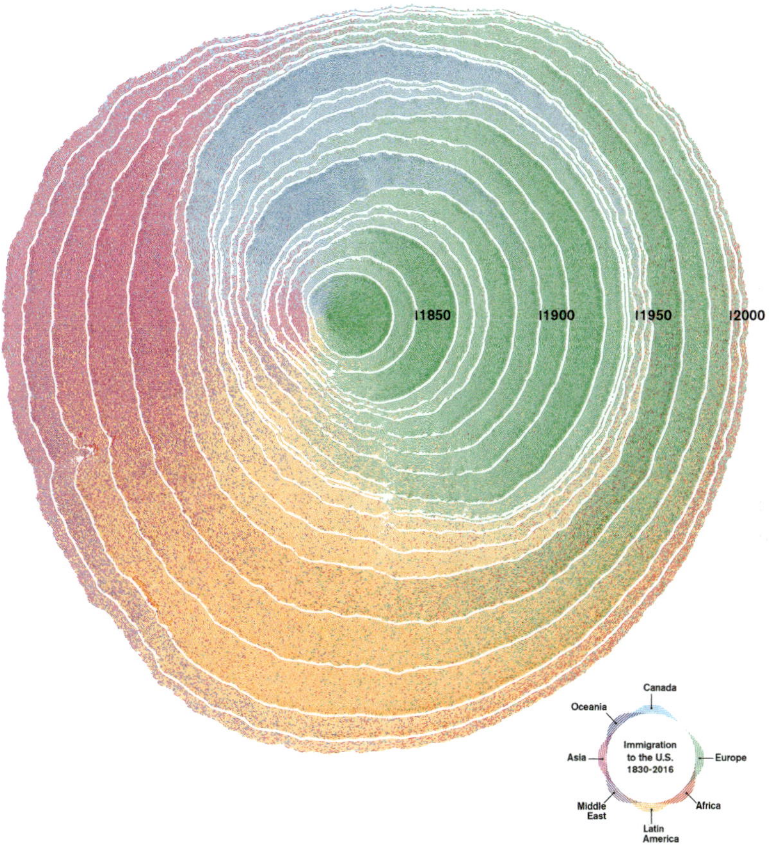

Figure 22.5. Visualization of US immigration as metaphorical rings in a growing tree trunk, with each dash representing 100 immigrants and each ring representing one decade. The image is based on Census data relating to persons' origin at birth, 1830-2015.

The video version of the visualization, some six minutes long, produces perhaps an even more powerful effect than the still images, as it allows the viewer to experience the full growth of the tree rings. That video can be found at: https://vimeo.com/276140430.

The video shows the simulation of the system: as data are injected into the visualization, new cells spawn that represent incoming immigrants in a given period in time. The specific places of origin of immigration for a certain decade are displayed as a list on the left side of the canvas, sorted by descending number of immigrants. As times passes, the tree registers every immigrant who arrived according to the dataset. One can observe the tree's state at six points in time in (Figure 22.5): 1880, 1910, 1940, 1970, 2000, and 2015.

Figure 22.6. Evolution over time of visual simulation of US immigration as metaphorical rings in a growing tree trunk.

The biological metaphor that inspires this visualization was chosen for its connotations and the form of discourse it evokes and produces. Trees in their natural setting have annual growth rings that reflect varying environmental conditions; the rings' forms are neither perfect circles nor ellipses. Our algorithm is inspired by this variation and accordingly deposits immigrant cells in specific directions, depending on the geographic origin of the immigrant. Rings that are more skewed toward the country's East, for example, show more immigration from Europe, while rings skewed South show more immigration from Latin America. With this, it is possible to observe the quantity of immigration through the thickness of the rings. As mentioned, the colour of the cells corresponds to specific cultural-geographical regions, which the key and labels indicate.

Like countries, trees can be hundreds, even thousands, of years old. The cells grow slowly, and their pattern of growth influences the shape of the tree's trunk. They are all part and parcel of the organism's growth. This idea lends itself to the representation of history, as it shows a sequence of events that have left a mark and shaped the present. Just as cells leave an informational mark in the tree, so too can incoming immigrants be seen as natural contributors. Our visualization suggests that these marks of the past are immutable and cannot be erased, regardless of how one reads them—or how one might prefer to shape the marks of the future.

Our data story uses an algorithmic 'physics engine' to simulate how cells interact with each other while creating a visualization with emergent patterns as one watches. This means that as new cells grow in the system, they are simulated as physical bodies that push and compress nearby bodies. As new cells grow, the physical constraints are solved in a certain number of steps, enabling the system to reach to near equilibrium states before advancing to the next data injection (Jakobsen, 2001; Press, Teukolsky, Vetterling, & Flannery, 2007). This causes a cascade of actions-reactions that result in the visual organization of our data, by simulating natural phenomena and obtaining a visual resemblance with organic forms (Cruz, 2017).

The cells and rings in a tree are nature's own way of organizing information. The United States is, of course, currently organized into fifty distinct states. Each state has grown at different rates and with varying immigration profiles. Some will be larger, some will be smaller, some will have complex shapes that represent waves of immigrants, and others will be perfectly circular due to the absence of immigration. Each state has its own signature and can be characterized individually. The country can therefore also be envisioned as a forest of trees, providing additional layers of complexity that tell the evolving story of American diversity. The visualization of such trees are cross-sections of their trunks that reveal the tree rings inside. In fact, when one looks at a set of tree rings, what is presented is a sample of all of the tree's cells. This dynamic can be observed in our project's context as well, in the sense that the visualized data are a sample of the universe of study.

US immigrants come from multiple geographical directions, so it makes sense that a tree can grow more in the direction where immigration is coming from. In order to do this, the seven cultural-geographical groups were attributed to specific directions (e.g. Canada → North, Europe → East, Latin America → South, and so on). With these directions, a Gaussian (normal) distribution can be created for each immigration group, with the average centred on the corresponding direction. This results in each state having its own form derived from data. Rings that are more skewed toward East, for example, show more immigration from Europe, while rings skewed South show more immigration from Latin America. Fifty sets of tree rings were simulated to show different profiles of growth and immigration for each US state (see Figure 22.7 for individual examples). An algorithm was devised in order to attain a resemblance with tree rings while carrying the semantic context that has been described. This algorithm was the result of an iterative design process that is described elsewhere (Cruz et al., 2018).

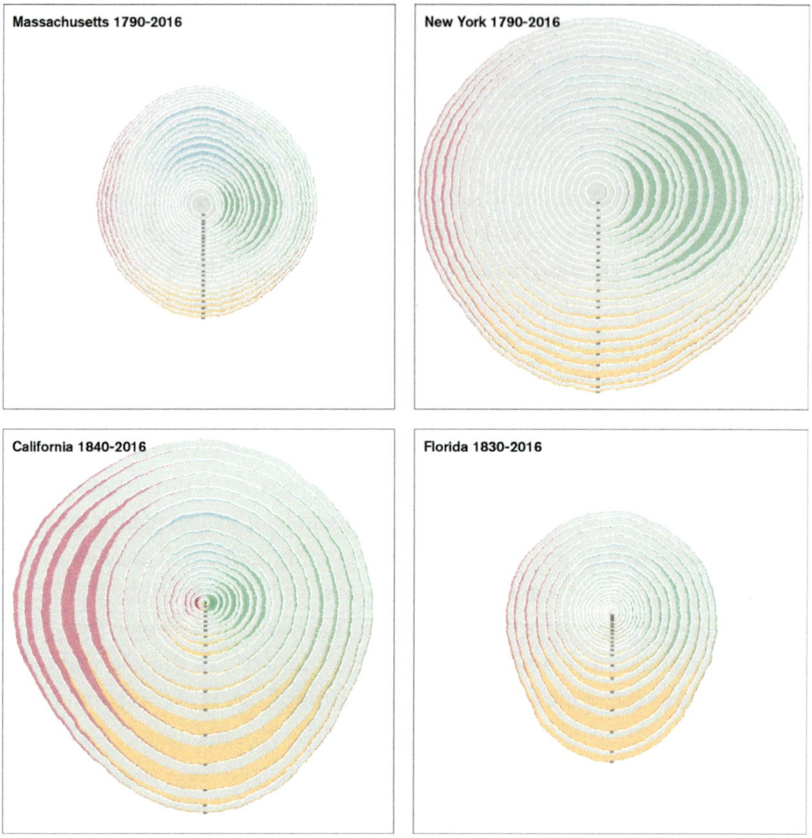

Figure 22.7. Examples of visual patterns in specific US states relating to immigration as metaphorical rings in a growing tree trunk. White cells represent native-born persons, while coloured cells represent immigrants.

Transcending literal representation

No visual form could, of course, capture the diversity of the US population, which in a very real sense mirrors the diversity of the globe itself. In this way, the project case study we present is inherently an artificial exercise. Yet, by moving from traditional conventions of precise correspondence to visual metaphor, we believe data visualization might indeed *more accurately* capture this sprawling and endlessly nuanced historical pattern and phenomenon, as compared with less artistic data forms. We use design elements to amplify and communicate messages of interrelatedness, cultural accumulation and accretion, and complex evolution; this might be contrasted with more literal visual expressions of diversity, where precise racial and origins categories might be directly compared in a more clinical and purely scientific way. Our visualization aims to embed more meaning, and to produce both more emotion and curiosity in the viewer, than would a more standard, minimal, and sterile depiction of a dataset. The use of a physics engine algorithm that introduces indeterminacy and produces emergent phenomena—rendering simulations novel and unpredictable to some degree—is also faithful to the phenomenon of immigration itself, which is a function of a non-linear, complex global system of push and pull factors.

Through the visual metaphor employed here, and the figurative representations, interesting and important historical patterns can be discerned. Through the visualization, the viewer may note that the origins of US immigrant populations transform from era to era. In the 1840s and 1880s, European immigrants came mainly from northern and western Europe, whereas the famous influx of the early 1900s, symbolized by Ellis Island's gateway, emanated mostly from southern and eastern Europe. Immigration from Asia rose between 1970 and 2000, while large-scale immigration from Latin America began in 1950 and lasted for half a century. Immigration from Africa only becomes visible in the 21st century.

As discussed, no data picture of diversity in the United States can fully account for the lack of data for certain marginalized groups. In addition, the categories in each successive decade after the country's founding may have gotten more inclusive and precise over time, but they still often suffer from certain historical biases. Undercounting was inevitable, and still is, with regard to newly arrived, and thus highly transient and vulnerable, populations.

There can be no neutral rendering in this domain, no purely objective point of view, and thus no representational act that avoids questions of exclusivity. Knowing this, we choose through the creation of this case study to accept the burden of fallibility in the service of trying to convey a higher set of insights. Imperfect and tragically flawed in its origins as well as its current history, the country is becoming more diverse ethnically and racially each year now, and the percentage of the population made up of foreign born persons is approaching historic levels (Zong & Batalova, 2017), even as a policy backlash and anti-immigration sentiment continue to simmer.

The counting of populations and the rendering of pictures based on those data are an inherently political act. Controversies continue to grow over how resources will be used in future census counts; there remain grave concerns that persons of colour and marginalized groups, in particular, will not be sufficiently represented in official statistics (Chevat & Lowenthal, 2015). Data visualization that is rigorously rooted from a computational and statistical perspective, while at the same time innovative in generating ideational and compositional meanings, can help transcend limitations of administrative data and produce new discourses about diversity and its importance in society.

References

Amaro, H., & Zambrana, R. E. (2000). Criollo, mestizo, mulato, LatiNegro, indígena, white, or black? The US Hispanic/Latino population and multiple responses in the 2000 census. *American Journal of Public Health, 90*(11), 1724-1727.

Aranti, L. (2018, April 6). Secret use of Census info helped send Japanese Americans to internment camps in WWII. *The Washington Post.* Retrieved from https://www.washingtonpost.com/news/retropolis/wp/2018/04/03/secret-use-of-census-info-helped-send-japanese-americans-to-internment-camps-in-wwii/

Armstrong Dunbar, E. (2008). *A fragile freedom: African American women and emancipation in the antebellum city.* New Haven: Yale University Press.

Borkin, M. A., Vo, A. A., Bylinskii, Z., Isola, P., Sunkavalli, S., Oliva, A., & Pfister, H. (2013). What makes a visualization memorable? *IEEE Transactions on Visualization and Computer Graphics, 19*(12), 2306-2315. http://doi.org/10.1109/TVCG.2013.234

Chandler, D. (2017). *Semiotics: The basics.* New York: Taylor & Francis.

Chang, A. (2017, August 3). *Watch how immigration in America has changed in the last 200 years.* Vox. Retrieved from https://www.vox.com/2016/1/4/10709366/immigration-america-200-years

Chevat, B., & Lowenthal, T. A. (2015, March 3). Meeting the challenges of the 2020 Census. *Center for American Progress.* Retrieved from https://www.american-progress.org/issues/race/reports/2015/03/03/106224/meeting-the-challenges-of-the-2020-census/

Cohn, D. (2010, March 3). Census history: Counting Hispanics. *Pew Research Center.* Retrieved from http://www.pewsocialtrends.org/2010/03/03/census-history-counting-hispanics-2/

Cox, D. (2006). Metaphoric mappings: The art of visualization. In: P. A. Fishwick (Ed.), *Aesthetic Computing.* (pp. 89-114). Cambridge: MIT Press.

Cruz, P. (2015). Wrongfully right: Applications of semantic figurative metaphors in information visualization. *Proceedings of IEEE VIS 2015 Arts Program, VISAP'15: Data Improvisations*, Chicago, Illinois, 14-21.

Cruz, P. (2017). Adapted Dorling cartogram on wage inequality in Portugal. *Proceedings of the IEEE VIS Arts Program (VISAP)*, Phoenix, Arizona.

Cruz, P.M., Wihbey, J., Ghael, A., Costa, S., Chao, R., & Shibuya, F. (2018). Process of simulating tree rings for immigration in the US. *IEEE VIS Arts Program Annotated Projects.*

Daniels, R. (1989). *History of Indian immigration to the United States: An interpretive essay.* New York: Asia Society.

Du Bois, W. E. B. (1900). *The Georgia Negro: A social study* [Map]. Library of Congress Prints and Photographs Division Washington, D.C.

Gibson, C., & Jung, K. (2006). *Historical Census Statistics on the Foreign-Born Population of the United States: 1850 to 2000—U.S. Census Bureau Working Paper No. 81.* US Census Bureau. Retrieved October 11, 2018 from https://www.census.gov/population/www/documentation/twps0081/twps0081.html

Handlin, O. (1959). *Immigration as a factor in American history.* Englewood Cliffs, NJ: Prentice-Hall.

Harding, S. (1993). *The 'racial' economy of science: Toward a democratic future.* Bloomington: Indiana University Press.

Hart, P. (2009, February 5). Census Bureau tracks changing face of US University Times. Retrieved from https://www.utimes.pitt.edu/archives/?p=8642

Horton, L. E. (1985). Community organization and social activism: Black Boston and the antislavery movement. *Sociological Inquiry, 55*(2), 182-199. http://doi.org/10.1111/j.1475-682x.1985.tb00858.x

Ignatiev, N. (2012). *How the Irish became white.* New York: Routledge.

IPUMS USA. (n.d.). 1890 Census: Instructions to enumerators. Retrieved June 1, 2018 from https://usa.ipums.org/usa/voliii/inst1890.shtml

Jakobsen, T. (2001). Advanced character physics. In: Proceedings of Game Developers Conference, San Francisco, pp. 383-401.

Jarvis, B. (2017, January 18). Who decides who counts as Native American? *The New York Times Magazine.* Retrieved from https://www.nytimes.com/2017/01/18/magazine/who-decides-who-counts-as-native-american.html

Kertzer, D. I., & Arel, D. (2002). *Census and identity: The politics of race, ethnicity, and language in national censuses.* Cambridge: Cambridge University Press.

Lakoff, G., & Johnson, M. (1980). *Metaphors we live by.* Chicago: The University of Chicago Press.

Lakoff, G. (1993). The contemporary theory of metaphor. In: A. Ortony (ed.), *Metaphor and thought.* (pp. 202-251). Cambridge: Cambridge University Press.

Lakota, H. (1902). *Winter Count on cloth by Long Soldier* [Painting]. Smithsonian, National Museum of the American Indian. Retrieved October 1, 2018 from https://americanindian.si.edu/exhibitions/horsenation/wintercount.html

Portes, A., & Rumbaut, R. G. (1990). *Immigrant America: A portrait.* Berkeley and Los Angeles, CA: University of California Press.

Press, W. H., Teukolsky, S. A., Vetterling, W. T., & Flannery, B. P. (2007). *Numerical Recipes* (3rd ed.). Cambridge: Cambridge University Press.

Nicholson, F. (1724/1900). *Map of the several nations of Indians to the northwest of South Carolina* [Map]. Library of Congress. Retrieved from http://hdl.loc.gov/loc.gmd/g3860.ct000734

Nugent, W. T. K. (1995). *Crossings: The great transatlantic migrations 1870-1914*. Bloomington: Indiana University Press.

Ruggles, S., Genadek, K., Goeken, R., Grover, J., & Sobek, M. (2017). *Integrated public use microdata series: Version 7.0* [Data Set]. Minneapolis: University of Minnesota. https://doi.org/10.18128/D010.V7.0.

Schmidt, R. W. (2011). American Indian identity and blood quantum in the 21st century: A critical review. *Journal of Anthropology,* 2011. http://dx.doi.org/10.1155/2011/549521

Steinmetz, K. (2018, March 27). The debate over a new citizenship question isn't the first Census fight. Here's why the count is controversial. *Time*. Retrieved from http://time.com/5217151/census-questions-citizenship-controversy/

Takaki, R. T. (2012). *Strangers from a different shore a history of Asian Americans* (Updated and Revised). Boston, New York, & London: Little, Brown and Company.

Truesdell, L. E. (1930). 1930 Census: Volume 6. [Supplement] Special report on foreign-born white families by country of birth of head with an appendix giving statistics for Mexican, Indian, Chinese, and Japanese families. US Census Bureau. Retrieved from https://catalog.hathitrust.org/Record/001310758

Tufte, E. R. (1983). *The visual display of quantitative information*. Cheshire, CT: Graphics Press.

US Census Bureau. (n.d.). *Measuring Race and Ethnicity Across the Decades: 1790-2010*. Retrieved June 15, 2018 from https://www.census.gov/data-tools/demo/race/MREAD_1790_2010.html

US Census Bureau. (n.d.). *Statistical abstracts of the United States*. Retrieved August 15, 2018 from https://www.census.gov/library/publications/time-series/statistical_abstracts.html

US Census Bureau. (1896). *Statistical abstract of the United States 1897—Part 2*. Retrieved August 15, 2018 from https://www.census.gov/library/publications/1898/compendia/statab/20ed.html

US Census Bureau. (1983). *Statistical abstract of the United States: 1984—Section 1 Population*. Retrieved August 15, 2018 from https://www.census.gov/library/publications/1983/compendia/statab/104ed.html

Zinn, H. (2015). *A people's history of the United States: 1492-Present* (3rd ed.). London: Routledge.

Zong, J., & Batalova, J. (2017, March 8). Frequently requested statistics on immigrants and immigration in the United States. *Migration Policy Institute*. Retrieved from https://www.migrationpolicy.org/article/frequently-requested-statistics-immigrants-and-immigration-united-states-6

About the authors

John Wihbey is an Assistant Professor of journalism and media innovation at Northeastern University in Boston, where he heads the graduate programmes in the School of Journalism. He is a faculty director of the Co-Laboratory for Data Impact, based at Northeastern's College of Arts, Media and Design.

Pedro Cruz is an Assistant Professor in the Department of Art + Design at Northeastern University. He is a faculty director of the Co-Laboratory for Data Impact.

Sarah J. Jackson is an Associate Professor of Communication Studies at Northeastern University, where she is also a faculty affiliate of the Department of Cultures, Societies and Global Studies.

Brooke Foucault Welles is an Associate Professor of Communication Studies at Northeastern University and a core faculty member at the Network Science Institute.

23. What is at stake in data visualization? A feminist critique of the rhetorical power of data visualizations in the media

Rosemary Lucy Hill

Abstract

Data visualizations are powerful semiotic resources, which, it is sometimes claimed, have the power to change the world. This chapter argues that to understand this power we need to consider the uses to which visualizations have been put. Using visualizations relating to abortion as a case study alongside Klein and D'Ignazio's notion of a 'Bring Back the Bodies' in data visualization, I argue that visualizations tell a narrow story, removing contextual detail and omitting to ask questions important to women's health. To grasp the significance of this I propose a new body issue: the neglect of the viewer and those affected by decisions taken based on visualized data. Far from being a simple device to graphically display numerical data, therefore, there are important social and ethical issues at stake in data visualization.

Keywords: Abortion; Data visualization; Feminism; Bodies

Introduction

What is data visualization for? Data visualizations in the media are not just about giving people easy or pretty access to information. They are about telling stories and they therefore work within the narrative frames of their designers and disseminators. When influential data visualizers write that data visualization can 'change the world' (Kosara, Cohen, Cukier, & Wattenberg, 2009), implicitly *for the better,* we therefore need to ask questions

Engebretsen, M. and H. Kennedy (eds.), *Data Visualization in Society.*
Taylor & Francis Group, 2020
DOI 10.5117/9789463722902_CH23

of what they mean. If data visualization can change the world, then there is much at stake in the form. The assumption is that access to more data can enable us to make more rational decisions (Dur, 2014). This idea is in part built on the belief in the power of *big* data to tell us something new about the world (cf. the famous claim that the data themselves are enough and we don't need theories to help us understand them anymore in Anderson, 2008). However, the idea that more data can enable more rational decision-making is deeply problematic. Feminist methodological arguments problematize the idea that research data have intrinsic objectivity (Ramazanoglu & Holland, 2002). Dorothy E. Smith (1974) argues that researchers' claims to objectivity position the researcher as apart from society, as able to take a completely objective viewpoint—what Haraway would call a 'god trick' (1988, p. 581). But of course it is not possible to be outside society, and those producing data make decisions which are fundamentally informed by their social positions. This therefore raises important questions about the data that are produced and who is producing them. What assumptions are built in? Who and what is counted? Who and what left out? How do gendered power relations impact the processes of data creation? When we consider 'big' data, it is not enough to assume that the data will speak for themselves. We must ask questions of the data (boyd & Crawford, 2012). When it comes to data visualization some research queries the form's objectivity (e.g. Ambrosio, 2015; Bowie & Reyburn, 2014; Kennedy, Hill, Aiello, & Allen, 2016, and others), but consideration of the political and rhetorical work of data visualizations has been more muted.

In this chapter I draw on my research into online visualizations relating to abortion. On the *Persuasive Data* project I examine visualizations made by campaigning individuals and groups, and consider how visualizations work in situ as rhetorical devices which attempt to persuade viewers about the rectitude of abortion. My position is pro-choice: I believe that women should have access to safe, legal abortion as a necessary part of healthcare and reproductive rights. For this reason, in analysing these visualizations, it is necessary to think about who is being counted and who is left out, who is doing the data creation and visualization. Drawing on feminist methodological ideas, D'Ignazio and Klein (forthcoming) argue that data visualization has an issue with bodies. They determine that there are four ways in which bodies are missing from data visualization:

1. 'Bodies are extracted' (D'Ignazio, Thylstrup, & Veel, 2017, p. 69). States, institutions, and companies have the power to collect data, which means they extract data from people, leaving the 'bodies' behind. Institutions determine what kinds of data are collected and what it will be used for, but not all institutions handle sensitive data in a safe, just, or ethical way.

2. 'Bodies are absent' (p. 69). The standpoints (including the privileges) of the people doing the work of data creation or extraction and visualization are unacknowledged. This matters because when data are posited to be objective, the privileges and biases of data producers and visualizers are encoded into them without recognition of this fact. D'Ignazio et al. highlight the overrepresentation of white men in tech and STEM: 'humans might make computers dumber by encoding our age-old biases and structural inequalities into the system' (p. 69, drawing on Kate Crawford).
3. 'Bodies go uncounted' (p. 69). There are differential amounts of data created about things that are important to men and things that are important to women, since, the authors argue, most data scientists are men. For example, there are much more data on erectile dysfunction than on 'the composition of breastmilk' (p. 69). The result is that those things on which there are data are seen to be important, whilst those things which are not quantified are not. This produces a very uneven view of the world.
4. 'Bodies are rendered invisible' (p. 69). Building on the idea that s/he who makes data and visualizations has an impact on how we see the world, D'Ignazio and Klein argue that visualizations give the appearance of neutrality and objectivity to the data within, whereas, as noted above, they often represent the viewpoints of those who are dominant (see body issues 2 and 3); thus the dominant viewpoints are presented as offering the normal view of the world. Visualizations therefore have discursive power.

Furthermore, O'Riordan (2016) posits that there is a risk of dehumanizing people through the processes of turning us into data and data points, arguing that we need to 'bring up the bodies', to re-embody disembodied data. This idea of the missing body is crucial for thinking about what is at stake in data visualization, especially if we want data visualizations to do good work in the world, to change people's minds, to spur people to action towards making Earth a more just, safe, and beautiful place to be. Fundamentally, data visualization is a creation of people. People—embodied, emotional, enmeshed, messy people—therefore must be at the heart of our critical thinking about data visualization.

My aim in this chapter is to use the four body issues listed above to critically address the work that data visualizations about abortion in the media do in the world. The four issues are not distinct: they interact and overlap with particular results. I also propose that a fifth body issue—that of the viewer—needs to be taken into account if we are to understand what is at stake in data visualization in the media. Thinking about abortion visualizations in the light of bringing back the bodies enables us to understand why the absence of bodies is a serious problem and how the abstraction of data has the potential to undermine fundamental human rights.

Methodology

First of all, a word on the methodology underpinning the research. In order to understand how visualizations relating to abortion are used by campaigning groups, I used the University of Amsterdam's Google Image Scraper to scrape Google Images for data visualizations about abortion, whilst also harvesting their URLs for deeper examination. Google Image Search is likely to be a common method for people seeking visual data about abortion. It can therefore be viewed as a valuable tool for groups wanting to influence minds about the rectitude of abortion. In order to gain a sense of what other people may see when using Google Image Search, I cleared my search history to ensure that the results were unaffected by Google's personalized results system.

The term 'abortion data visuali*ation' is most likely to be used by data specialists, but I wanted to get a sense of what kinds of graphical representations of data are available online without being restricted by specialist terms. I therefore also used the everyday alternatives 'abortion chart' and 'abortion graph'. The three terms provided slightly different images, but there was significant overlap, with a number of the same visualizations and the same webpages being returned for each term. I focused on the top 20 search results in each search. These 60 search results are just a snapshot of abortion-related visualizations, but a snapshot has meaning when we acknowledge that few people look beyond a first page of search results. These are the kinds of visualizations that will typically by found and viewed. I paid particular attention to the kinds of data being used, the claims being made in the surrounding texts, and the discourses employed in both written and visual texts. Using these close readings of visualizations in my dataset, I now explore how the body issues can be seen in three of the top visualizations in the results.

Body issue 1: Data are extracted from bodies

One of the major concerns with data visualizations is that, whilst sources of data may be in evidence (i.e. we know who created the data), very little

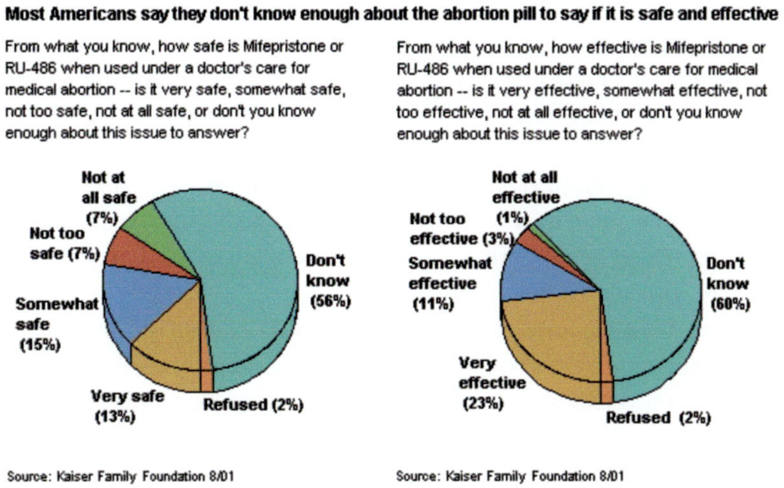

Figure 23.1. Most Americans say they don't know enough about the abortion pill to say if it is safe and effective. By S. Terzo, 2012. (http://clinicquotes.com/abortionvisual-aids-graphs-and-charts/). Copyright 2012 by S. Terzo. Reprinted with permission.

information about how the dataset was created is usually available. As Bowker (2005) argues, data are never raw, they are always 'cooked'—datasets bear the impression of those who made them. Knowing little about this process is problematic, as organizations may display specific data in particular ways, in order to suit their own agenda. This is the case with the visualizations in my dataset from ClinicQuotes, a US anti-choice blog which gathers together images and stories about the perceived ills of abortion. A large number of the images in the search results come from one page on the blog, 'Abortion Visual Aids, Graphs and Charts', which brings together many visualizations and presents them with minimal information about data generation or how data were visualized. One example is the visualization 'Most Americans say they don't know enough about the abortion pill to say if it is safe and effective', shown in Figure 23.1.

The visualization contains two 3D pie charts which show responses to polling about people's opinions about the medical abortion drug mifepristone, undertaken by the Kaiser Family Foundation (KFF). The largest segment of both charts is 'don't know'. The main message of the visualization is that people do not know what to think about mifepristone; they feel ill-informed. Whilst KFF may be supportive of abortion, ClinicQuotes definitely is not, and this visualization used on the site implies support for anti-abortion arguments. The fact of asking this particular question suggests that people ought to be well-informed about mifepristone. But other questions regarding mifepristone could have been asked. The drug's safety is arguably not in question since it is approved by the FDA and is regarded as 95 percent effective. In the UK a number of women's health organizations are calling for medical abortion to be conducted in women's homes to ensure that they are in a safe environment when they begin to miscarry, rather than travelling from clinic to home. Thus the kinds of questions that could have been asked about medical abortion could relate to the effects of needing to travel to and from clinics and experiencing abortion whilst in transit, for example.

We need to ask questions about the people being polled too: how much is the general population likely to know about the safety and efficacy of *any* drug? Who was polled? It is likely that the only people qualified to make judgements on the topic are those who are medically trained to evaluate the evidence. Yet the visualization notes only that 'Americans' were polled. If the organization were aiming for a representative sample then around half of those polled would be men and a significant number of the women would be post-menopausal, sterilized, infertile, using long-term contraception, or not in heterosexual relationships (Goldstein, 2010). In other words, it is possible that many people polled are unlikely to have much awareness of mifespristone because they are unlikely to come into contact with it. It therefore should not be surprising that more than half the sample said they did not know about the safety and effectiveness of the drug. Ordinary people's opinions say little about the actual safety or effectiveness of the drug. These polling data should not, therefore, be taken as indicating that it is a problem that people know little about mifespristone, but the visualization shows how particular data questions can be used in order to produce visualizations which reinforce particular political agendas.

Body issue 2: Visualizers are subject to their own situated knowledge

Just as the data extraction process is usually opaque in finished visualizations, so is the visualization process. Visualizations are provided as finished products, their clean lines, space, and flat colours drawing on their origins in modernist art (Kennedy et al., 2016). However, like all text producers, visualizers tend to let the beliefs, assumptions, and perspectives characteristic of their own social group—that is, their 'situated knowledge' (Haraway, 1988)—influence on the choices they make during the process of production. And data scientists and visualizers tend to be members of privileged groups.

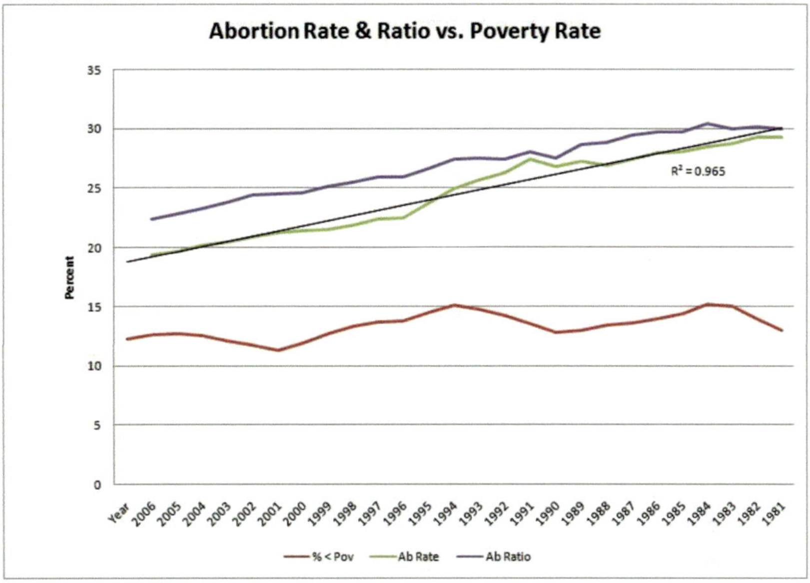

Figure 23.2. Abortion Rate & Ratio vs. Poverty Rate. By Darwin, 2008 (http://darwincatholic.blogspot.co.uk/2008/03/poverty-and-abortion-new-analysis.html). Copyright 2008 by Darwin. Reprinted with permission.

They are often, as D'Ignazio and Klein note (forthcoming), white men. The visualization 'Abortion Rate & Ratio vs. Poverty Rate' (Figure 23.2) forms part of a long article about abortion and poverty, and the author, Darwin, presents his views as scientifically based.

The visualization and article encourage the viewer to understand for themselves—to *see* and *know*—that there is no correlation between abortion and poverty, and to view these data as the facts of the matter. But there is a relationship between poverty and abortion rates, with poorer women obtaining abortion at higher rates (Jones, Darroch, & Henshaw, 2002) and to deny this obscures the structural reasons for abortion decisions, as I discuss further below, and the continued need for safe access to reproductive healthcare. Darwin is anti-abortion and seeks to bring a scientific examination of data to religious discussions. The article uses the language of statistics, although, note that the timeline on this graph runs backwards, which somewhat undermines the author's authority when it comes to statistical literacy. The visualization therefore gives a sense of rationality and contributing to informed debate, although there is very little information here. Neither the visualization nor the article discusses why women have abortions, access to contraception, or what it means to be a mother on the breadline, that is, what the actual relationship between poverty and abortion might be.

Both poverty and abortion are taken out of the context of women's lives and decision-making about their families. Darwin suggests that the high rates of recorded pregnancies in 1973 (when Roe v. Wade was passed) represent a euphoric moment: women could now easily get abortions—and so they did. He goes on to argue that numbers of abortions in the US are falling of their own accord due to people realizing that there is a personal cost to terminating a pregnancy. According to Darwin, the fall is therefore a natural decline. Darwin does not take into account that *reporting* of abortions would have increased post-1973, since abortion was no longer criminalized. No evidence is presented for the claim that the fall in numbers of abortions is due to 'a build-up of painful experience, which has overcome the initial impression that the costs of getting pregnant (and getting out of getting pregnant) are not as high as they were before 1973' (Darwin, 2008). Indeed, it is disputed by the UK Royal College of Obstetricians and Gynaecologists (2016), which found that continuing an unwanted pregnancy has a more detrimental impact on women than terminating one. The reasons for a fall in the abortion rate is actually unknown. Thus, Darwin's contribution can be seen as an effort to mobilize data visualization's rhetorical objectivity to support a subjective point of view.

Body issue 3: Data important to women are missing

Very few of the visualizations in my dataset centre pro-choice arguments. Of the 60 visualizations, 28 sit on anti-abortion websites and only 9 are located on pro-choice sites. Others are on news, health, educational (such as university), and visualization critique sites. A large number of anti-abortion visualizations across the dataset (14) are hosted by one site: ClinicQuotes. Anti-abortion campaigning sites use more data visualizations than pro-choice groups, and there is a difference in the kinds of data being visualized. Anti-abortion groups tend to use polling figures relating to opinions on abortion, statistics on numbers of abortions, who has them at which point in their lives and at which point in their pregnancies. On the other hand, the few pro-choice visualizations present charts relating to threats against abortion providers and restrictions on abortions in different states, data on misinformation in state-mandated documents given to women seeking terminations, and visualizations about women's fertility choices over their lifetimes. These offer a different perspective from the anti-abortion statistics. They focus on the tactics of anti-abortion groups and laws in an effort to protect access to abortion. These different topics of pro-choice campaigning visualizations suggest that the kinds of data relating to abortion that might be useful to women are quite different from the data on numbers of abortions presented by anti-abortion groups. For example, being aware that the information about abortion presented to you by your state has been judged to be misleading (Daniels, Ferguson, Howard, & Roberti, 2016) may enable women to counter the emotive arguments of anti-choice campaigning at the point of decision-making, or it may lessen the emotive impact of such information. However, there remains a gap here in offering data that might be helpful, for example data about how to access abortion in the US (e.g. how far people have to travel to attend a clinic, or how much it costs, or length of waiting times—all things that could be quantified), or the social 'push' factors that lead women to conclude that an abortion is the only realistic option. The question of who is socially supported and financially able to raise a child reveals that 'choosing' an abortion is not a free choice; it can be a forced decision based on a lack of necessary resources to raise a child, an issue of reproductive justice that has significant intersections with class, race, disability, age, and citizenship status (Lonergan, 2012; Ross, 2017). Data on these aspects of abortion are missing from the examples discussed here.

Since much of the data being visualized by anti-choice groups comes from large statistical polling organizations (e.g. Guttmacher, Gallup), we also encounter the first body issue as well: data are extracted from female bodies for purposes which are not fundamentally about sustaining or extending women's rights. This becomes more problematic when we think about what the data are that are being visualized, i.e. numbers of abortions, question about the safety of mifepristone. Those data which *are* visualized come to be seen as important, and those data which are not, to be of no value. That the datasets visualized are created by large well-respected organizations deepens this valuing of particular kinds of data. It raises a further issue of how minimal visualizations strip contextual detail out of issues where such detail is important.

Body issue 4: Data are abstracted in the visualization

In her investigation into the use of sonogram images (technical representations of ultrasound data used in examining the foetus inside the womb) by anti-abortion campaigners, Julie Palmer (2009) argues that sonograms have proven highly emotive and powerful tools. In part this is because seeing a sonogram image is confused with knowing the foetus, as if the sonogram provides a real, objective photograph-like image, rather than being a technological creation. This 'knowledge' is then used to further the aim of reducing the time in which women can legally have abortions by making scientific arguments about the viability of foetuses, for example during debate in the UK House of Commons Science and Technology Committee. Those who are experts in interpreting sonogram images acknowledge their 'beauty' and emotional power, but contest their ability to tell a truth. They argue that the emotion is in the viewer, not the foetus, and that sonogram images do not produce scientific knowledge in themselves (Palmer, 2009). This conflation of 'seeing' with 'knowing' is evident in the 'Abortion Rate & Ratio vs. Poverty Rate' visualization (Figure 23.2), but presenting data in minimal visualizations as in Figure 23.2 further abstracts both the woman and the foetus, and provides a new layer of perceived objectivity. Using data visualizations could be argued to be a step away from the emotionally arresting images previously used by campaigning groups, e.g. powerfully affective photographs of babies and foetuses (Hopkins, Zeedyk, & Raitt, 2005). However, to see visualizations as only rational, neutral artefacts is to fail to recognize the rhetorical and emotional work that they do. This matters because the abstraction takes abortion out of the context of women's lives, out of the context of women making decisions that affect them and their families, and that are part of a wider landscape of reproductive decision-making.

This is particularly evident in the Live Citizen visualization, 'Abortion in the United States' (see Figure 23.3), which appears on a number of visualization critique sites (the original Live Citizen website has been taken down). What is striking is that bodies are in evidence, but the isotypes and area charts representing data use a widely understood icon for women to tell a political story about women's place in the world.

The visualization shows statistics about abortion rates worldwide and in the US. It uses metaphors in which the birth rate is represented through visuals of mothering and nursing newborns (women holding babies, prams), and the abortion rate is represented through visuals of women discarding newborn babies into dustbins. Actually, most terminations happen within the first three months of pregnancy when the foetus is not baby-like and could not survive outside the womb. The equation of the foetus with a baby is a common representational tactic in anti-abortion campaigning (Daniels et al., 2016). Blue and pink icons divide the population into equal parts male and female, using the common convention of gendered colour associations. The visualization thus makes use of some common discourses: the gender binary is natural; babies are nursed by women; women are responsible for birth rates and abortion rates; abortion is casually done (the most common reason for

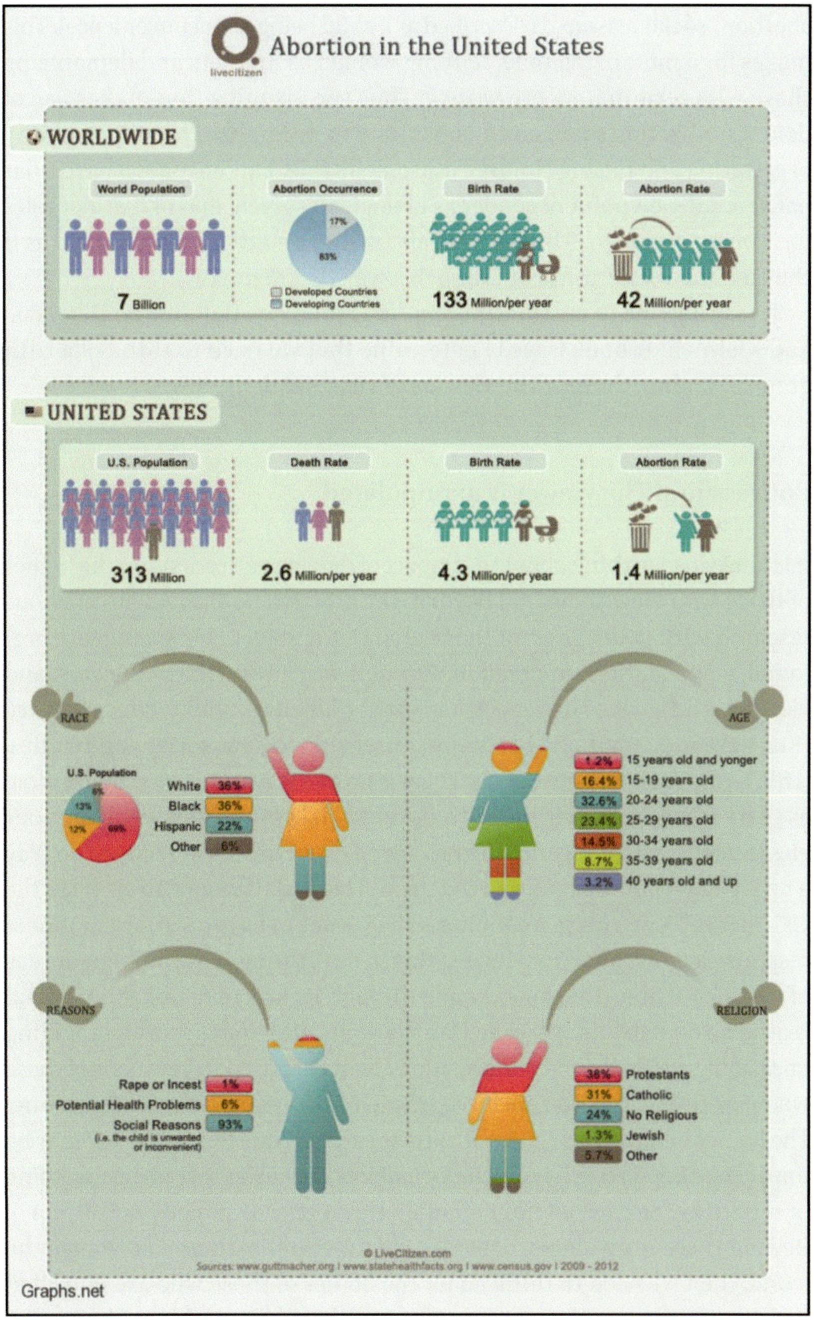

Figure 23.3. Abortion in the United States. By Live Citizen, n.d. (http://schoolofdata.metamorphosis.org.mk/category/data-journalism/page/3/). No copyright information available. Permission sought.

abortion, 'social reasons' is described as a child being an inconvenience). This makes for a moralizing tone, reifying women as mothers and demonizing those who terminate a pregnancy. Thus the visualization makes use of data visualization's perceived objectivity to normalize the responsibility of gestating and raising children as women's role. This encodes a particular patriarchal viewpoint of gender as biologically given, and of distinct roles for women and men. The data are abstracted and then re-embodied as if they tell the whole truth, but in such a way that distorts.

This brings me to my final missing body problem. Building on D'Ignazio and Klein's four body issues I determine that we need to think of a fifth group of bodies: those of the viewers of visualizations.

Body issue 5: The viewer is manipulated

It is vital that we think about the impact of data visualizations on the bodies of those who view them and beyond: the affected bodies. As my previous research with colleagues on the Seeing Data project (see seeingdata.org) found, visualizations are read in different ways by different viewers, and viewing is influenced by gender, nationality, language ability, education, age (Kennedy et al., 2016), and by the discourses around data, society, and culture (Hill, Kennedy, & Gerrard, 2016). There is no one way to view a visualization, as there is no single way to read a novel: social circumstances change our engagements with culture (Barthes, 1977). However, visualizations do play a role in determining how we read them. The visualization 'Abortion in the US' (Figure 23.3) tries to manipulate the viewer to have a strong emotional response against abortion. It does this in part through the ambiguous use of data about abortion, for example through its lack of detail about 'social reasons' and baby imagery. As Daniels et al. (2016) have found, providing misleading and inaccurate information about abortion is a key tactic of those who seek to ban abortion, including those who form part of state legislature. The bodies of those seeing visualizations such as 'Abortion in the US' may be impacted directly by viewing the visualization; they may find it convincing or upsetting, or have another emotional response (Kennedy & Hill, 2017). Beyond these individuals, however, if data visualizations *can* change the world, then we need to think about the bodies of those who are impacted at more of a remove. If data visualizations like these are able to change the world, then the direct impact of them may be on women's ability to access healthcare: working class women who have less money to travel and pay for a procedure; younger women and girls who may be unable to travel; those who cannot take time from caring, family, or work commitments; black, ethnic minority, and poor women who have less access to contraception; migrant women whose citizenship status means they find it harder to access reproductive healthcare (Lonergan, 2012). Banning abortion does not prevent all abortions, but rather forces women to seek illegal, unregulated, expensive, and often unsafe reproductive healthcare. Combined with the severe restrictions on abortion in some US states, and in the UK a lack of free abortion services for migrant women and those in Northern Ireland (up until 2018), these data suggest that those who are more able to gain abortion services are middle and upper class white women living in the right geographical area and holding the right citizenship. Working class, black and minority ethnic, and migrant groups are disadvantaged through lack of funds and other resources needed to seek out abortion, whether legal, private, or 'backstreet' providers. The missing bodies of viewers, and of those who may be affected by decisions informed by visualizations, need to be brought into discussions of the power of data visualizations.

Conclusion

Visualizations about abortion matter when we think about what is at stake in visualizations in the media. These visualizations were in the top portion of Google Image Search results. They have a part to play in changing the world—but not for the better and certainly not because they are providing useful information for making rational decisions. They are offering misleading interpretations of small amounts of data on particular aspects of abortion, leaving out contexts of data creation and visualization, and ignoring significant aspects of factors that affect the experience of abortion. They are leaving the bodies of those from whom data have been extracted, of the visualizers, of women managing their fertility, of women terminating pregnancies, and of those viewing and making decisions based on visualizations out of the frame. These absent bodies mean that it is important to rethink what it means to argue that visualizations can change the world. Abortion is a complex issue and these visualizations and others in my dataset show that simple statistical graphics are unlikely to capture that complexity. But also, more worryingly, such graphics reveal that visualizations are being used as a tool to argue for limits on access to reproductive healthcare. Visualizations can indeed then play a role in changing the world, but it is utopian to imagine that the changes they bring about are always of a positive kind.

References

Ambrosio, C. (2015). Objectivity and representative practices across artistic and scientific visualization. In: A. Carusi, A. S. Hoel, T. Webmoor, & S. Woolgar (Eds.), *Visualization in the age of computerization.* (pp. 118-144). London: Routledge.

Anderson, C. (2008, June 23). The end of theory: The data deluge makes the scientific method obsolete. *Wired.* Retrieved from http://www.wired.com/science/discoveries/magazine/16-07/pb_theory

Barthes, R. (1977). *Image, music, text.* (S. Heath, Trans.). London: Fontana Press.

Bowie, A., & Reyburn, D. (2014). Expanding the terministic screen: A Burkean critique of information visualization in the context of design education. *KB Journal, 10*(1).

Bowker, G. C. (2005). *Memory practices in the sciences.* London: MIT Press.

boyd, d., & Crawford, K. (2012). Critical questions for big data: Provocations for a cultural, technological, and scholarly phenomenon. *Information, Communication and Society, 15*(5), 662-679. https://doi.org/10.1080/1369118X.2012.678878

D'Ignazio, C., & Klein, L. F. (forthcoming). *Data feminism.* Cambridge, MA: MIT Press.

D'Ignazio, C., Thylstrup, N., & Veel, K. (2017). Data visualization from a feminist perspective: interview with Catherine D'Ignazio. *Women, Gender & Research, 17*(1), 67-71.

Daniels, C. R., Ferguson, J., Howard, G., & Roberti, A. (2016). Informed or misinformed consent? Abortion policy in the United States. *Journal of Health Politics, Policy and Law, 41*(2), 181-209. http://doi.org/10.1215/03616878-3476105

Darwin. (2008, March 6). Poverty and abortion: A new analysis [Blog post]. Retrieved December 19, 2016 from http://darwincatholic.blogspot.co.uk/2008/03/poverty-and-abortion-new-analysis.html

Dur, B. U. (2014). Data visualization and infographics in visual communication design education at the age of information. *Journal of Arts and Humanities (JAH), 3*(5), 39-50.

Goldstein, R. D. (2010). Picturing the life course of procreative choice. *UCLA Law Review, 58*(Discourse 5), 5-25.

Haraway, D. (1988). Situated knowledges: The science question in feminism and the privilege of partial perspective. *Feminist studies, 14*(3), 575-599.

Hill, R. L., Kennedy, H., & Gerrard, Y. (2016). Visualizing junk: Big data visualizations and the need for feminist data studies. *Journal of Communication Inquiry, 40*(4), 331-350. https://doi.org/10.1177/0196859916666041

Hopkins, N., Zeedyk, S., & Raitt, F. (2005). Visualising abortion: Emotion discourse and fetal imagery in a contemporary abortion debate. *Social Science & Medicine, 61*(2), 393-403. http://dx.doi.org/10.1016/j.socscimed.2004.11.049

Jones, R. K., Darroch, J. E., & Henshaw, S. K. (2002). Patterns in the socioeconomic characteristics of women obtaining abortions in 2000–2001. *Perspectives on Sexual and Reproductive Health, 34*(5), 226-235.

Kennedy, H., & Hill, R. L. (2017). The feeling of numbers: Emotions in everyday engagements with data and their visualisation. *Sociology, 52*(4), 830-848. http://doi.org/10.1177/0038038516674675

Kennedy, H., Hill, R. L., Aiello, G., & Allen, W. (2016). The work that visualisation conventions do. *Information, Communication and Society, 19*(6), 715-735. https://doi.org/10.1080/1369118X.2016.1153126

Kennedy, H., Hill, R. L., Allen, W., & Kirk, A. (2016). Engaging with (big) data visualizations: Factors that affect engagement and resulting new definitions of effectiveness. *First Monday, 21*(11). https://doi.org/10.5210/fm.v21i11.6389

Kosara, R., Cohen, S., Cukier, J., & Wattenberg, M. (2009). *Panel: Changing the world with visualization.* Paper presented at the IEEE Visualization Conference Compendium.

Lonergan, G. (2012). Reproductive justice and migrant women in Great Britain. *Women: A Cultural Review, 23*(1), 26-45.

O'Riordan, K. (2016, October). *Feminist data visualisation.* Paper presented at the AoIR conference, Berlin.

Palmer, J. (2009). Seeing and knowing: Ultrasound images in the contemporary abortion debate. *Feminist Theory, 10*(2), 173-189. http://doi.org/10.1177/1464700109104923

Ramazanoglu, C., & Holland, J. (2002). *Feminist methodology: Challenges and choices.* London: Sage.

Ross, L. J. (2017). Reproductive justice as intersectional feminist activism. *Souls, 19*(3), 286-314. https://doi.org/10.1080/10999949.2017.1389634

Royal College of Obstetricians and Gynaecologists. (2016). *Q&A: Abortion and mental health.* Retrieved December 19, 2016 from https://www.rcog.org.uk/en/news/campaigns-and-opinions/human-fertilisation-and-embryology-bill/qa-abortion-and-mental-health/

Smith, D. E. (1974). Women's perspective as a radical critique of sociology. *Sociological Inquiry, 44*(1), 7-13. https://doi.org/10.1111/j.1475-682X.1974.tb00718.x

Terzo, S. (2012). Abortion visual aids, graphs and charts. Retrieved January 5, 2017 from http://clinicquotes.com/abortion-visual-aids-graphs-and-charts/

About the author

Rosemary Lucy Hill researches gender, popular music, and big data. She is the author of *Gender, Metal and the Media: Women Fans and the Gendered Experience of Music* (Palgrave) and numerous articles on the politics of data visualizations. She is currently investigating sexual violence at live music events.

24. The power of visualization choices: Different images of patterns in space

Britta Ricker, Menno-Jan Kraak, and Yuri Engelhardt

Abstract

Maps are representations of the world. They offer summaries or simplifications of data that are collected, attempt to reveal unknowns, to simplify and communicate complex spatial phenomena. Numerous decisions are made in the process of creating a map. Seemingly inconsequential variations of cartographic design decisions offer many ways to illustrate this process. We use an open dataset related to the United Nations Gender Inequality Index to demonstrate design decision points and their output. As governments are increasingly making data open to the public, and map-making tools and software are now more accessible online, these considerations are important both for those making and reading maps online.

Keywords: Cartography; SDGs; Open data,

Introduction

Cartography is defined as the art, science, and technology of making and using maps. It requires both qualitative and quantitative methodologies associated with data handling and information communication design. Maps are often seen as authoritative representations of truth (Pickles, 1995; Wood, 1992). There are interrelated processes, interactions, and negotiations during the design and the data collection phases of map-making (Pavlovskaya, 2018). When designed effectively, maps and diagrams can tell stories, offer interactive, dynamic insights into geographical patterns at multiple scales, and show trends over various temporal scales. A map should help the user to quickly grasp a concept or an idea.

Engebretsen, M. and H. Kennedy (eds.), *Data Visualization in Society.*
Taylor & Francis Group, 2020
DOI 10.5117/9789463722902_CH24

Maps can be powerful communication tools to convey the distribution and the magnitude of challenges society faces. Increasingly, people with access to the internet also have access to data that are collected by governments and online data visualization. These data could be used to make maps to lead advocacy efforts linked to gender inequality and feminist discourse, but the creation and use of maps for such efforts are not straightforward. Design defaults or visualization constraints in common mapping software packages and Application Programming Interfaces (API) can impose unintended communication consequences.

In this chapter, we pose the following research questions: What are significant cartographic design decision points? What are various cartographic output possibilities resulting at design decision points? To answer these research questions, we have selected the United Nations Gender Inequality Index (GII). This dataset is used to walk through the cartographic design process, reveal different design decisions, and discuss their possible communication outcomes.

The United Nations (UN) has identified seventeen sustainable development goals (SDGs) in an attempt 'to end all forms of poverty, fight inequalities and tackle climate change while ensuring that no one is left behind' by globally measuring and monitoring a consistent set of variables (United Nations, 2016). These SDGs range from eradicating hunger, to access to clean water, to improving the health of the oceans. The aim of the SDGs is to transform how we live today on a global level, shifting policy and practice, with local authorities held accountable to make changes (Fukuda-Parr, 2016). We fully acknowledge that the SDG indicators and indices are far from perfect, but they are useful for the discussion of map-making. SDG indicators that are calculated as part of the GII include maternal health outcomes, adolescent birth rate, a population with at least secondary education, female and male shares of parliamentary seats, and female and male labour force (United Nations Development Programme, 2016).

In an attempt to make a step towards a feminist geography workflow, we reveal problems in the existing dataset, such as missing relevant data, and we question who is and is not included (Moore, 2018; Pavlovskaya & St. Martin, 2007). We unpack an authoritative dataset assembled by the UN, question it, and offer alternative visualizations thereof, using maps as a tool for fostering discussion. Mapping these data can help elucidate its problems. A map can reveal data inadequacies and support improvements or changes in data collection. It is the first step in a longer process.

We first show how the cartographic process can be blended with a feminist approach to design. As a response to the call for 'doing feminist data visualization' (Hill, Kennedy, & Gerrard, 2016), our aim is to demystify the process of cartography and to address points where a designer makes pivotal design decisions that may alter the narrative of the visualization. By identifying these decision points, we encourage critical thinking and visual literacy associated with cartography whilst drawing attention to data representing women and challenges unique to women. Here we present ways in which the cartographic process, both for a mapmaker and a map user, can adhere to feminist epistemologies. In this chapter, we provide illustrated examples covering data processing, transformation, and visualization to display how design decisions may influence meaning-making. We aim to identify cartographic visualization design and interpretation techniques that foster understanding and inspire action to combat inequities. This responsibility is not only for the mapmaker but also the map-reader.

Gender inequality and the power of cartography

Gender inequity is a type of injustice that affects and is affected by all and is a significant problem in all regions of the world. When women are supported, so are men, when women have educational opportunities, economic improvement advances for everyone (Duflo, 1997). The causes of social inequities are deeply rooted in complex historic socio-economic challenges that plague entire regions and influence daily life as well as policy and leadership. Maps could help reveal the complex structures that cause inequity by parsing out individual variables, revealing patterns that may otherwise go unnoticed. This conversion from data to information may lead to new insights that can be used to identify appropriate and localized solutions.

Maps and visualizations are thought to be factual, objective, and transparent, but maps are made by people with individual positionalities and epistemology (Elwood, 2009; Kennedy, Hill, Aiello, & Allen, 2016). Maps often offer both direct and indirect messages with unintended and implicit meanings (MacEachren, 2004). These meanings are conveyed through representations and symbols. Depending on the nature of the data, symbols can vary in appearance. Symbols can vary in size depending on quantities to be represented, or have different colours depending on the quality of the data elements. Colours evoke emotions and hold a different meaning in different cultures. Feelings that emerge from interacting with specific visualizations influence what is learned from them (Kennedy & Hill, 2017). The use of visual variables, particularly when representing sensitive and important messages associated with underrepresented populations, will influence how and what a map user interprets and/or learns from a map.

Feminist geography, GIS, and cartography

Feminist geography is a way of knowing and influencing choices of research methodologies, including how they are implemented (England, 2006). Cartography and GIS have been critiqued by feminist scholars who report that this external vantage point, the all-knowing 'God's eye view' of most maps, that seems to offer the ability to see everything from nowhere (Haraway, 1991) is actually situated (Kwan, 2002b). This God's eye view is seen as authoritative, unquestionable, and objective, yet cartographers know well that the map-making process is filled with decision-making and uncertainty.

Cartography has historically been associated with positivist forms of knowledge production and may help answer questions about what and where. Feminist methodologies are associated with qualitative research and can best answer questions about why and how. Mixed methods approaches to research are recognized as an asset of geographic inquiry (England, 2006). Cartography and information visualization can be used to push a feminist agenda by revealing inequity through the use of the same tools that have (inadvertently or not) perpetuated inequity by omitting women's voices and needs from the map, to begin with (Pavlovskaya & St. Martin, 2007). It is important for feminist geographers to participate in the development of strategies to give a physical map form to feminist discourse by utilizing GIS to advance feminist practices (Kwan, 2002b). We recognize that using tools associated with scientific cartography alone is not enough; it is still critical to acknowledge what is not in the data, and what is missing from the data and the map (Moore, 2018). We take a seemingly traditional cartographic approach but identify points in this process where ideas from the feminist data visualization framework of D'Ignazio and Klein (2016) can be inserted, as a step towards feminist cartography.

Hill et al. (2016) call for more examination of data visualization to uncover hidden biases and sexist discourses within them. Few have attempted to actually 'do' feminist cartography. Feminist cartographies or GIS are practices using tools associated with quantitative methods to further a feminist presence on the map (Pavlovskaya & St. Martin, 2007). One example of feminist GIS is visualizing the spatially limited, daily trajectory of a subset of women reflected in a space-time cube map (Kwan, 2002a; Kwan, 1999). Kwan (1999) used a unique network-based GIS method to illuminate that conventional accessibility measures have gender bias because they do not take into account gender roles and expectations. Another example of applied feminist GIS is Stephens's (2013) study of OpenStreetMap (OSM) which reflects the gendered priorities in points of interest contributed through

Volunteered Geographic Information (VGI). OSM is made up entirely of VGI that form the base map used in many popular location-based services (useful for finding a business or service based on your current location). OSM offers several different classifications and options for brothels yet only one for childcare, but adding or changing the classification of points of interest in OSM requires collective consensus (Stephens, 2013). Positionality and interests are reflected in modern forms of map-making in many ways. Those with the power and technical background to make maps often are unaware of specific injustices to represent on a map. Thus, there is a risk of reproducing and reinforcing inequality that occurs in reality, on the map (Elwood & Leszczynski, 2013; Leszczynski, 2015).

Open data

In line with the call for equity in the map-making process, there has been a significant push for transparency in governmental data, particularly spatial data. Open data initiatives aim to have a positive impact on both governance and economic sectors and potentially create an 'open innovation economy through software developers, civic society and participation from residents' (Ojo, Curry, & Zeleti, 2015, p. 2333). This availability presents new opportunities for civic engagement with data. When raw data are publicly available, skills and specialized training are still required to generate effective maps. Technical skills, hardware, software, and time are required to interact with data. These are resources that those suffering inequities often lack.

Open data offer opportunities to analyse data that those in positions of power also use, to produce maps that display data in different ways, to speak to those in power in a visual language that they already understand. However, 'the master's tools will never dismantle the master's house' (Lorde, 1983, p. 27). In other words, open data could be seen as an attempt to preoccupy the oppressed with the master's concerns (Lorde, 1983) as the data that the government collects and makes available may reflect the government's interests which do not always match the people's. Nonetheless, having open data available, as well as tools to interact with data, presents an exciting opportunity for feminist scholars who wish to do advocacy research. Here we argue that cartographic skills and know-how are more important than ever before. There is an opportunity to inspect open data, to create visualizations, illuminate social disparities, to reflect on whose representation of reality is being revealed, and who is benefiting and who is not, according to a particular data visualization. We will demonstrate that fundamental cartographic principles do not change based on the purpose of a map.

Visualization choices: Points of reflection during the cartographic workflow

Maps are produced by following a cartographic workflow (Kraak & Ormeling, 2010). Figure 24.1 depicts a version of this flow, which is usually split into two connected processes: data analysis and design. Even before the cartographer looks at the data in detail several questions need to be considered. The most important are: What is the purpose of the map, is there a story to be told? Who are the target users and/or other potential users (and can they be involved in the design process)? What is the map use environment? Which type of medium will be used to make the map—static or interactive, paper or digital?

With answers to these questions in mind, the cartographer must consider what is possible to communicate with the available data? Are the data qualitative or quantitative? Quantitative data symbology requires the use of specific perceptual properties to encourage the map-reader to immediately comprehend if the map is representing quantities or numbers. Quantitative data may require proportional perceptual properties that can be communicated by applying visual variables based on size to represent amounts. During each step of the process, the cartographer needs to remain critical and reflective to consider each design and alternative options. This typically requires design understanding and skills. The map user should also be 'skilled' and critical, asking, 'What do I really see? And why does the map look as it does?' Does it privilege a certain viewpoint over another?

Answers to these questions are influenced significantly by both the mapmaker and the map user's positionality. Individual experiences and worldviews will influence what is understood from the visualization made and/or offered. In qualitative research, rigour is tightly connected to reflexivity and positionality (England, 1994; Rose, 1997) because these deeply influence how media are produced and understood. Reflexivity needs to be inserted into cartographic design, questioning each design and interpretation choice along the way.

In Figure 24.1 we summarize a set of questions, which we present to cartography students in classrooms. These questions have to be asked because the tools available to make maps encourage the user to simply push a few buttons using default settings and operations which often result in maps that do not meet the intended communication goals. Not all of the best cartographic design processes or practices are available in popular mapping APIs and software, and to date, machines do not spot the bias that may be caused by using inappropriate visualization choices or that could be found in the data themselves.

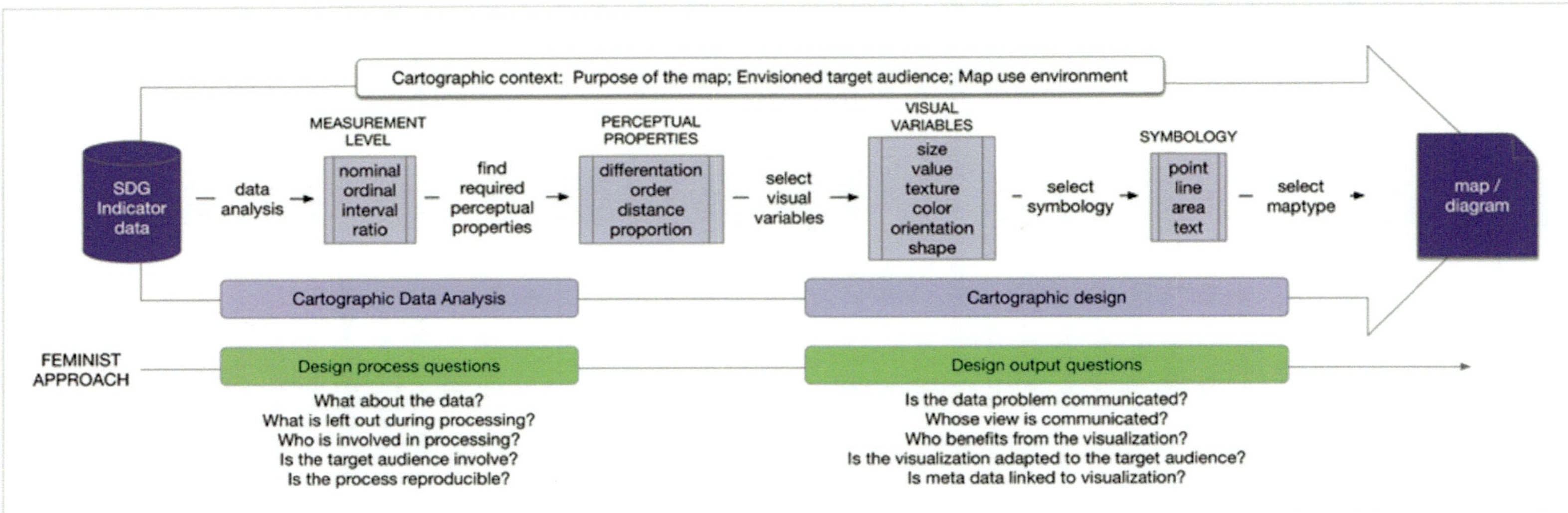

Figure 24.1. The traditional cartographic workflow with special attention to feminist cartography. The diagram shows the steps needed to go from data to map or diagram, split between the cartographic data analysis and the cartographic design. Parallel to this workflow is the feminist data visualization workflow. Illustration by B. Ricker, M.-J. Kraak, and Y. Engelhardt.

For the purposes of illuminating this workflow, we refer back to the research questions that inform the design process related to the maps we would make using the GII. What is the purpose of the map, is there a story to be told? The purpose of the map is to display the spatial trend of gender inequality as measured and defined by the UN. What is the communication aim? For map-readers to understand what it represents and to critically reflect where GII is a significant problem. Who is the target audience? We see the target users and/or other potential users to be policymakers. What is the map use environment? Which type of medium will be used to make the map—static or interactive, paper or digital? Since this is a book chapter, we decided to make static paper maps. The map use environment is book reading. Next, we start to interact with the data.

Figure 24.2a presents a table with the numbers associated with the GII for each country in alphabetical order. This makes it easy to find a country of interest in the table. Each value is a number between 0 and 1. In this index 0 represents no inequality and 1 is maximum inequality. Note that the table does not have an index value for every country. Data in this format are often considered the 'pure' data, but these data are the result of choices that have been made. Here the data have been organized in alphabetical order in English, the number of columns has been chosen, and the index itself is based on choices highlighted in the inset above the bar graph.

In Figure 24.2b, a bar chart offers the same data found in the table, but this time ordered from the highest inequality (left) to lowest (right) for the reader to quickly compare values and countries. Country names are only given every other bar due to space limitations and the countries with no data are omitted. Again, these are examples of choices that have been made; countries have been left out. This does not mean that gender inequality does not exist in these places that lack data.

Figure 24.2c shows a thematic map, a choropleth map that reveals the spatial distribution of GII; in this map the index values have been grouped in four categories based on a particular classification method (natural breaks, which divide classes where there are dramatic differences between low and high points in the data). The visual variable of value has been used to help the user quickly perceive inequality from high (dark) to low inequality (light). Now the countries with the highest inequality are noticed first. This common map type has certain disadvantages. When mapped phenomena are related to the population it is important for the reader to consider that people are seldom homogeneously distributed over the area of the geographic unit (e.g. most of Canada's population lives in a small corridor in the south).

Name	GI index	Name	GI index	Name	GI index	Name	GI index	Name	GI index	Name	GI index
Afghanistan	0,67	Cote d'Ivoire	0,67	Guinea	-	Kuwait	0,33	Norway	0,05	Serbia	0,18
Angola	-	Cameroon	0,57	Gambia	0,64	Laos	0,47	Nepal	0,50	Suriname	0,45
Albania	0,27	Dem. Rep. Congo	0,66	Guinea-Bissau	-	Lebanon	0,38	New Zealand	0,16	Slovakia	0,18
United Arab Emirates	0,23	Congo	0,59	Eq. Guinea	-	Liberia	0,65	Oman	0,28	Slovenia	0,05
Argentina	0,36	Colombia	0,39	Greece	0,12	Libya	0,17	Pakistan	0,55	Sweden	0,05
Armenia	0,29	Costa Rica	0,31	Greenland	-	Sri Lanka	0,39	Panama	0,46	Swaziland	0,57
Australia	0,12	Cuba	0,30	Guatemala	0,49	Lesotho	0,55	Peru	0,39	Syria	0,55
Austria	0,08	N. Cyprus	-	Guyana	0,51	Lithuania	0,12	Philippines	0,44	Chad	0,69
Azerbaijan	0,33	Cyprus	0,12	Honduras	0,46	Luxembourg	0,07	Papua New Guinea	0,59	Togo	0,56
Burundi	0,47	Czechia	0,13	Croatia	0,14	Latvia	0,19	Poland	0,14	Thailand	0,37
Belgium	0,07	Germany	0,07	Haiti	0,59	Morocco	0,49	Puerto Rico	-	Tajikistan	0,32
Benin	0,61	Djibouti	-	Hungary	0,25	Moldova	0,23	North Korea	-	Turkmenistan	-
Burkina Faso	0,62	Denmark	0,04	Indonesia	0,47	Madagascar	-	Portugal	0,09	Timor-Leste	-
Bangladesh	0,52	Dominican Rep.	0,47	India	0,53	Mexico	0,35	Paraguay	0,46	Trinidad and Tobago	0,32
Bulgaria	0,22	Algeria	0,43	Ireland	0,13	Macedonia	0,16	Palestine	-	Tunisia	0,29
Bahamas	0,36	Ecuador	0,39	Iran	0,51	Mali	0,69	Qatar	0,54	Turkey	0,33
Bosnia and Herz.	0,16	Egypt	0,57	Iraq	0,53	Myanmar	0,37	Romania	0,34	Taiwan	-
Belarus	0,14	Eritrea	-	Iceland	0,05	Montenegro	0,16	Russia	0,27	Tanzania	0,54
Belize	0,38	Spain	0,08	Israel	0,10	Mongolia	0,28	Rwanda	0,38	Uganda	0,52
Bolivia	0,45	Estonia	0,13	Italy	0,08	Mozambique	0,57	W. Sahara	-	Ukraine	0,28
Brazil	0,41	Ethiopia	0,50	Jamaica	0,42	Mauritania	0,63	Saudi Arabia	0,26	Uruguay	0,28
Brunei	-	Finland	0,06	Jordan	0,48	Malawi	0,61	Sudan	0,57	United States of America	0,20
Bhutan	0,48	Fiji	0,36	Japan	0,12	Malaysia	0,29	S. Sudan	-	Uzbekistan	0,29
Botswana	0,44	Falkland Is.	-	Kazakhstan	0,20	Namibia	0,47	Senegal	0,52	Venezuela	0,46
Central African Rep.	0,65	France	0,10	Kenya	0,56	New Caledonia	-	Solomon Is.	-	Vietnam	0,34
Canada	0,10	Gabon	0,54	Kyrgyzstan	0,39	Niger	0,70	Sierra Leone	0,65	Vanuatu	-
Switzerland	0,04	United Kingdom	0,13	Cambodia	0,48	Nigeria	-	El Salvador	0,38	Yemen	0,77
Chile	0,32	Georgia	0,36	South Korea	0,07	Nicaragua	0,46	Somaliland	-	South Africa	0,39
China	0,16	Ghana	0,55	Kosovo	-	Netherlands	0,04	Somalia	-	Zambia	0,53
										Zimbabwe	0,54

a)

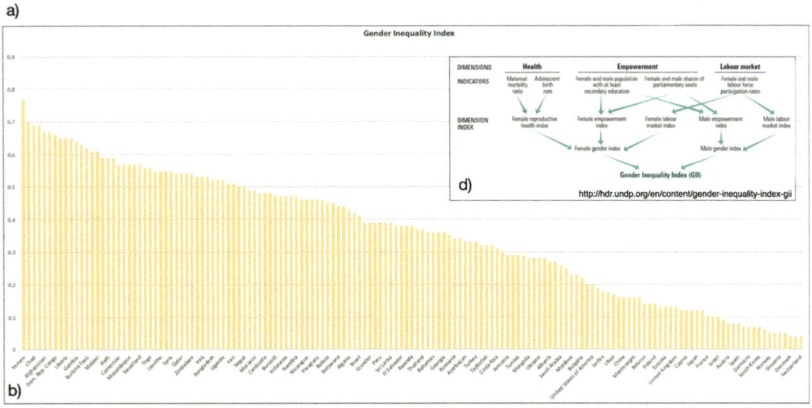

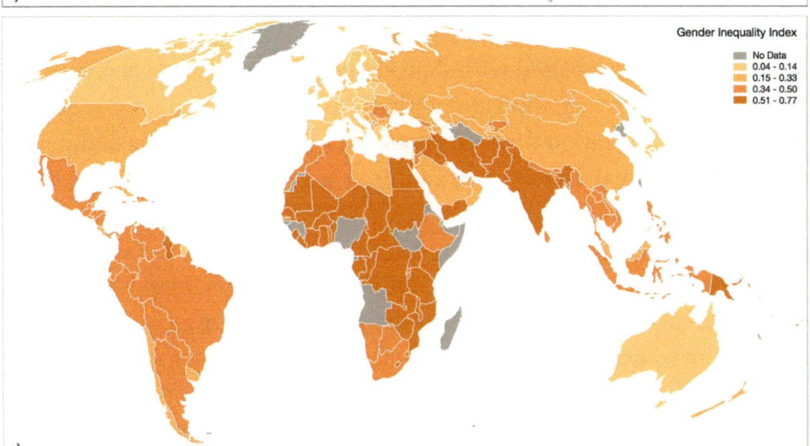

Figure 24.2. The Gender Inequality Index: a) a table with the index value for each country; b) a bar graph with the index order from high to low inequality; c) a map with the geographic distribution of inequality emphasizing high inequality; d) the variables included in the Gender Inequality Index. Source: a), b), c) designed by Ricker, Kraak, and Engelhardt, d) retrieved from http://hdr.undp.org/en/content/gender-inequality-index-gii.

If we take the feminist geographer's call to read between the lines, and to think about what is missing (Moore, 2018), then gender inequality is considered, valued, and measured differently between individuals, countries, regions, and offices. D'Ignazio and Klein (2016) argue that labour should be made visible. This variable measured in the GII as unpaid labour is a variable within this index, but it is completely unclear how these data are collected. How was the time of unpaid labour measured? Where was it measured? In the context of this research, labour is not visible in terms of who made the maps. According to D'Ignazio & Klein (2016), this metadata should be recorded and visualized.

Presenting all three graphics in a linked interactive environment would be ideal. The user could select a country in the table, graph, or map, and at the same time see it highlighted in the other graphics.

Following the workflow presented in Figure 24.1 means making many choices with the map objective and targeted audiences in mind. Paying close attention to colour and data classification, we reveal different outcomes of these choices in Figure 24.3 a-d. Each of the four maps presented in Figure 24.3 shows a detailed view of the same data in Africa and Europe. The colour will influence the audience's conclusions. All four maps are choropleth maps. Both maps on the top use a red colour ramp and both maps on the bottom use a green colour ramp. The meaning associated with each of these colours will vary based on the cultural context of the map-reader. Red is typically associated with danger/bad and green is associated with safe/good. With this in mind and with the assumption that low inequality is the optimal situation different patterns emerge. By emphasizing high inequality via the dark red map, Figure 24.3a sends an alert message: this is not good, it is hoped someone will act. If the colour ramp is reversed as in map Figure 24.3b the problem seems less urgent since the emphasis is on those areas where inequality is not as big a problem. Map Figure 24.3c uses the same approach to the colour ramp as map Figure 24.3a, only now in green. Green does not immediately send an alert, and the mapmaker might downplay the phenomenon. Map Figure 24.3d emphasizes the countries with a low index value. If the maps are there to encourage governments to act, we believe that map Figure 24.3a sends an urgent message. The other maps tell the same story but one has to see through the inverted colour ramps and the colour. The legend shows four classes based on the natural breaks classification. If the number of classes changes or a different classification method were used, the pattern in the map would change dramatically. The audience of any map may not ask, What do I really see? Often a first impression of the map pushes the user to make a particular conclusion.

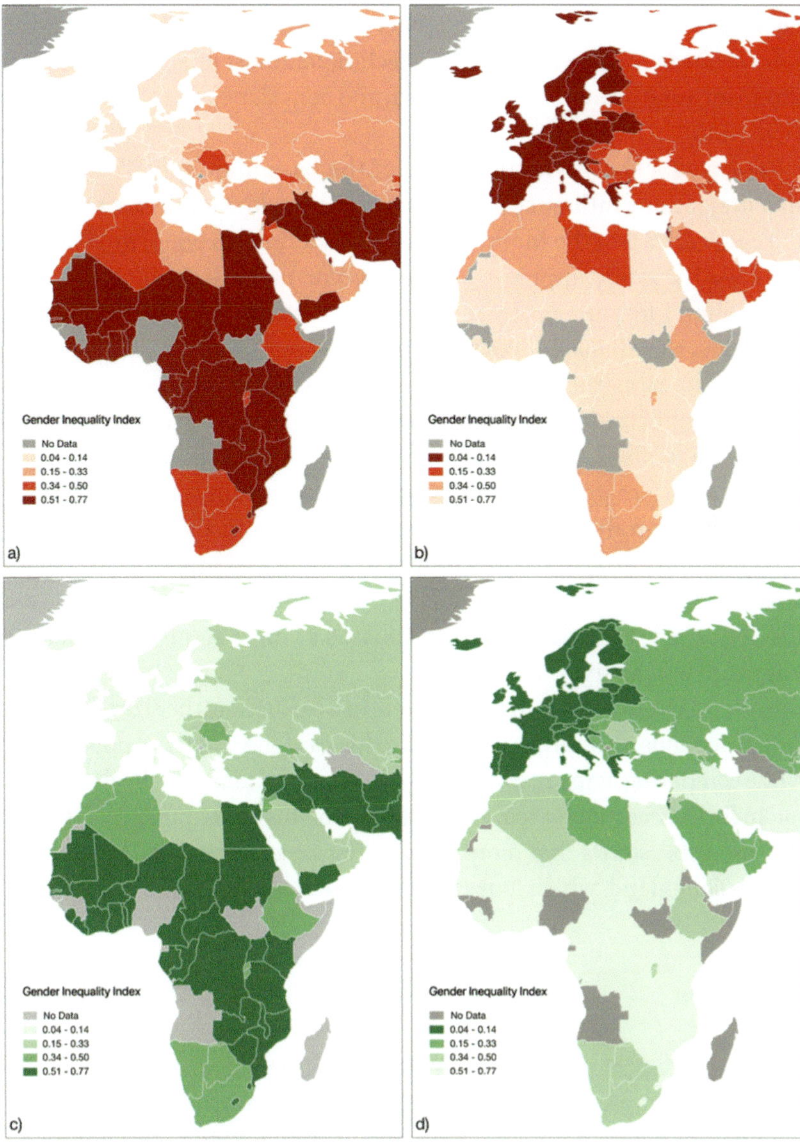

Figure 24.3. Choosing colours and colour ramps for a choropleth map: a) red, emphasizing high inequality; b) red, emphasizing low inequality; c) green, emphasizing high inequality; d) green, emphasizing low inequality. Illustration by Ricker, Kraak, and Engelhardt.

Other key considerations related to the base map selected are the political boundaries and the projection. Borders used in maps convey a specific worldview; lines on maps may cause controversy. Different cartographic techniques, such as blurred or dotted lines, can visualize uncertain and disputed boundaries, but no single map will be able to honour all viewpoints regarding any political issue or boundary dispute at a global scale. A decision must be made about which of these techniques most effectively embrace pluralism, multiple views of the world, and consider context (D'Ignazio & Klein, 2016).

Figure 24.4 shows the influence of the base map, specifically how map projections influence depictions of the world. Figures 24.2, 24.3a-d and 24.4a are all in the Mollweide map projection, an equal area projection. Saudi Arabia (approx. 2.15 million km^2) and Greenland (approx. 2.16 million km^2) appear to be roughly the same size in the Mollweide projection. Not all map projections preserve area. In map 23.4b the world is projected using the Mercator projection, used by the majority of web mapping applications. Mercator's strength is preserving angles, but not size. Saudi Arabia appears to be far smaller in area than Greenland in map 23.4b. Mercator is an inappropriate projection for a global scale choropleth map since a map user will consider both the representation of colour and the size of the countries. Canada is only slightly bigger than Australia in area (see top map 23.4a), but map 23.4b may suggest a different story: Canada might be evaluated as more significant because it appears much larger in size.

In map 23.4c a cartogram is presented. In a cartogram, the size of the area shown depends on another variable, in this case the country's population. While comparing maps 23.4a and 23.4c the size of Canada has been reduced and India has been increased. The advantage of this cartogram is that India's increased size shows that more people are being affected by the problem than in Greenland, which looks much bigger in 23.4b.

Conclusion

Maps evoke emotion, and design choices influence meaning or understanding of maps. Incorporating ideas from feminist GIS, we offered examples of how seemingly simple and arbitrary design decisions may alter what is learned from a map. We displayed common design decisions that cartographers have to make, and in Figure 24.1 we synthesized reflective practice associated with the cartographic process. Based on Figures 24.2-24.4, and connecting them with the design output questions offered in Figure 24.1, we asked: Do we communicate the data problems? Whose view is communicated? Who benefits from the visualization? Do we adapt the visualization to the target audience? Is metadata linked to the visualization?

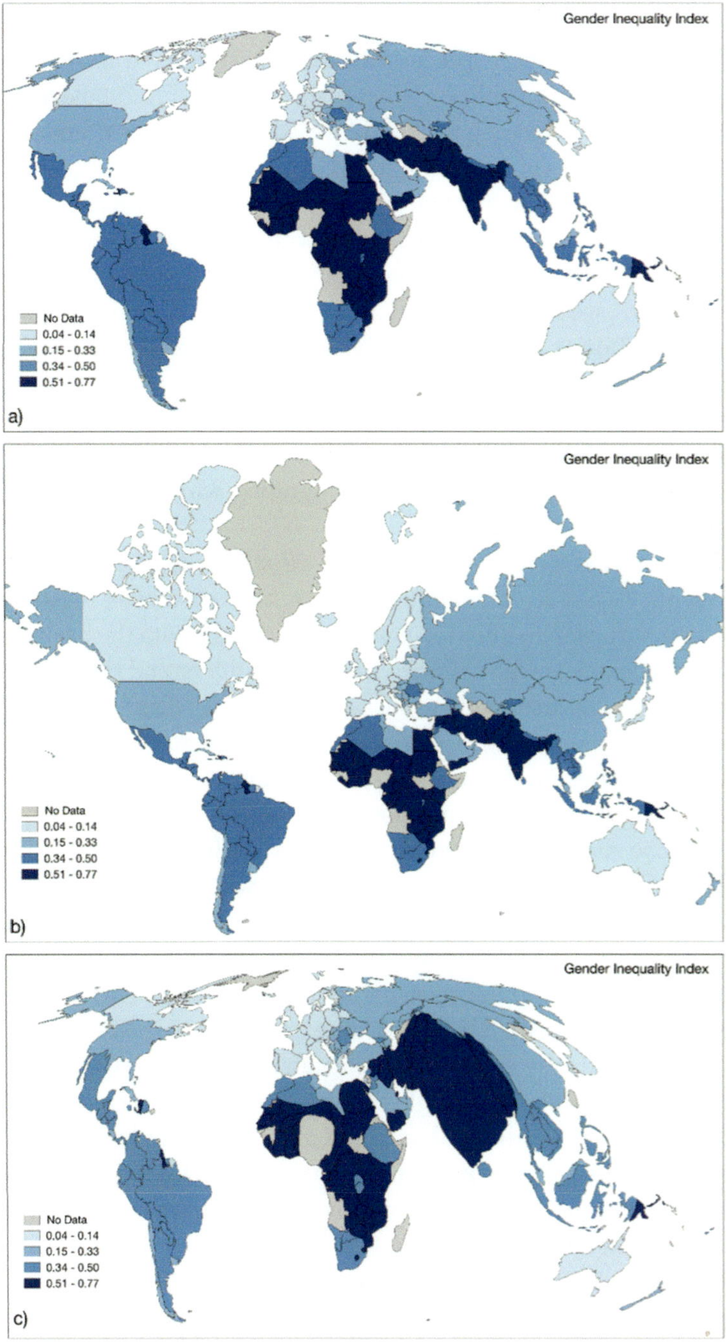

Figure 24.4. The influence of the map projection: a) the equal-area Mollweide projection; b) the conformal Mercator projection; c) a cartogram based on the country population. Illustration by Ricker, Kraak, and Engelhardt.

It is well known that cartography is not an objective science. The positionality of the cartographer influences the data interpretation and design decisions, and their subject position will influence the meaning of the map. Maps can be a tool for advocacy, and for this reason, we encourage feminist geographers to use maps as a praxis and part of advocacy efforts to combat and mitigate gender inequality. Our aim is to encourage more people to create visualizations that reflect and reveal women's subject position in an effort to combat and reduce gender inequity. Using a feminist lens, like Kwan (2002a), we see the methods associated with GIS, cartography, and geovisualization as means to identify what type of gender inequity is happening and where it is happening. Future research is needed to then implement qualitative methods to explore the why of these phenomena. As England (2006) points out, qualitative and quantitative methods are not mutually exclusive, particularly in cartography, which is an art and a science.

We acknowledge the critiques of GIS, data visualization, and the SDGs themselves, but we also see an opportunity to use all of these tools together for subversive purposes, to illuminate problems with the data themselves. These could be activists who would like to better understand the spatial distribution of gender inequality as measured by the UN, and challenge these measures. By identifying specific points in the cartographic workflow where design decisions could change outcomes, we aim to encourage critical thinking about and questioning of maps.

Here we echo the call to 'do' more feminist visualization (Hill et al., 2016). Feminist cartography is not a straightforward task. We do not claim to have achieved this end, but we hope these are steps that could precede or complement additional efforts that are taken by feminist geographers. Through mapping open data, what is missing from the data can more easily be identified by a critical map-reader. This can be seen in parallel with 'embrace pluralism' as suggested by D'Ignazio & Klein (2016) to be open for a different angle on map design and interpretation. Taking this seemingly traditional mapping practice of an authoritative data source could be a step in problematizing the data. This step could be seen as one in feminist geography workflows, to read between the lines and consider what is missing from the data and the map (Moore, 2018; Pavlovskaya & St. Martin, 2007). We encourage others to utilize open data in an effort to improve data collection, to become a more inclusive practice. We invite more participants to share what they value, to map their worldview.

References

D'Ignazio, C., & Klein, L. (2016, October). *Feminist data visualization*. Paper presented at and published in the workshop proceedings from the Workshop on Visualization for the Digital Humanities at IEEE VIS Conference, Baltimore, Maryland.

Duflo, E. (1997). Women empowerment and economic development. *Journal of Economic Literature, 50*(4), 1051-1079. Retrieved September 18, 2019 from http://www.jstor.org/stable/1049571

Elwood, S. (2009). Geographic Information Science: New geovisualization technologies—emerging questions and linkages with GIScience research. *Progress in Human Geography, 33*(2), 256-263. https://doi.org/10.1177/0309132508094076

Elwood, S., & Leszczynski, A. (2013). New spatial media, new knowledge politics. *Transactions of the Institute of British Geographers, 38*(4), 544-559. https://doi.org/10.1111/j.1475-5661.2012.00543.x

England, K. (1994). Getting personal: Reflexivity, positionality, and feminist research. *Professional Geographer, 46*(1), 80-89. https://doi.org/10.1111/j.0033-0124.1994.00080.x

England, K. (2006). Producing feminist geographies: Theory, methodologies, and research strategies. In: S. Aitken & G. Valentine (Eds.), *Approaches to human geography.* (pp. 286-297). London: Thousand Oaks; New Delhi: SAGE Publications.

Fukuda-Parr, S. (2016). From the Millennium Development Goals to the Sustainable Development Goals: Shifts in purpose, concept, and politics of global goal setting for development. *Gender & Development, 24*(1), 43-52. https://doi.org/10.1080/13552074.2016.1145895

Haraway, D. (1991). *Simians, cyborgs, and women: The reinvention of nature.* New York: Routledge.

Hill, R. L., Kennedy, H., & Gerrard, Y. (2016). Visualizing junk: Big data visualizations and the need for feminist data studies. *Journal of Communication Inquiry, 40*(4), 331-350. https://doi.org/10.1177/0196859916666041

Kennedy, H., & Hill, R. L. (2017). The feeling of numbers: Emotions in everyday engagements with data and their visualisation. *Sociology, 52*(4), 830-848. https://doi.org/10.1177/0038038516674675

Kennedy, H., Hill, R. L., Aiello, G., & Allen, W. (2016). The work that visualisation conventions do. *Information, Communication and Society, 19*(6), 715-735. https://doi.org/10.1080/1369118X.2016.1153126

Kraak, M. J., Ormeling, F. (2010). *Cartography: Visualization of geospatial data* (3rd ed.). Abingdon: Routledge.

Kwan, M.-P. (2002a). Feminist visualization: Re-envisioning GIS as a method in feminist geographic research. *Annals of the Association of American Geographers, 92*(4), 645-661. https://doi.org/10.1111/1467-8306.00309

Kwan, M.-P. (2002b). Is GIS for women? Reflections on the critical discourse in the 1990s. *Gender, Place, and Culture: A Journal of Feminist Geography, 9*(3), 271-279.

Kwan, M. P. (1999). Gender and individual access to urban opportunities: A study using space-time measures. *The Professional Geographer, 51*(2), 210-227. https://doi.org/10.1111/0033-0124.00158

Leszczynski, A. (2015). Spatial media/tion. *Progress in Human Geography, 39*(6), 729-751. https://doi.org/10.1177/0309132514558443

Lorde, A. (1983). The Master's Tools Will never dismantle the Master's house. In: C. Moraga, G. Anzaldúa, T. C. Bambara (eds.), *This bridge called my back: Writings by radical women of color.* (pp. 94-101). New York: Kitchen Table Press.

MacEachren, A. M. (2004). *How maps work: Representation, visualization, and design.* New York: Guilford Press.

Moore, F. (2018). Historical geography, feminist research and the gender politics of the present. *Geography Compass, 12*(9), 1-8. https://doi.org/10.1111/gec3.12398

United Nations. (2018). Sustainable Development Goal indicators website. Retrieved June 28, 2018 from https://unstats.un.org/sdgs/

Ojo, A., Curry, E., & Zeleti, F. A. (2015). A tale of open data innovations in five smart cities. *Proceedings of the Annual Hawaii International Conference on System Sciences, 2015–March*, 2326-2335. https://doi.org/10.1109/HICSS.2015.280

Pavlovskaya, M. (2018). Critical GIS as a tool for social transformation. *The Canadian Geographer / Le Géographe Canadien, 62*(1), 40-54. https://doi.org/10.1111/cag.12438

Pavlovskaya, M., & St. Martin, K. (2007). Feminism and Geographic Information Systems: From a missing object to a mapping subject. *Geography Compass, 1*(3), 583-606. https://doi.org/10.1111/j.1749-8198.2007.00028.x

Pickles, J. (1995). *Ground truth: The social implications of geographic information systems.* New York: Guilford Press.

United Nations Development Programme [UNDP] (2016). Gender Inequality Index. Retrieved June 28, 2018 from http://hdr.undp.org/en/content/gender-inequality-index-gii

Rose, G. (1997). Situating knowledges: Positionality, reflexivities, and other tactics. *Progress in Human Geography, 21*(3), 305-320. https://doi.org/10.1191/030913297673302122

Stephens, M. (2013). Gender and the GeoWeb: Divisions in the production of user-generated cartographic information. *GeoJournal, 78*, 981-996.

United Nations website (2016). (Retrieved in 2018).

https://www.un.org/sustainabledevelopment/development-agenda/

Wood, D. (1992). *The power of maps.* New York: Guilford Press.

About the authors

Britta Ricker is an Assistant Professor at the University of Twente in the Netherlands. Her research interests focus on accessible spatial technologies, particularly the use of open data and mobile map interfaces.

Menno-Jan Kraak is a Professor of Geovisual Analytics and Cartography and head of the Geo-Information Processing Department at the University of Twente/ITC. He was President of the International Cartographic Association from 2015-2019. He has written more than 200 publications about cartography, geovisualization, and mapping time.

Yuri Engelhardt is an Assistant Professor of Data Visualization at the University of Twente. He is passionate about how visualizing information can contribute to sustainable human well-being. He has written about 'The language of graphics' and 'Diagrams for the masses'.

25. Making visible politically masked risks: Inspecting unconventional data visualization of the Southeast Asian haze

Anna Berti Suman

Abstract
This contribution investigates the potential and challenges of data visualization in stimulating a socially and legally accountable governance of environmental risk affecting public health. The visualization analysed results from 'middle-up' mapping efforts of the Southeast Asian haze performed by environmental NGOs and civil society. It is argued that haze governance failures are associated with both a lack of reliable evidence on the haze risk and a denied access to existing information. In response to this informational gap, unconventional solutions to state haze mapping were generated by non-governmental actors. The aim of the chapter is to explore to what extent such counter-mapping succeeded in making *visible* politically masked risks, triggering human agency at the individual and collective levels, and enabling a more accountable governance of the haze risk.

Keywords: Counter-mapping; Environmental risk; Public health; Accountability; Environmental information; Information access.

Introduction

This contribution investigates the societal response to a pressing environmental risk to public health, the Southeast Asian haze, manifested through data visualization practices. A socio-legal analysis is conducted to understand the implications of haze counter-mapping by non-state actors

Engebretsen, M. and H. Kennedy (eds.), *Data Visualization in Society.*
Taylor & Francis Group, 2020
DOI 10.5117/9789463722902_CH25

along three lines: firstly, with regard to the filling of informational gaps in haze mapping, making *visible* 'politically masked' risks; secondly, as a source of social accountability and a trigger for personal and social agency; and thirdly, as a catalyst for legal accountability.

The digital display of the haze phenomenon on online maps appears particularly powerful as it shows the extent of the risk, its rapid movement from one land to another, its volatility but also its origin, shedding light on possible responsibility links. This data visualization therefore communicates complex risk patterns and dynamics in a direct and accessible way. Throughout such dissemination of haze data a process of risk *sense-making, learning,* and *engaging* is triggered (see Kennedy & Engebretsen, this volume). The data visualization at issue has the power to evoke emotions, democratic participation, and other forms of engagement. The emotions awoken are extremely contextual (Kennedy & Hill, 2017) and intertwine with a pre-existing sense of injustice, distrust, and feelings of anger and insecurity. In line with the approach adopted in this book, the engagement with the haze maps is here understood as layered. In other words, people *engaged with* the mapping and got *engaged by* it. Audience responses, identified in the influence on haze-affected individuals and on haze policymakers, are considered of particular interest. The type of communication inspected is *unconventional* in the sense of conveying non-institutional and non-institutionalized meaning, being an unexpected and deviating response to dominant patterns of haze risk governance. Eventually, the displayed haze data fill informational gaps and can become influential for decision-making. The *social power* of data visualizations is inspected in its reactive and transformative potential.

Some of the questions raised by this chapter have been investigated by critical cartography scholars that focused on 'mapping the unmapped' and mapping in situations of crisis. Discussing the accountability potential of counter-mapping, I build on the reflections of Georgiadou et al. (2011) on accountability in the specific context of East Africa; Hohenthal, Minoia and Pellikka (2017) on the potential of critical cartography for co-governing resources; Milan and Gutiérrez (2017) on data activism in Latin America; and my earlier exploratory and more targeted analysis on the subject (Berti Suman, 2017; 2019).

Before answering the key questions of this contribution, the haze phenomenon is contextualized and the response to the haze from relevant institutions and from the local community is described. Subsequently, I discuss the potential and challenges of the haze mapping to fill informational gaps, to foster social accountability, personal and social agency, and legal accountability. In the conclusion, recommendations are sketched on how to release the full potential of this unconventional mapping. Overall, this chapter contributes to the still scarce understanding of the social utility and legal admissibility of non-institutional maps.

The methodology that shaped this study is based on a combination between desk research and empirical qualitative research. The empirical research has been deployed through observations performed at Greenpeace International, Legal Unit, Amsterdam, over the period comprised between January and April 2017; web analysis of haze mapping platforms such as Greenpeace's Kepo Hutan and Global Forest Watch Fires; (virtual) participation in topical meetings of Greenpeace Southeast Asia on the haze matter; and targeted communications with stakeholders (e.g. ministries) and organizations (e.g. action groups) involved in the haze issue both in Southeast Asia and elsewhere. In conducting the empirical research, I signed a collaboration agreement with Greenpeace International and have disclosed the research findings to the organization. Nonetheless, the present piece represents exclusively my view and opinion and in no way should be regarded as expressing Greenpeace's position.

The risk at issue: The Southeast Asian Haze

The Southeast Asian haze is an aggregation in the atmosphere of fine, dispersed, solid or liquid particles or smoke which gives the air an opalescent appearance and may cause harm to human health when inhaled. The haze mostly derives from illegal burning of forests and peatlands. These illegal activities are mainly located in Indonesia and are aimed at preparing land for agricultural use, burning agricultural residue, clearing forest for land acquisition, but they can also be the result of vandalism, accidental ignition, or a mechanism to force inhabitants off the land (Simorangkir, 2007).

The haze is toxic for human health as, during the combustion of forests and peatlands, high amounts of noxious fine particulate matter are released. In addition, the substance is highly volatile, being transported by winds to densely populated areas (Koplitz et al., 2016). Exposure to the haze generates respiratory, heart, and eye-related diseases (Stephen & Low, 2002) and can hinder the health status of the population in the long term (Quah & Johnston, 2001).

Digital maps as a response to the haze problem

The haze challenge has not been properly tackled due to a number of reasons, such as economic and political interests, institutional failures, but also a lack of (access to) information. The problem is two-sided: on the one hand, data on fires locations and on the stakeholders responsible are scarce; on the other hand, when the information is available, it is rarely openly accessible to *the public* (civil society at large). Access to strategic information is here regarded as a crucial 'gatekeeper' that prevents environmental NGOs, local civil society organizations, and individuals from exercising their agency in reacting to the haze. However, when the right to access environmental information is violated, people may find creative ways to get the information they need through *unconventional* ways, for example making use of mobile technologies and sensors operated by non-state actors.

This non-institutional way of tracking (environmental) risks and displaying them through data visualization can be regarded as 'critical mapping'. Specifically with regard to the Indonesian context, critical mapping practices have been discussed by Peluso (1995), arguably before the data visualization 'hype'. The author inspects the politics of land and forest rights in Kalimantan, its representation in official forest mapping, and the indigenous people's response through 'sketch maps' aimed at reclaiming territories. This chapter builds on this discussion, inserting it in today's reality of massive data visualization. Yet the underlying claims and anger of local people do not differ substantially from the reality described by Peluso. Only the means differ, arguably reinforcing the potential of this critical mapping.

The topic of a tech-enhanced counter-mapping has been discussed by Radjawali, Pye and Flitner (2017), again from a specific Indonesian angle. The authors inspect the use of drones to produce 'high-quality community controlled maps' aimed at challenging state spatial planning. In the case, the use of drones boosted the demands of the organized (NGOs) and unorganized (the Dayak communities) civil society and facilitated the institutional recognition of the territorial claims. However, the authors also stressed possible adverse social and political consequences of counter-mapping.

This chapter does not argue that such non-institutional mapping can substitute government's action, but rather that it can push policymakers towards action, in addition to providing useful information for the government itself. Such an outcome has been recently illustrated by Gutiérrez (2018) who discusses maps as forms of political counter-power strengthened by data's new potential. The author conceptualizes a 'digital' critical cartography as a 'new paradigm in activism and humanitarianism' through a discussion of two platforms, Ushahidi in Kenya (on the topic see also Bailard et al., 2012) and InfoAmazonia in the Amazon region (the latter platform specifically aimed at tracking environmental issues).

Throughout the next subsections, the following three aspects of the potential of haze counter-mapping initiatives are investigated: the maps as instruments to fill institutional gaps and make visible politically masked risks; the initiatives as a trigger for social accountability and for personal and social agency; and the data as a source of legal accountability to be possibly used in courts for law enforcement.

Counter-mapping as an instrument to fill institutional gaps

At the institutional level, governmental actors have engaged in a series of initiatives tackling the haze problem, such as the creation of a haze-targeted Geographical Information System on the Indonesian Jambi province and the launch of a Sub-Regional Haze Monitoring System supported by the Association of Southeast Asian Nations (ASEAN) (see http://haze.asean.org/). In addition, it is worth mentioning the 'One Map initiative' by the Indonesian government, aimed at creating a unified database on Indonesian land use, land tenure, and other spatial data. Despite these efforts, however, most concession maps are not digitalized at present and inter-governmental cooperation against the haze is still weak (Shah, 2016).

In response to these weaknesses, non-governmental actors and civil society have mobilized their energy. In the first group are Indonesian organizations such as 'Walhi' (or Friends of the Earth Indonesia), 'Jatam' (the local Mining Advocacy Network), 'Jakarta Legal Aid' (a local non-profit for legal support), and the Indonesia Centre for Environmental Law. At the international level, the World Resources Institute (WRI) and Greenpeace International together with Greenpeace Southeast Asia intervened. For the second group, the civic response assumed either the form of 'passive' data visualization (accessing maps) or 'active' contribution to visualization (feeding data into existing maps). By launching maps, feeding, or visualizing them, both groups aimed at making visible the 'politically masked' risk posed by the haze (on this concept see Berti Suman, 2018).

Two counter-maps are specifically investigated here, the 'Global Forest Watch Fires Map' and 'Kepo Hutan Map'. Both maps have been launched by international organizations but have generated an actual impact on the countries affected and on the local inhabitants. The choice to focus on these two platforms has been guided by a series of reasons: first, accessibility (I could access both maps remotely and information was available in English although this is only partially true for the Kepo Hutan Map); second, opportunity (I had the opportunity to observe from the Greenpeace International premises the functioning of the platforms, the actors operating them, and the response from the local community); third, scope (these mapping initiatives, differently from smaller ones, have a considerable data visualization potential and thus trigger important questions on the effect thereof). Yet, these justifications do not exempt me from acknowledging that my own 'visualization choices' have consequences for the shape of this chapter and that more local initiatives also deserve attention. The focus here on Western-oriented international organizations such as the WRI and Greenpeace is not intended to devalorise the work of local organizations. Future research needs to explore the haze counter-mapping from a local perspective.

The first map, the Global Forest Watch (GFW) Fires was launched by the WRI in 2014 (see Figure 25.1; for GFW maps, see http://fires.globalforestwatch.org/home/, and http://www.wri-indonesia.org/en/resources/maps). The map combines 'real-time satellite data from NASA [...], detailed maps of land cover and concessions [...], weather conditions and air quality data' in order to show fire activity and related effects in the region (Global Forest Watch Fires, n.d.). The platform cooperates with national and local governments, NGOs, corporations, and individuals. The website announces that the map 'offers on-the-fly analysis to show where fires occur' (thus filling institutional gaps) and 'who might be responsible', in order to ensure that 'those who are illegally burning are held accountable' (the legal accountability component) (Global Forest Watch Fires, n.d.). Lastly, the potential for social accountability and personal and social agency References evident in the statement: 'GFW Fires is free to use and follows an open data approach in putting decision-relevant information *in the hands of all* who want to minimize the impacts of fires [...]' (Global Forest Watch Fires, n.d., emphasis added).

Greenpeace Southeast Asia joined the 'visualization' efforts creating the 'Kepo Hutan' interactive map, produced using open source technology provided by GFW (for Kepo Hutan maps, see http://greenpeace.org/seasia/id/Global/seasia/Indonesia/Code/Forest-Map/index.html (Indonesian) and http://greenpeace.org/seasia/id/Global/seasia/Indonesia/Code/Forest-Map/en/index.html (English)). The website affirms that the map is 'a tool to help anyone working on land use [...] and conservation in Indonesia' (personal and social agency) and also 'provides a benchmark for the government's One Map initiative' discussed above (contribution to policymaking) (Greenpeace Southeast Asia, n.d.). The map's 'primary function is to provide *greater transparency* about who controls areas of land and what happens within it. Previously, this information has not been publicly available to this extent [...]' (Greenpeace Southeast Asia, n.d., emphasis added). From such a statement, the social and legal accountability potential of the initiative emerges. The beneficiaries are both civil society and government bodies. Specific

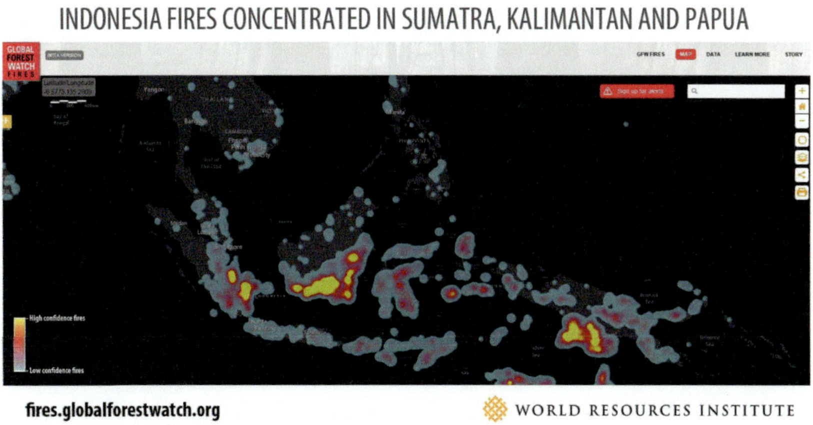

Figure 25.1. The Global Forest Watch Fires Map by the World Resource Institute. From the Global Forest Watch Fires website (https://fires.globalforestwatch.org/). Creative Commons license. Reprinted with permission.

shortcomings of the maps are also acknowledged, such as the diversity of the sources and the obsolete concession data, hindering the digital visualization process. Remarkably, Greenpeace 'invite[s] all stakeholders to help [them] improve [the map]' (Greenpeace Southeast Asia, n.d.). The main obstacle is identified in the government's reluctance to make recent data on concessions 'freely accessible as shapefiles' (Greenpeace Southeast Asia, n.d.), which is an easily analysable format. The website stresses this inaccessibility: '[The] maps presented on this platform are unofficial copies from various sources [...]. Official data are not currently available due to restrictions imposed by the Government of Indonesia' (Greenpeace Southeast Asia, n.d.).

From the combined efforts of Greenpeace and Global Forest Watch Fires, an integrated multilayered platform was generated, based on a vast array of services (e.g. NASA Fire Information System, fire data from the MODIS and VIIRS satellites, and maps from Google Earth) and on a network of infrared sensors capturing heat signatures of fires from the infrared spectral band (see https://fires.globalforestwatch.org/).

The accuracy of fire detection is considered very high: fire data from the MODIS satellite are approximately 1 km resolution and VIIRS satellite data have a resolution of 375m, with a very low rate of false positives. In addition, the algorithm used by the maps to detect fires can eliminate sources of false positives and it will send a fire alert only if the system has enough information. Remarkably, the platform can be fed with information gathered by individuals who can upload data from their mobile phones and, in turn, receive information in real time from the platform.

Figure 25.2. Kepo Hutan Map by Greenpeace in collaboration with the World Resource Institute. From Greenpeace Southeast Asia website (http://greenpeace.org/seasia/id/Global/seasia/Indonesia/Code/Forest-Map/en/index.html). Copyright by Greenpeace. Reprinted with permission.

The discussed map system could conceivably fill informational gaps and enhance transparency in haze governance. In addition, it could allow haze-affected people to be alerted about and react to haze events. Eventually, policymakers could also use the system as a tool to understand fire dynamics and plan more informed haze-tackling strategies. Yet non-institutional mapping efforts can contribute to haze policymaking only if they are recognized as valid by state actors.

Non-institutional mapping as a source of social and legal accountability

Maps as a source of social accountability and human agency

Individuals exposed to data visualization may experience strong emotional reactions, identified by Kennedy and Hill (2017) as, among others, anger, sadness, and offence. Both maps here inspected can trigger intense feelings as, first, they resonate with haze daily experiences and the related sense of fear, concern, and even injustice. Second, being dynamic and often near-real-time, they give the observer an understanding of a phenomenon which is often perceived as obscure by the local inhabitants, as information on its causation and extent is rarely accessible.

The maps discussed, although being unconventional in the sense that they deviate from the governmental approach, cannot be considered expressions of the grassroots or 'bottom-up'. Both the WRI and Greenpeace are non-governmental yet *institutionalized* actors that have a (more or less conflictive) relationship with local governments and hold a specific power, which differentiate them from the *pure* civil society. These maps are initiated by an 'organized' civil society whose organizational structure may shape and orient the counter-mapping agenda. The hope is that such an organizational structure reflects as much as possible the needs, desires, and the expectations of the civil society which the data visualization aims to benefit. This hope seems met as the maps became a platform for the grassroots to gather and communicate haze information. In a sense, these international organizations may be seen as 'middle-up gatekeepers' (as discussed for the case of the Ushahidi crisis mapping by Gutiérrez, 2018). They are intermediary entities which connect the 'bottom' with the 'top' and in some instances are necessary shields for local activists. The truly *lay people* in the haze discourse would be the individuals who experience risk and who seek/feed information online on a daily basis. Future research should inspect the extent to which the agenda of local organizations and civil society is the same as that of the international organizations involved in haze counter-mapping.

In addition, although the use of the words 'bottom-up' and 'top-down' is dominant in the literature (see, for example, Hai-Ying et al., 2014, p. 1), ongoing discussions suggest that this division may fail to capture the blurred reality of social interactions where the boundaries between the two categories are often confused. Authors discussing participatory environmental governance argue that 'bottom-up' and 'top-down' approaches often coexist (Liu et al.,2014, p. 6) and have the same goals but 'different paths'. Along this line, Rey-Mazón et al. stress that the understanding of the term 'veillance' should go 'beyond mere "sur"- or "sous"-topologies', similarly to what should happen for 'bottom-up' and 'top-down' (2018, p. 24). The dichotomy may require more dynamic terms, reflecting processes of 'closing down' and 'opening up' a field to a narrower or broader number of actors (Stirling, 2008, p. 262). Lastly, these unconventional maps stimulate a discussion on how to categorize them. The options range from counter-mapping, participatory mapping, grassroots-driven mapping (Dosemagen, Warren & Wylie, 2011), radical mapping (Denil, 2011), volunteer geographic information (Gutiérrez, this volume), or 'citizen sensing' as argued with regard to noise pollution critical-mapping (Berti Suman, 2018). A combination of counter-mapping, participatory mapping, and citizen sensing may apply.

The preceding discussion suggests that the two maps investigated can be considered *external* factors to the local context eventually activating a response from the local civil society. Such a response is identified in feelings but also actions. Both feelings and actions configure forms of social accountability or 'accountability through engagement' (see, for example, Hughes and Mellado, 2016; Bonner, 2009).

The maps discussed triggered a change in traditional risk mapping, suggesting that not only are public institutions responsible for tracking risks endangering public health, but also non-governmental and civil society actors. With regard to civil society engagement, it should be stressed that the question of *who these people are* matters. In consideration of uneven access to technology, there is still the fear that individuals in remote areas—often the most exposed to haze pollution—will be excluded from haze counter-mapping. Rather, well-educated and wealthy activists may be the fulcrum of such initiatives. Future research should tackle this issue by questioning which groups with which interests generate and visualize the data, to which aims, and—importantly—which groups are *missing*?

Despite the limits of the discussed initiatives, it is worth underlining that the parallel development of (at least) two systems, the institutional and the non-institutional, stimulated a constructive discourse on the appropriateness of the government's approach to the haze. The 'cross-checking' potential of these maps, when state mapping is undermined by a lack of trust, could trigger a social accountability outcome in the sense of multiplying the 'eyes' watching over the government. Although there are clear technical challenges to the success of the unconventional maps, these platforms enabled local actors to gain awareness on the risk they were exposed to and to react to it through adaptation of their personal behaviour and through collective action (the personal and social agency discussed in Poell et al., 2015). Both at the individual and at the collective levels, a transition from passively tracked and profiled individuals and communities to 'active trackers' emerges, configuring a shift from a 'quantified self' to a 'quantified surrounding' of the self (Berti Suman, 2018). The discourse of 'empowering', 'consulting', or 'including' the people in governance processes is arguably replaced by a self-empowerment and auto-inclusion, although through the mediation of pre-existing organizational structures (e.g. Greenpeace). These unconventional mapping initiatives resist top-down imposed means for 'engaging' the people. When communities (more or less supported by organizations) start mapping, they are not 'being consulted', but they actually organize themselves in the first place to carry on their own mapping (see Chambers, 1994, for map-making in rural contexts). They are not only 'data seekers', but actually generate data ('data providers') and become able to exercise their power as concerned stakeholders and as *aware* societal agents that demand more accountability ('critical map-makers'). Traditional patterns of dominance facilitated by restricted access to strategic information are challenged. Technology becomes not only a tool in the hands of corporate and government actors for massive surveillance (Crampton, 2003) but also an opportunity for proactive data activism (Gutiérrez, 2018). A 'reverse' surveillance may even take place as detailed by Radjawali, Pye and Flitner (2017), who showed how community drones have been used to 'watch over' the actions of corporations and the government in areas interdicted to public access.

Maps as a source of legal accountability

Having reflected on the maps as a source of social accountability and human agency, the focus is now shifted on the possibility that such non-institutional data visualization efforts may 'find their way' to courts. The potential of such maps to provide legally acceptable evidence that could enable civil society to make governors and companies accountable for the haze before judicial bodies is inspected, as a way to bring even further human agency. This analysis resonates with Gutiérrez's (2018) discussion of the evidence-generation potential of geoweb technologies.

The underlying assumption to this discussion lays in the acknowledgments that these mapping efforts need to be somehow 'institutionalized' to succeed in court. For example, the maps could find legitimization in the Transboundary Haze Pollution Act (THPA) of 2014, aimed at preventing and punishing the causation of transboundary haze pollution. The THPA recognizes the use of digital maps to enforce justice against haze-causing actors. Specifically, Part II—Liability for Transboundary Haze Pollution, Subsection 8—Presumptions, suggests that the haze presumptions can also be based on *satellite information*. It is made clear that *any* satellite information applies, thus arguably also including unofficial maps. Part II affirms that ownership/occupation of the land shall be presumed on the basis of *maps* which can derive from governmental sources but also from *any prescribed person* through *any prescribed means*. This very open clause in terms of maps' admissibility leaves room for unconventional haze mapping evidence to be considered valid before courts.

The local academic discussion has devoted attention to the recent developments in the use of electronic evidence in court. For example, Low (2012) discussed Art 116A of the Singaporean Evidence (Amendment) Bill 2012, which states that the Minister of Justice may define a *certified process* for generating digital evidence from e.g. tracking tools. Pursuant to this provision, the unconventional maps would need to be recognized as resulting from a certified process in order to be used in court. Yet the admission of electronic evidence may also be dependent on the technology involved

(Seng & Chakravarthi, 2003), and unconventional maps' validity could be undermined by alleged measuring bias and inaccuracies.

The need to rely on lay-produced maps in haze court proceedings has become pressing due to the reluctance of institutional stakeholders to release official maps, often supported by the judiciary. For example, in February 2017, the Indonesian Administrative High Court ruled in favour of the Environment and Forestry Ministry, judging as lawful the government's decision *not to disclose* forest and concessions maps, as requested by Greenpeace Indonesia. This ruling clearly opposes a transparent and accountable governance of the haze and was highly criticized by local and international organizations (Jong, 2017). The local environmental organization Walhi was instead successful in having the Indonesian government condemned for negligence in the management of the 2015 haze crisis. The evidence presented during the trial in part derives from the GFW Fires maps, which suggests that unofficial mapping may eventually succeed in court.

Conclusion

The preceding discussion shows that unconventional data visualization can do 'good' to the affected communities, to the extent that they arise from informational/governance failures. Yet it has been stressed that the reality is more nuanced, as it is unclear how accessible this non-institutional mapping is to the local civil society, and to what extent the agenda of the international organizations behind the maps coincides with that of local actors. Nonetheless, I have argued that it is time to go beyond a polarized 'top versus bottom' debate, as it has been shown that the two layers (and the many more existing layers) often overlap and are blurred. The question of how all these actors and levels of governance differently relate to the challenge of making haze data open to the public has been partially discussed in this chapter but deserves further attention in future research. The relation between data accessibility and the legal framework of different countries, enabling or hindering use of counter-mapping in court, should also be inspected in future research. The extent to which these critical maps can be considered valid (or 'just good enough'; Gabrys, Pritchard, & Barratt, 2016) to show correlations between concessions' ownership, illegal fires, and haze events should also be further analysed.

Despite the need for future research, this discussion contributes to the ongoing debate on the need for more transparent and accountable haze governance in Southeast Asia. It has been demonstrated that unconventional mapping efforts can effectively track and visualize the haze. In addition, they can be a tool enabling local dwellers to react to haze events. Furthermore, such instruments bring the promise to fill gaps generated by institutional failures and make visible politically masked risks. Moreover, their potential to stimulate personal and social agency has been stressed. The rise of a 'cross-checking' mechanism has been identified as a potential source of social accountability. Lastly, even the possibility of a legal accountability outcome has been advanced.

Considering the outlined opportunities and challenges, the following recommendations can be made for the release of the maps' full potential:

- These initiatives should respond to informational/governance failures, which justify the need for an additional system;
- The organizational structure behind these mapping initiatives should reflect as much as possible the needs, desires, and expectations of the civil society for whose benefit the data visualization is aimed, in order not to undermine the social accountability and human agency potential;
- To be a source of legal accountability, the maps should be recognized as valid by state actors, for example through 'certification' mechanisms (which, however, may compromise their non-institutional nature);
- Measuring bias and inaccuracies have to be addressed in order to facilitate their admissibility before courts.

The present reflections and recommendations could guide organizations and civil society actors in developing (haze) data visualization initiatives in a way that ensures their effectiveness and impact, while preserving their non-institutional nature and potential for human agency.

References

Bailard, C., Baker, R., Hindman, M., Livingston, S., & Meier, P. (2012). *Mapping the maps: A meta-level analysis of Ushahidi and Crowdmap*. Washington: Crowdglobe.

Berti Suman, A. (2017). Towards a map-based accountability for environmental health risk? The case of the Equatorial Asian Haze. *Proceedings of the 13th International Conference on Internet, Law & Politics, Managing Risk In the Digital Society*. ISBN: 978-84-697-4474-1.

Berti Suman, A. (2018). Challenging risk governance patterns through citizen sensing: the Schiphol Airport case. *International Review of Law, Computers & Technology*, 32(1), 155-173. http://doi.org/10.1080/13600869.2018.1429186

Berti Suman, A. (2019). The role of information in multilateral governance of environmental health risk: lessons from the Equatorial Asian Haze case". *Journal of Risk Research* – Special Issue on 'Multilateral Governance of Technological Risk'. https://doi.org/10.1080/13669877.2019.1617338

Bonner, M. D. (2009). Media as social accountability: The case of police violence in Argentina. *International Journal of Press/Politics, 14*(3), 296-312.

Chambers, R. (1994). Participatory Rural Appraisal (PRA): Analysis of experience. *World Development, 22*(9), 1253-1268.

Crampton, J. W. (2003). Cartographic rationality and the politics of geosurveillance and security. *Cartography and Geographic Information Science, 30*(2), 135-148.

Denil, M. (2011). The search for a radical cartography. *Cartographic Perspectives, 68*, 7-28. http://doi.org/0.14714/CP68.6

Dosemagen, S., Warren, J., & Wylie, S. (2011). Grassroots mapping: Creating a participatory map-making process centered on discourse. *Journal of Aesthetics and Protest* (8). Retrieved March 9, 2018 from https://www.joaap.org/issue8/GrassrootsMapping.htm.

Gabrys, J., Pritchard, H., and Barratt, B. (2016). Just good enough data: Figuring data citizenships through air pollution sensing and data stories. *Big Data & Society, 3*(2). http://doi.org/10.1177/2053951716679677

Georgiadou, Y., Baba, B., Becht, R., Hoppe, R., Ikingura, J., Kraak, M.-J., Lance, K., ... Verplanke, J. (2011). Sensors, empowerment, and accountability: A Digital Earth view from East Africa. *International Journal of Digital Earth, 4*(4), 285-304. https://doi.org/10.1080/17538947.2011.585184

Global Forest Watch Fires. (n.d.). *About.* Retrieved November 28, 2018 from https://fires.globalforestwatch.org/about/

Greenpeace Southeast Asia. (n.d.). *Protecting forests & peatlands in Indonesia.* Retrieved November 28, 2018 from https://greenpeace.org/seasia/id/Global/seasia/Indonesia/Code/Forest-Map/en/about.html

Gutiérrez, M. (2018). Maputopias: Cartographies of knowledge, communication and action in the big data society—The cases of Ushahidi and InfoAmazonia. *GeoJournal,* 1-20. http://doi.org/10.1007/s10708-018-9853-8

Hai-Ying, L., Kobernus, M., Broday, D.M., & Bartonova, A. (2014). A conceptual approach to a citizens' observatory: Supporting community-based environmental governance. *Environmental Health, 13*. http://doi.org/10.1186/1476-069X-13-107

Hohenthal, M., Minoia, P., & Pellikka, P. (2017). Mapping meaning: Critical cartographies for participatory water management in Taita Hills, Kenya. *The Professional Geographer, 69*(3), 383-395. http://doi.org/10.1177/1940161215614565.

Hughes, S., & Mellado, C. (2016). Protest and accountability without the press: The press, politicians, and civil society in Chile. *The International Journal of Press/Politics, 21*(1), 48-67.

Jong, H.N. (2017, February 16). Forest cover maps to remain confidential: Court. *The Jakarta Post*. Retrieved from http://www.thejakartapost.com/news/2017/02/16/forest-cover-maps-to-remain-confidential-court.html

Kennedy, H., & Hill, R. L. (2017). The feeling of numbers: Emotions in everyday engagements with data and their visualisation. *Sociology, 52*(4), 830-848. https://doi.org/10.1177/0038038516674675.

Koplitz, S. N., Mickley, L. J., Marlier, M. E., Buonocore, J. J., Kim, P. S., Liu, T., Sulprizio, M. P., DeFries, R. S., Jacob, D. J., Schwartz, J. (2016). Public health impacts of the severe haze in Equatorial Asia in September-October 2015: Demonstration of a new framework for informing fire management strategies to reduce downwind smoke exposure. *Environmental Research Letter, 11*. http://doi.org/10.1088/1748-9326/11/9/094023

Liu, H. Y., Kobernus, M., Broday, D. M., & Bartonova, A. (2014). A conceptual approach to a citizens' observatory – supporting community-based environmental governance. *Environmental Health,* (13), 107. http://doi.org/10.1186/1476-069X-13-107

Low, W. (2012). A Commentary on the Amendments to the Electronic Evidence Provisions in the Singapore Evidence Act. *Singapore Law Gazette* 191, 11-23.

Milan, S., & Gutiérrez, M. (2017). Technopolitics in the age of big data: The rise of proactive data activism in Latin America. In: F. Sierra Caballero & T. Gravante (Eds.), *Networks, movements and technopolitics in Latin America: Critical analysis and current challenges*. (pp. 95-109). London: Palgrave

Peluso, N. L. (1995). Whose woods are these? Counter-mapping forest territories in Kalimantan, Indonesia. *Antipode, 27*(4). http://doi.org/10.1111/j.1467-8330.1995.tb00286.x.

Poell, T., Kennedy, H., & van Dijck, J. (2015, December 7). Special theme: Data & agency [Blog post]. Retrieved from http://bigdatasoc.blogspot.com.es/2015/12/special-theme-data-agency.html.

Quah, E., & Johnston, D. (2001). Forest fires and environmental haze in Southeast Asia: Using the 'stakeholder' approach to assign costs and responsibilities. *Journal of Environmental Management,* 63, 181-191.

Radjawali, I., Pye, O., & Flitner, M. (2017). Recognition through reconnaissance? Using drones for counter-mapping in Indonesia. *The Journal of Peasant Studies, 44*(4), 817-833. http://doi.org/10.1080/03066150.2016.1264937

Rey-Mazón, P., Keysar, H., Dosemagen, S., D'Ignazio, C., & Blair, D. (2018). Public lab: Community-based approaches to urban and environmental health and justice. *Science and Engineering Ethics, 24*(3), 971-997. http://doi.org/10.1007/s11948-018-0059-8

Seng, D., & Chakravarthi, S. (2003). *Computer output as evidence: Final report*. Singapore: Technology Law Development Group, Singapore Academy of Law. Retrieved April 10, 2018 from https://www.sal.org.sg/Portals/0/PDF%20Files/Law%20Reform/TLDG-2003-09%20-%20Computer%20Output%20as%20Evidence.pdf

Shah, V. (2016). New map launched on company concessions in Indonesia. Press release of 22 March 2016. Ecobusiness. Retrieved February 14, 2018. http://www.eco-business.com/news/new-map-launched-on-company-concessions-in-indonesia/

Simorangkir, D. (2007). Fire use: Is it really the cheaper land preparation method for large-scale plantations? *Mitigation and Adaptation Strategies for Global Change*, 12, 147-164.

Stephen, A., & Low, L. G. (2002). Impact of haze on health, mortality and mitigation programme. *World Conference on Land and Forest Fire Hazards 2002*, Kuala Lumpur, 319-323.

Stirling, A. (2008). 'Opening Up' and 'Closing Down': Power, participation, and pluralism in the social appraisal of technology. *Science, Technology, & Human Values*, 33(2), 262-294. http://doi.org/10.1177/0162243907311265

About the author

Anna Berti Suman is a PhD researcher at the Tilburg Institute for Law, Technology, and Society. Her PhD project aims at investigating the influence of 'Citizen Sensing' practices on environmental risk governance. She has work and research experience in environmental health and technology, environmental law, and community engagement in environmental decision-making.

26. How interactive maps mobilize people in geoactivism

Miren Gutiérrez

Abstract

Thus far little has been said about how maps are employed in activism to unleash sentiments. Employing as a lens the emotional turn currently influencing geography, this article looks at a 15M map, a cartographic animation that shows a 'connected multitude' of indignad@s as they demonstrated in Spain in 2011; the 'Left-to-die boat' map, tracing the course of a ship in which 63 refugees lost their lives; and the 'Western Africa missing fish' map, which shows foreign fishing vessels operating irregularly in African waters. Interviews, fieldwork, and participatory observation are employed to understand how maps are designed to activate people through emotions. Based on DeSoto (2014) and Muehlenhaus (2013), the chapter also offers a taxonomy as a heuristic tool.

Keywords: Emotions; Critical cartography; Data activism; Maps

Emotions in mobilization and maps

In *Poststructuralist Geographies*, Doel (1999) challenged readers to envisage a cartography that shimmers and think about ways in which flows, relations, and change can be mapped. Two decades ago, imagining a sparkling map entailed a leap of imagination; today, maps can glitter thanks to the geoweb—combining geographic, geospatial, and geotag overlay systems (Scharl & Tochtermann, 2007)—and other technologies. The 15M map included in this study is an example. The 15M or indignad@s movement was a citizen uprising formed in the wake of a sit-down protest on May 15, 2011 in Madrid to demand a more representative democracy. Figure 26.1 shows an instance of the animated chart, which starts with a few sparkles

Engebretsen, M. and H. Kennedy (eds.), *Data Visualization in Society*.
Taylor & Francis Group, 2020
DOI 10.5117/9789463722902_CH26

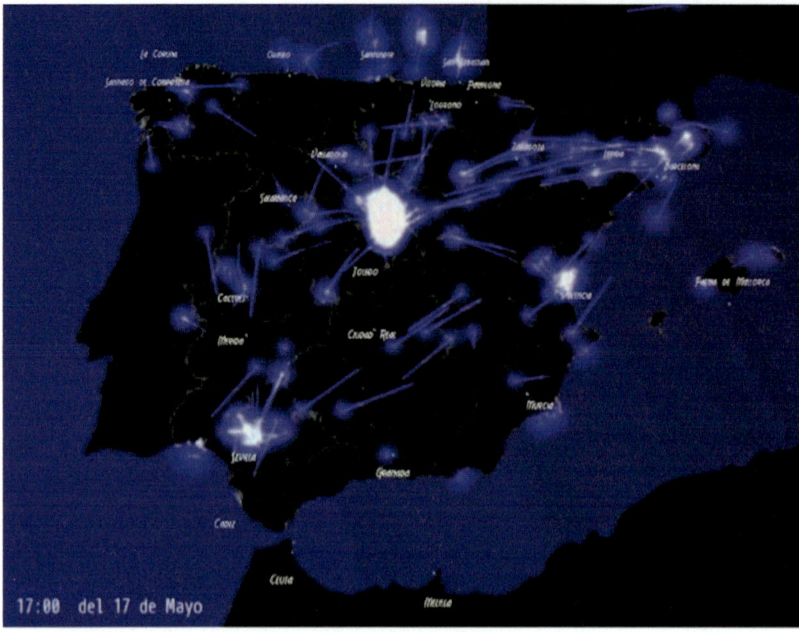

Figure 26.1. A moment in the 15M map. By Instituto de Biocomputación y Física de Sistemas Complejos, Universidad de Zaragoza. Reprinted with permission.

developing into a landscape of increasingly bright jumping lights indicating interactions—tweets—among the 15M supporters as they joined in. The 15M map is Doel's vision come true.

This chapter is positioned at the juncture between approaches that reckon emotions central in social mobilization (della Porta & Diani, 2006; Goodwin, Jasper, & Polletta, 2001, 2004; Melucci, 1996) and the 'emotional turn' in critical cartography, produced by the need to integrate affects in the study of places (Griffin & Mcquoid, 2012; Maddrell, 2016). On the one hand, there is no protest without strong emotions (Jasper, 1998), which can include 'anger and indignation, fear and disgust, joy, and love' (Goodwin et al., 2001, p. 2). Anger can spur participation, but it cannot sustain it for long. Hope, which Goodwin, Jasper, and Polletta deem 'crucial to sustaining movements', breeds an 'anticipation of improvement' (pp. 19, 66). Namely, mobilization can start with anger, but is sustained by hopefulness. On the other hand, Kennedy and Hill (2017) discuss strong emotional reactions—including 'pleasure, anger, sadness, guilt, shame, relief, worry, love, empathy, excitement, offence'—amongst participants in focus groups exposed to data visualizations. Maps—as particular visualizations—can spur sentiment too (Fabrikant, Christophe, Papastefanou, & Maggi, 2012; Griffin & Mcquoid, 2012). In his interview, Panek speaks about how making maps can spark strong emotions and feelings of belonging among cartographers.

There is no consensus about what differentiates emotions from other affective states, such as feelings and sentiments (Klettner et al., 2013, p. 66); I am using these terms as synonyms. The focus here is how maps are designed to mobilize people in geoactivism, understood as activism that relies on digital cartography (Gutiérrez, 2018a), not the people being mobilized by charts. Likewise, I make no distinction between 'mobilization', 'action', and 'protest' since these words all indicate 'doing'.

In-depth interviews and empirical observations of relevant cases, as well as participatory observation of one of 'Western Africa's missing fish' maps, which I co-led with colleagues at the Overseas Development Institute (ODI), are employed to observe geoactivist maps. Mapmakers were questioned about how they design maps. The interviewees include Lorenzo Pezzani, researcher at Forensic Oceanography; Juan Carlos Alonso, a designer at Vizzuality (currently at satellitestud.io), which offers cartography for global campaigns on issues such as climate change; Jiri Panek, an expert on emotional mapping; and Daniel Huffman, an independent mapmaker. They have been selected not only for their experience in generating *maps with a cause*, but also because they are vocal about their strategies as mapmakers. They have given their permission to be named in this article. Empirical observation is employed to contextualize the maps studied here and collect data about the campaigns to which they belong. No causal relationship can be established between these maps and these campaigns' outcomes. The idea instead is to examine how maps are devised to generate reactions, and to determine what elements make them effective. As the lead mapmaker in the 'Western Africa missing fish' project, I employed participant observation to collect data from the processes, meetings, decisions, and internal documents behind the initiative, which are used to enrich this study. The conclusions respond to the initial research question and reflect on the implications of this research for contemporary activism. The main research question in this study is: 'What is it in maps than can make them successfully mobilize people?' A taxonomy of maps is offered at the end of this article as a heuristic tool to create new maps or to examine new cases.

A landscape of emotions

Some emotions in social movements are shaped by collective action around concrete events and issues, while others can exist in people before they connect to campaigning groups (Jasper, 1998, p. 397). The first perspective is employed in this study to establish how cartography is used in advocacy as an instrument of what Muehlenhaus calls 'persuasive geocommunication' (2013). For example, in his interview Alonso talks about how he tries to empower people to act on climate change by using local impact maps to make this global phenomenon more approachable. The 'emotional turn' has influenced cartography to study the links between maps and sentiment from different viewpoints. One is based on the use of technologies to collect and chart emotions spawned by locations (Hauthal & Burghardt, 2013; Klettner et al., 2013). A second approach focuses on the exploration of the feelings engendered by cartography (Fabrikant et al., 2012; Griffin & Mcquoid, 2012). Nold (2009) combines both viewpoints. His emotional cartography captures individual biometric data and then explores its emotional implications. This study uses the second approach. Visualizations can present particular standpoints more convincingly than others (Kennedy et al., 2016). As rhetorical artefacts serving someone's interests (Harley, 1989), maps are often fashioned in a way that they can evoke an emotional response and persuade users to believe or do something (Griffin & Mcquoid, 2012). In his interview, Alonso hypothesizes that maps can situate the observer in remote places by generating vivid emotions connected with places. 'Maps further enhance this recall effect by providing a geographical context that transports the receiver to the place of events', says Alonso. Meanwhile, Huffman explains that when a map is being created with an intent to arouse, this purpose becomes part of its functionality. But what are maps in geoactivism?

A map is a map

First, geoactivist maps comply with the rules of *mapicity,* that is, the properties that make a map recognized by users as useful and believable as maps (Denil, 2011). Second, maps can enable people to plan, coordinate, and mobilize. The data infrastructure, information and communication technologies, the geoweb, and other technologies such as data crowdsourcing platforms, satellite data and imagery, sensors and drones, augment the map's capacities. In the hands of activists, as I have written elsewhere, cartography becomes 'action-oriented, participatory, and production tools signifying complex social, political, or technological processes, and mutable interactions and networks for action' (Gutiérrez, 2018a, p. 15). Third, maps are emotion-producing machineries. Kent (2005) focuses on how aesthetic features in cartography can revive memories, while Preston (2008) examines how map symbols can trigger emotional responses. The maps inspected in

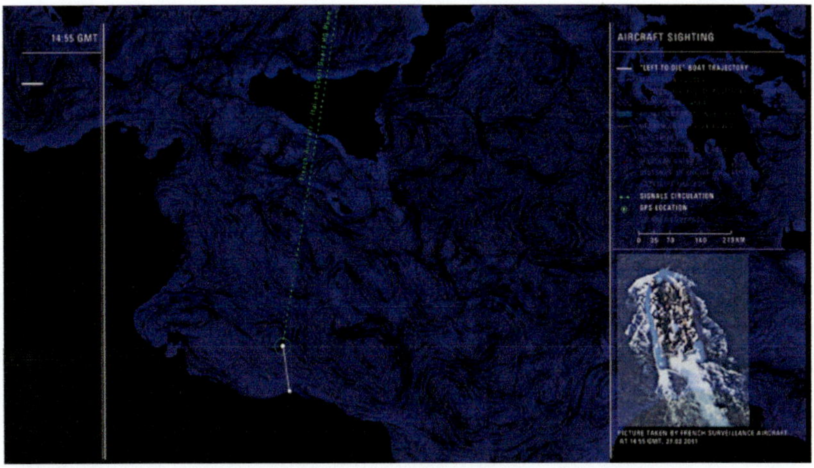

Figure 26.2 A moment of 'Left-to-die boat' map. From *Liquid Traces—The Left-to-Die Boat case* (https://vimeo.com/89790770). Copyright 2014 by C. Heller & L. Pezzani. Reprinted with permission.

this study contain impactful elements that are designed to activate reactions. The 'Left-to-die boat' map includes a narrator, videos, and a soundtrack. The 'Western Africa missing fish' map challenges users to detect irregular fishing activity. Meanwhile, the 15M map evolved in real time, gathering sentiment and pouring it back into the map in the form of light. Figures 26.1, 26.2, and 26.3 show static snapshots of these dynamic maps.

Technologies seem to augment the map's ability to wake up emotions. Van Lammeren et al. conclude in a study comparing 2D and 3D visualizations that the latter generate stronger affective assessments in people (2010, p. 465). Finally, the practice of critical cartography—understood as 'counter-mapping' (Peluso, 1995) and 'radical cartography' (Denil, 2011)—transforms maps into action. For Doel (1999) and DeSoto (2014), among others, maps are in a constant state of becoming. With the geoweb and other technologies, maps incorporate the dimension of sequential time, sometimes even real time, becoming dynamic (Gutiérrez, 2018b). The properties of digital maps allow civil society organizations to engender 'ways of organising collective life' (Gray, Bounegru, Milan, & Ciuccarelli, 2016). This is an appropriate idea to understand maps in activism, since social mobilization is never static.

See-think/feel-do

The maps studied here demand more than just watching or interpretation; they respond to the formula 'see-think-do', coined by Netek and Panek (2016), an awareness-based approach employed to look at crisis mapping.

I propose a variation, 'see-think/feel-do', since maps can generate meanings and feelings, which jointly shape reactions to the visualizations (Lemke, 2015). Crisis mapping—or the geoactivist practice of charting emergency reports and channelling them to responders on the ground so that they can act (Gutiérrez, 2018a)—deserves closer examination as an example of a 'see-think/feel-do' mechanism. It cannot occur without convincing mapmakers (deployers), enthusiastic witnesses (reporters), and cooperating humanitarian workers: the three communities that converge around the map (Gutiérrez, 2018a). The challenge in mapping citizen data does not derive from its technical complexity, but from the summoning power of the map and the mapmakers' capacity to sustain the effort. An example is Ushahidi, a visualization tool that is widely employed in mapping humanitarian emergencies. Hundreds of Ushahidi deployments have flopped due to the lack of crowds transmitting reports (Vota, 2012). Fatigue is an issue in Ushahidi deployments as some volunteering mapmakers can exhaust themselves in the effort to help disaster victims (Gutiérrez, 2018a).

This study is focused on how activist maps that comply with the rules of mapicity are devised to generate emotions and ultimately mobilize. The next section considers three examples.

Three geoactivist maps

The 'Left-to-die boat' map

On March 27, 2011, a group of 72 people were forced by armed Libyan soldiers on-board an inflatable craft, which headed in the direction of the island of Lampedusa (BBC, 2012). Only nine would survive. The 'Left-to-die boat' map shows that the failure to save them was due to callousness, since satellite imagery and data, testimonies, and other evidence substantiate that their dire situation had been detected and ignored. What made this case different from that of the 1,500 people who died attempting to cross the Mediterranean in 2011 was that the boat's calls 'would appear to have been ignored by a range of fishing vessels, a military helicopter and a large naval vessel' (Committee on Migration, Refugees and Displaced Persons, 2012). The people on the boat launched distress signals transmitting their location, and sustained repeated interactions with others (Heller & Pezzani, 2014). As part of its operations in Libya, an arms embargo was enforced in the Mediterranean Sea, making the area 'the most highly surveilled section of the sea in the entire world' (Heller & Pezzani, 2014). The map illustrates how other ships come across the boat's path but never pause to try a rescue. However, the Convention on the Law of the Sea calls for ships to 'render assistance to any person found at sea in danger of being lost' (United Nations, 1994, p. 60). Figure 26.2 is a static snapshot of the online version of the map.

The map was crafted by Forensic Oceanography, a team based within the Forensic Architecture agency, which specializes in techniques to scrutinize deaths and human rights abuses (Forensic Architecture, 2016). It employs a charged visual, textual, and acoustic language. It darkens when night falls, and lightens when the sun rises, making the journey realistic. Looking at emotional response to map design, Fabrikant et al. conclude the maps that use 'semantically correct colour assignments', for example blue for water, receive better responses (2012, p. 3). The ominous twirling cobalt shades that surround the Left-to-die boat signify a threat too. The animation is accompanied by the voice of a narrator, who uses loaded terms to recount the trip. The people on-board were escaping 'violent repression' in Libya and, without food, water, or fuel, ended up 'chained to the seas' open expanse', the narrator claims. The map shows that other ships in the boat's vicinity did not respond to its calls; the storyteller says that the refugees were 'denied minimal assistance' (Heller & Pezzani, 2014). A timeline induces a sense of urgency as the account proceeds. The soundtrack—a recording by the Laboratory of Applied Bioacoustics (Listening to the Deep Ocean Environment, 2017)—feels like a threatening marine roar. Consistent with Edsall's explorations of the use of music to convey the emotional context of geospatial data (2011), the ocean's thunder employed by the 'Left-to-die boat' map suits the grim facts as they unfold.

In his interview, Pezzani, who made the map, explains these formal elements were the result of a search for an adequate language 'to engage with the migrant crisis', an attempt to offer a different view than the conventional 'spectacle' of migrants as either invaders or victims. He adds: 'We did not want to risk being unwillingly complicit with the *border regime*'. The idea was to cast 'a disobedient gaze over this situation' because the migrant crisis is also a 'visual struggle', yet there is a 'lack of certain images'. The migrants encountered several ships, and these engagements were photographed by both sides, but the pictures have been kept secret, he recounts, as the survivors were stripped of their belongings. Every detail is carefully orchestrated in this map, but this is done without sentimentalism, as the pivotal element of the project is the map and the data. The complete meaning of the map is begot by the interrelation of the signs and words working together. The map arouses emotions partly in response to the mapmaker's 'new approach to mapping', paraphrasing Field and Demaj (2012, p. 73). 'Distress is probably one of the most fitting reactions' to the map, argues Pezzani.

The map supported a report to the European Parliamentary Assembly, which concludes that 'too many persons have lost their lives in circumstances similar to the 63 persons on board the "left-to-die boat"' (Committee on Migration, Refugees and Displaced Persons, 2012). This report includes only one image: that of the 'Left-to-die boat' map (p. 23). A coalition of organizations led a series of legal actions in the national courts of each of the states participating in the military operations against Libya (Pezzani & Heller, 2011). Should these states fail to investigate the incident, a case may be brought to the European Court of Human Rights, says Pezzani.

The 'Western Africa missing fish' map

Western Africa has some of the world's most abundant fishery resources, which are under threat from illegal or irregular fishing, which puts at risk the food security of millions of people. An industrialized fleet is catching too many fish, and much of the activity falls into irregular fishing, which is difficult to deter. The interactive 'Western Africa missing fish' map used by the ODI provides comprehensive visual evidence, for the first time, about foreign fleets engaging in irregular operations in developing countries.

The static snapshot in Figure 26.3 is based on the interactive 'Western Africa missing fish' map, which tracks 35 fish cargo vessels for a year. Figure 26.3 visualizes Automatic Identification System (AIS) signals emitted by one of these vessels, Sierra Loba, from 6 to 23 August 2013, as the ship operates in Senegal's waters. Vessels of a certain size must regularly launch AIS signals to avoid a collision. Fish cargo vessels such as Sierra Loba are specialized in gathering, processing, and deep-freezing fish for transportation. The zigzagging movements in Figure 26.3 are typical of a cargo ship in search of fishing vessels willing to empty their holds (Daniels et al., 2016, p. 18). The snapshot of the map in Figure 26.3 exposes the boat searching and stopping to transfer fish, an operation that is marked by the red circles indicating that the vessel is stationary for a number of hours (p. 18). However, Senegal forbids fish transfers in its waters as it lacks the resources to monitor whether fish caught illegally are involved in the operation.

The map and its accompanying report are unadorned, as they display no human suffering, but it unleashed a wave of disapproval. News media coverage played a key role. The project had been designed by my colleagues and I as a full-blown data activist endeavour, complete with a media outreach

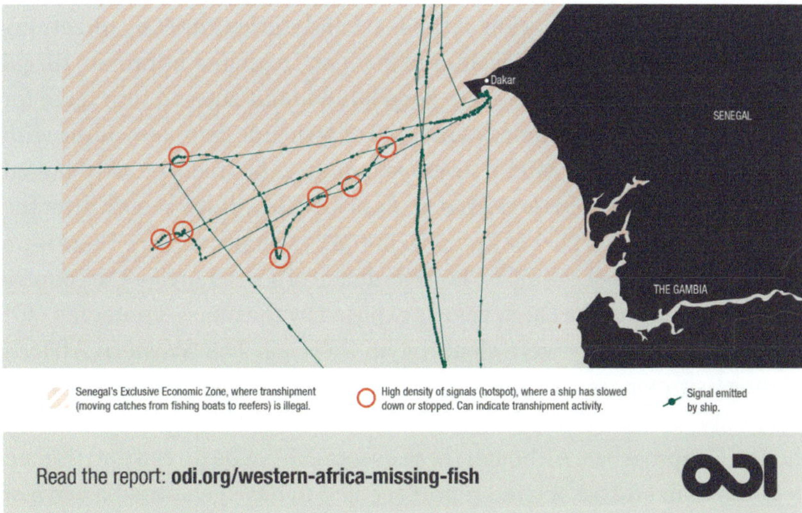

Figure 26.3. Sierra Loba, as it engages in irregular operations in Senegalese coastal waters. From 'Western Africa's missing fish: The impacts of illegal, unreported and unregulated fishing and under-reporting catches by foreign fleets', by A. Daniels, M. Gutiérrez, G. Fanjul, A. Guereña, I. Matheson, & K. Watkins, 2016 (https://www.odi.org/publications/10459-western-africas-missing-fish-impacts-illegal-unreported-and-unregulated-fishing-and-under-reporting). Copyright 2016 by Overseas Development Institute. Reprinted with permission.

plan. As a result, more than 150 media outlets from 35 countries had covered the story in several languages only weeks after its publication. Headlines including 'Illegal fishing "killing" livelihoods across in West Africa' (Bahati, 2016), 'UE, accomplice in the pillaging of African waters' (Caballero, 2016) and 'To end the looting of African waters' (Jarrett, 2017) stressed the loss that illegal fishing means for Africa.[1] These stories included poignant words such as 'pillage', 'looting', 'fight', 'killing', 'starving', and 'plundering'. That is, the loaded language was not offered by the map, but by the interpretations it triggered within the different news media. Journalistic coverage was especially intense in Western Africa; dozens of local media outlets in Congo, Morocco, Ghana, South Africa, Mauritania, Nigeria, Burkina Faso, Senegal, Ivory Coast, and Guinea picked up the news, which is not usual with ODI reports. Right after the publication of the map in June 2016, Guinea banned all international fishing activities in its waters, referring to the ODI investigation as the trigger for this decision (BBC, 2016). Consequently, Guinea was removed from the list of non-cooperative countries in the fight against illegal fishing (Karuri, 2016). The report accompanying the map

1 Orginal titles are *'La Unión Europea, cómplice del saqueo de los mares africanos'* and *'Pour en finir avec le pillage des eaux africaines'*. All translations by the author.

endorsed the implementation of the Food and Agriculture Organization's Agreement on Port State Measures to Prevent, Deter and Eliminate Illegal Fishing, which was concluded in 2009 but had not entered into force for lack of signatories. It became operational only in June 2016, coinciding with the launch of the map. After a seven-year hiatus, the treaty quickly gained traction and more countries joined in. As of July 2016, only a month after the launch, it had 34 state parties (from the initial 25 a month earlier). There were other aftershocks. For example, the government of South Korea—the country of origin of some of the cargo vessels exposed by the map—contacted ODI to proclaim that they were cleaning up their act. The 'Western Africa's missing fish' map was the first published study on illegal fishing behaviour using data visualizations. Since 2016, new maps have been produced using the same approaches. Although these events cannot be directly attributed to it, the map and associated report seemed to have generated a wave of annoyance that put illegal fishing in Africa on the table.

The 15M map

The 15M map is one in a series that guided the indignad@s movement in Spain. DeSoto (2014) notes they incorporate network visualizations, conceptual maps, alert systems, databases, and georeferenced wikis at a scale never seen before 2011, illustrating 'the art of cartography by connected multitudes'. The map in Figure 26.1, made by Instituto de Biocomputación y Física de Sistemas Complejos, Universidad de Zaragoza, stands out among the series because of its resemblance to Doel's simmering map. However, all the 15M charts are central to a broader *tekné* that allowed a great number of 'brains and bodies' to connect through time, space, emotion, and behaviour (DeSoto, 2014, p. 360). DeSoto classifies the 15M maps into two groups: diagnostic 'discomforting maps', responding to an initial phase of indignation, and performative 'empowering maps', conducive to action.[2] Figure 26.1—an empowering map—shows the contagious 'emotional climate' of 'joy' that powered the indignad@s protests (p. 359).

Are the three maps equality stimulating? The maps included in this chapter share some commonalities, but they are basically different. To understand what makes them effective, next they are appraised based on DeSoto's dual taxonomy (discomforting/empowering) and Muehlenhaus's classification of persuasive geocommunication. The idea is to produce a taxonomy that can serve as a heuristic tool for further analysis.

2 *'Mapas del malestar'* and *'mapas de la potencia'. Idem.*

Taxonomy

Muehlenhaus divides maps into four categories depending on the amount of data they assimilate and whether they are 'rationalist' or 'emotive' (2013). Maps can employ an 'authoritative style', which is data rich and magisterial looking; an 'understated style', relying on small datasets and minimalist presentations; a 'propagandist style', data-light and 'rhetorical in nature'; and a 'sensationalist style', resorting to rich datasets and making 'heavy use of rhetorical styling' (Muehlenhaus, 2013, pp. 6-10). Although Muehlenhaus acknowledges that this classification may not include all varieties, it is a useful tool to examine geoactivist maps. Table 26.1 combines DeSoto's and Muehlenhaus's classifications.

Table 26.1 Three maps seen from DeSoto's and Muehlenhaus's categorizations

	'Left-to-die boat'	'Western Africa missing fish'	15M map
Objective	Discomforting	Discomforting	Empowering
Data volume	Rich	Rich	Rich
Appearance	Partly rationalist/ partly emotive	Mostly rationalist/slightly emotive	Slightly rationalist/ mostly emotive

Note. Elaboration by the author based on Muehlenhaus (2013) and DeSoto (2014).

These maps differ in the type of emotions that they were designed to trigger. By denouncing a wrong, the 'Western Africa missing fish' and the 'Left-to-die boat' maps generate distress; they are discomforting maps. Meanwhile, the 15M is an empowering map, which feeds and grows on enthusiasm.

Maps can integrate massive amounts of data without incorporating different types of data. Therefore, the origins of the data are critical to determining data richness in maps. The 'Left-to-die boat' visualizes public data from satellite AIS data providers, heat signatures, radar signals, and data from other surveillance technologies recording the movement of nearby ships (Heller & Pezzani, 2014). It also counts on the testimonies from the survivors, a soundtrack and music, and other content. The mixture of types of data makes it complex when it integrates filmed interviews, sounds mimicking what had happened on-board, and a narrator's voice from the perspective of the people on the boat. The 'Western Africa missing fish' map assimilates dynamic AIS data with a static database which describes each vessel, including physical information (e.g. length, carrying capacity), as well as information on its owners, operators, and registries. As with the previous example, this case not only includes massive amounts of data but also different data sources. Meanwhile, the 15M map shows big amounts of data, without the richness in variety displayed by the other two. In this animated chart, there are two basic data layers: geographic data and the timestamped data of the tweets and retweets. The difference is that this map evolves in quasi-real-time, producing an 'alternative, community-owned definition of a territory' (Dosemagen, Warren, & Wylie, 2011), which makes it complex. Namely, these maps are not characterized by their paucity of data; nor are they propagandistic (Muehlenhaus, 2013).

The 'Western Africa missing fish' chart is an example of 'authoritative' geocommunication, as paraphrasing Muehlenhaus (2013, p. 6), attempts to persuade by looking legitimate to make the spectator infer that scientific rigour is being observed. The 'Left-to-die boat' map shows a style halfway between authoritative and sensationalist; it can be said to include 'a variety of tricks to excite and engage map users' (p. 10). The narrator's voice in pseudo real-time, the sound of the threatening sea, the timeline, and the testimonials provide a charged atmosphere. However, the amount of data, as well as their variety, bestows the map a sense of reliability. The 15M map is more sensationalist than authoritative, but while it shows none of the data complexity of the other two, it cannot be said to be data poor. The dancing lights that weave the map's landscape, superimposed on a dark blue emphasizing them, are bursting with liveliness. The map does not look scientific or formal, but animated and exciting.

Discussion and conclusion

What is it in maps than can make them mobilize people in activism? The interviews and geoactivist maps observed in this study suggest that striking a balance between emotive elements and rich data, in terms of both quantity and complexity, is crucial. From the less emotive to the most emotive, these maps use expressive language too. The three of them generated strong reactions according to the 'see-think/feel-do' formula. Although it cannot be inferred from this exercise that all data-rich geoactivist maps that are rationalist or emotive in adequate proportions will be effective, these are characteristics found in three successful cases of maps which either mobilized people or sustained action. These samples illustrate how maps play a role in stimulating the two basic emotions that influence people to 'do' things: negative, motivating feelings during an early stage of the mobilization, and hope to sustain it.

References

Bahati, L. (2016, July 1). Illegal fishing 'killing' livelihoods across in West Africa. *AllAfrica News.* Retrieved from http://www.africanews.com/2016/07/01/illegal-fishing-in-western-africa/

BBC. (2012, October 28). The left to die boat. *BBC.* Retrieved from http://www.bbc.co.uk/programmes/p0101r27

BBC. (2016, June 30). Africa highlights: Guinea bans foreign fishing. *BBC.* Retrieved from http://www.bbc.com/news/live/world-africa-36452771

Caballero, C. (2016, July 5). La Unión Europea, cómplice del saqueo de los mares africanos. *El País.* Retrieved from https://elpais.com/elpais/2016/07/04/planeta_futuro/1467647004_670890.html

Committee on Migration, Refugees and Displaced Persons. (2012). *Lives lost in the Mediterranean Sea: Who is responsible?'* Parliamentary Assembly, Council of Europe. Retrieved September 12, 2018 from https://assembly.coe.int/CommitteeDocs/2012/20120329_mig_RPT.EN.pdf

Daniels, A., Gutiérrez, M., Fanjul, G., Guereña, A., Matheson, I., & Watkins, K. (2016). *Western Africa's missing fish.* London: Overseas Development Institute. Retrieved July 1, 2018 from https://www.odi.org/sites/odi.org.uk/files/resource-documents/10665.pdf

della Porta, D., & Diani, M. (2006). *Social movements: An introduction* (2nd ed.). Malden, Oxford, & Victoria: Blackwell Publishing.

Denil, M. (2011). The search for a radical cartography. *Cartographic Perspectives, 68.* Retrieved August 1, 2018 from http://cartographicperspectives.org/index.php/journal/article/view/cp68-denil/14

DeSoto, P. (2014). Los mapas del #15M: El arte de La cartografía multitud conectada'. *Universitat Oberta de Catalunya, 25.*

Doel, M. (1999). *Poststructualist geographies: The diabolical art of spatial science.* Edinburgh: University of Edinburgh Press.

Dosemagen, S., Warren, J., & Wylie, S. (2011). Grassroots mapping: Creating a participatory map-making process centered on discourse. *Journal of Aesthetics and Protest* (8). Retrieved August 16, 2018 from https://www.joaap.org/issue8/GrassrootsMapping.htm

Edsall, R. (2011). 'Sounds dangerous: Emotion, geovisual analytics, and music'. *Kartographische Nachrichten, 62*(6), 291-299.

Fabrikant, S. I., Christophe, S., Papastefanou, G., & Maggi, S. (2012). Emotional response to map design aesthetics. *GIScience.* Retrieved September 12, 2018 from http://www.zora.uzh.ch/71701/1/2012_FabrikantS_giscience2012_paper_64.pdf

Field, K., & Demaj, D. (2012). Reasserting design relevance in cartography: Some concepts. *The Cartographic Journal, 49*(1), 70-76. https://doi.org/10.1179/0008704112Z.00000000011

Forensic Architecture. (2016, November 29). *Left-to-die boat.* Retrieved November 29, 2018 from http://www.forensic-architecture.org/case/left-die-boat/

Goodwin, J., Jasper, J. M., & Polletta, F. (Eds.). (2001). *Passionate politics: Emotions and social movements.* Chicago: University of Chicago Press.

Goodwin, J., Jasper, J. M., & Polletta, F. (2004). Emotional dimensions of social movements. In: D. A. Snow, S. A. Soule, & H. Kriesi (Eds.), *The Blackwell companion to social movements.* (pp. 413-432). Malden, Oxford, & Carlton: Wiley-Blackwell.

Gray, J., Bounegru, L., Milan, S., & Ciuccarelli, P. (2016). Ways of seeing data: Toward a critical literacy for data visualizations as research objects and research devices. In: S. Kubitschko & A. Kaun (Eds.), *Innovative methods in media and communication research.* (pp. 290-325). Cham: Palgrave Macmillan.

Griffin, A., & Mcquoid, J. (2012). At the intersection of maps and emotion: The challenge of spatially representing experience. *Kartographische Nachrichten 62*(6), 291-99.

Gutiérrez, M. (2018a). *Data activism and social change.* London: Palgrave Pivot.

Gutiérrez, M. (2018b). Maputopias: cartographies of communication, coordination and action—the cases of Ushahidi and InfoAmazonia. *GeoJournal*, 1-20. https://doi.org/10.1007/s10708-018-9853-8

Harley, J. B. (1989). Deconstructing the map. *Cartographica: The International Journal for Geographic Information and Geovisualization 26*(2), 1-20. https://doi.org/10.3138/E635-7827-1757-9T53

Hauthal, E., & Burghardt, D. (2013). *Detection, analysis and visualisation of georeferenced emotions.* In: M. F. Buchroithner (Ed.). Proceedings of the 26th International Cartographic Conference, August 25-30, 2013, Dresden, Germany. https://icaci.org/files/documents/ICC_proceedings/ICC2013/ICC2013_Proceedings.pdf

Heller, C., & Pezzani, L. (2014). *Liquid traces—The Left-to-die boat case* [Video file]. Retrieved July 1, 2018 from https://vimeo.com/89790770

Jarrett, M. B. (2017, May 5). Pour en finir avec le pillage des eaux africaines. *Le Monde Afrique.* Retrieved from http://www.lemonde.fr/afrique/article/2017/05/05/pour-en-finir-avec-le-pillage-des-eaux-africaines_5123210_3212.html

Jasper, J. M. (1998). The emotions of protest: Affective and reactive emotions in and around social movements. *Sociological Forum, 13*(3), 397-424.

Karuri, K. (2016, October 12). Guinea removed from list of countries not cooperating with eliminating illegal fishing. *Africa News.* Retrieved from http://www.africanews.com/2016/10/12/guinea-removed-from-list-of-countries-not-cooperating-with-eliminating-illegal/

Kennedy, H., & Hill, R. L. (2017). The feeling of numbers: Emotions in everyday engagements with data and their visualisation. *Sociology, 52*(4), 830-848. https://doi.org/10.1177/0038038516674675

Kent, A. J. (2005). Aesthetics: A lost cause in cartographic theory?' *The Cartographic Journal, 42*(2), 182-188. https://doi.org/10.1179/000870405X61487

Klettner, S., Huang, H., Schmidt, M., & Gartner, G. (2013). Crowdsourcing affective responses to space. *Kartographische Nachrichten, 2*(3), 66-72.

Lemke, J. (2015). Feeling and meaning: A unitary bio-semiotic account. In: P. P. Trifonas (Ed.), *International handbook of semiotics.* (pp. 589-616). New York & London: Springer.

Listening to the Deep Ocean Environment. (2017). *Presentation.* Retrieved November 29, 2018 from http://www.listentothedeep.com/acoustics/index2.php?web=presentation&lang=en

Maddrell, A. (2016). Mapping grief: A conceptual framework for understanding the spatial dimensions of bereavement, mourning and remembrance. *Social & Cultural Geography, 17*(2), 166-88. https://doi.org/10.1080/14649365.2015.1075579

Melucci, A. (1996). *The playing self.* Cambridge: Cambridge University Press.

Muehlenhaus, I. (2013). *Four rhetorical styles of persuasive geocommunication: An initial taxonomy.* Dresden: International Cartographic Association. Retrieved August 1, 2018 from https://icaci.org/files/documents/ICC_proceedings/ICC2013/_extendedAbstract/355_proceeding.pdf

Netek, A., & Panek, J. (2016, July). *Framework See-Think-Do as a Tool for Crowdsourcing Support.* Paper presented at the XXIII ISPRS Congress. http://doi.org/10.5194/isprsarchives-XLI-B6-13-2016

Nold, C. (Ed.) (2009). *Emotional cartography: Technologies of the self.* Retrieved August 15, 2018 from http://www.emotionalcartography.net/

Peluso, N. L. (1995). Whose woods are these? Territories in Kalimantan'. *Antipode, 27*(4), 383-406. https://doi.org/10.1111/j.1467-8330.1995.tb00286.x

Pezzani, L., & Heller, C. (2011). Case: 'Left-to-die boat'. Retrieved July 1, 2018 from https://www.fidh.org/IMG/pdf/fo-report.pdf

Preston, D. (2008). *Off the map: An exploration of emotive cartography, power and place.* In: E. Tilley (Ed.), Power & place: Refereed proceedings of the Australian & New Zealand Communication Association Conference.

Scharl, A., & Tochtermann, K. (Eds.). (2007). *The geospatial web: How geobrowsers, social software and the web 2.0 are shaping the network society.* London: Springer-Verlag.

United Nations. (1994). *Convention on the law of the sea.* Retrieved July 28, 2018 from http://www.un.org/depts/los/convention_agreements/texts/unclos/unclos_e.pdf

van Lammeren, R., Houtkamp, J. M., Colijn, S., Hilferink, M., & Bouwman, A. (2010). Affective appraisal of 3D land use visualization. *Computers, Environment and Urban Systems, 34*(6), 465-475. https://doi.org/10.1016/j.compenvurbsys.2010.07.001
Vota, W. (2012, August 29). CrowdGlobe: The dead Ushahidi reality behind 12,795 Ushahidi crowdmaps. *ICTworks*. Retrieved from http://www.ictworks.org/2012/08/29/crowdglobe-dead-ushahidi-reality-behind-12795-crowdmaps/

About the author

Miren Gutiérrez's interests are on data activism, or how data infrastructure can be utilized for social change in development, climate change, and environmental issues. She is a researcher at the University of Deusto, and a Research Associate at Overseas Development Institute, London, and at Datactive, University of Amsterdam. She holds a PhD in Communication Sciences.

Index

References to data visualizations are abbreviated to DVs throughout the index, except for its own main entry where it is spelled out in full. References to illustrations and diagrams are in **bold**.